RESONANCE IONIZATION SPECTROSCOPY

RESONANCE IONIZATION SPECTROSCOPY

Ninth International Symposium

Manchester, United Kingdom   June 1998

EDITORS
John C. Vickerman
The University of Manchester
Institute of Science and Technology

Ian Lyon
The University of Manchester

Nick P. Lockyer
The University of Manchester
Institute of Science and Technology

James E. Parks
The University of Tennessee, Knoxville

AIP CONFERENCE
PROCEEDINGS 454

American Institute of Physics   Woodbury, New York

Editors:

John C. Vickerman and Nick P. Lockyer
The University of Manchester
Institute of Science and Technology (UMIST)
Department of Chemistry
Manchester M60 1QD
United Kingdom

E-mail: John.Vickerman@umist.ac.uk
E-mail: Nick.Lockyer@umist.ac.uk

Ian Lyon
Department of Earth Sciences
The University of Manchester
Williamson Building
Oxford Road
Manchester M13 9PL
United Kingdom

E-mail: Ian.Lyon@man.ac.uk

James E. Parks
Department of Physics and Astronomy
The University of Tennessee
401 Nielsen Physics Building
Knoxville, TN 37996-1200

E-mail: jeparks@utk.edu

Authorization to photocopy items for internal or personal use, beyond the free copying permitted under the 1978 U.S. Copyright Law (see statement below), is granted by the American Institute of Physics for users registered with the Copyright Clearance Center (CCC) Transactional Reporting Service, provided that the base fee of $15.00 per copy is paid directly to CCC, 222 Rosewood Drive, Danvers, MA 01923. For those organizations that have been granted a photocopy license by CCC, a separate system of payment has been arranged. The fee code for users of the Transactional Reporting Service is: 1-56396-810-X/ 98 /$15.00.

© 1998 American Institute of Physics

Individual readers of this volume and nonprofit libraries, acting for them, are permitted to make fair use of the material in it, such as copying an article for use in teaching or research. Permission is granted to quote from this volume in scientific work with the customary acknowledgment of the source. To reprint a figure, table, or other excerpt requires the consent of one of the original authors and notification to AIP. Republication or systematic or multiple reproduction of any material in this volume is permitted only under license from AIP. Address inquiries to Office of Rights and Permissions, 500 Sunnyside Boulevard, Woodbury, NY 11797-2999; phone: 516-576-2268; fax: 516-576-2499; e-mail: rights@aip.org.

L.C. Catalog Card No. 98-88679
ISBN 1-56396-810-X
ISSN 0094-243X
DOE CONF- 9806123

Printed in the United States of America

CONTENTS

Preface xi
Sponsors xiii
Committees xiv

SESSION I: RIS AND CLUSTERS

Infrared Resonance Enhanced Multiphoton Ionization of Fullerenes 3
    G. von Helden, I. Holleman, A. van Roij, G. M. H. Knippels,
    A. F. G. van der Meer, and G. Meijer

Growth of CdSe Clusters by Laser Ablation and Mass Spectroscopy
of Ion Flow 9
    A. Mazeikis, J. Miskinis, A. Petravicius, M. Robino, J. Vaitkus,
    and A. Zindulis

SESSION II: ATOMIC RIS

Experimental Studies of Quantum Control in Atoms 19
    D. Charalambidis, N. E. Karapanagioti, D. Xenakis, C. Fotakis, O. Faucher,
    E. Hertz, S. Cavalieri, R. Eramo, L. Fini, and M. Materazzi

Photoionization of Laser-Cooled Cesium 6 $^2P_{3/2}$ Atoms 27
    F. Fuso, O. Maragò, D. Ciampini, E. Arimondo, C. Gabbanini,
    and S. T. Manson

High-Lying Bound Rydberg States of Excited Hg($6s6p$ $^3P_1$) Atoms
from Two-Color Resonance Ionization Mass Spectroscopy 33
    P. Bisling, J. Dederichs, B. Neidhart, and C. Weitkamp

SESSION III: RIS AND ULTRA-TRACE ANALYSIS

Ultra-Trace Determination of Long-Lived Radioactive Isotopes 39
    K. Wendt

Ultra-Sensitive Detection of Hydrogen Isotopes by the Lyman-α RIS 47
    Y. Miyake, K. Shimomura, A. P. Mills, Jr., J. P. Marangos, and K. Nagamine

Multi-Element Trace Analysis of Solid Samples Using One-Photon
Two-Step RIMS 53
    H. H. Telle, C. J. Abraham, O. R. Jones, and T. Krustev

Resonance Enhanced Laser Mass Spectrometry for Process- and
Environmental-Analysis: Applications and Perspectives 59
    R. Zimmermann, H. J. Heger, R. Dorfner, U. Boesl, and A. Kettrup

SESSION IV: APPLICATIONS

Laser Mass Spectrometry of Chemical Warfare Agents Using Ultrashort Laser Pulses 67
    C. Weickhardt, C. Grun, and J. Grotemeyer

Ultra-Trace Determination of the Long-Lived Isotope ^{41}Ca by Narrowband CW-RIMS 73
    P. Müller, B. A. Bushaw, K. Blaum, W. Nörtershäuser, N. Trautmann, and K. Wendt

SESSION V: PHOTOELECTRONS AND PHOTODISSOCIATION

Photoelectron Spectroscopy of Reactive Intermediates 81
    J. M. Dyke, S. D. Gamblin, A. Morris, J. B. West, and T. G. Wright

Femtosecond Photodissociation Dynamics of $Cr(CO)_6$ 89
    T. Kono, V. Vorsa, K. F. Willey, and N. Winograd

Time-Resolved Photoelectron Emission from Rydberg Atoms 95
    L. D. Noordam

Formation of $O_2(b^1\Sigma_g^+)$ in the Ultraviolet Photolysis of O_3 103
    P. K. O'Keeffe, T. Ridley, S. Wang, K. P. Lawley, and R. J. Donovan

Multiphoton Ionization/Dissociation of Cyclopentanone at the Lower Rydberg States 109
    J. G. Philis, C. Kosmidis, and P. Tzallas

SESSION VI: MOLECULAR RIS

Charge Transfer and Charge Localization in Extended Radical Cations: Investigation of Model Molecules for Peptides 117
    R. Weinkauf and F. Lehrer

An Attempt to Study LiH and Li_2 Molecules by High Resolution Pulsed Laser Spectroscopy 125
    N. Bouloufa, L. Cabaret, P. Cacciani, P. Camus, B. Pitcheev, and R. Vetter

Dibenzo—p—dioxin and its Chlorinated Congeners: Molecular Geometry and Electronic Structure 131
    R. Zimmermann

Overtone Spectroscopy of Jet-Cooled Phenol Studied by Nonresonant Ionization Detected IR Spectroscopy 137
    S. Ishiuchi and M. Fujii

Resonant Ionization Mass Spectrometry of Ammonia 143
    T. Gibert-Legrand, T. Gonthiez, L. Vivet, and P. Brault

Translationally Cold Cs_2 Molecules Formation in a Magneto-Optical Trap 147
    A. Fioretti, D. Comparat, C. Drag, A. Crubellier, O. Dulieu, F. Masnou-Seeuws, C. Amiot, and P. Pillet

SESSION VII: FEMTO-SECOND AND HIGH INTENSITY RIS

The Dissociation Dynamics of Diatomic Molecules in Intense Laser Fields 155
    J. H. Posthumus and K. Codling
**Femtosecond Dynamics of Photoinduced Chemical Reactions Studied
by Intense-Field Dissociative Ionization** 163
    W. Fuß, K. L. Kompa, W. E. Schmid, and S. A. Trushin

SESSION VIII: APPLICATIONS II

**Lineshapes and Optical Selectivity in Double-Resonance RIMS
Measurements** 171
    W. Nörtershäuser, K. Blaum, B. A. Bushaw, P. Müller, N. Trautmann,
    and K. Wendt
Diode-Laser-Based RIMS Measurements of Strontium-90. 177
    B. A. Bushaw and B. D. Cannon
**Determination of the First Ionization Potential of Nine Actinide
Elements by Resonance Ionization Mass Spectroscopy.** 183
    G. Passler, M. Nunnemann, G. Huber, R. Deißenberger, N. Erdmann,
    S. Köhler, J. V. Kratz, N. Trautmann, A. Waldek, and J. R. Peterson

SESSION IX: NEW LASER DEVELOPMENTS AND APPLICATIONS

Laser Ion Source for Nuclear Spectroscopic Studies 191
    Y. Kudryavtsev, A. Andreyev, B. Bruyneel, S. Franchoo, J. Gentens,
    M. Huyse, K. Kruglov, W. Mueller, R. Raabe, I. Reusen, P. Van den Bergh,
    P. Van Duppen, J. Van Roosbroeck, L. Vermeeren, and L. Weissman

SESSION X: SURFACE APPLICATIONS AND SPUTTERING

**Resonance Ionization Spectroscopy Investigations of Electronic
Processes during Ion-Beam Sputtering of Metal Atoms** 197
    R. E. Silverans and P. Lievens
**Photoionization Studies of Small Biological Molecules Using Femtosecond
Laser Pulses.** 206
    V. Vorsa, K. F. Willey, T. Kono, and N. Winograd
**State-Selective Laser Photoionization of Neutral Benzene Molecules
Ejected from keV Ion Bombarded C_6H_6/Ag{111}** 210
    C. A. Meserole, E. Vandeweert, R. Chatterjee, B. R. Chakraborty,
    B. J. Garrison, N. Winograd, and Z. Postawa

SESSION XI: SURFACE APPLICATIONS AND SPUTTERING II

Detection of Large Neutral Clusters in Sputtering 217
    C. Staudt and A. Wucher

SESSION XII: RIS APPLIED TO NUCLEAR AND PARTICLE PHYSICS

Hyperfine Structure Studies with the COMPLIS Facility 225
    J. E. Crawford, J. K. P. Lee, F. Le Blanc, D. Lunney, J. Obert, J. Oms,
    J. C. Putaux, B. Roussière, J. Sauvage, S. Zemlyanoi, D. Verney, J. Pinard,
    L. Cabaret, H. T. Duong, G. Huber, M. Krieg, V. Sebastian, M. Girod,
    S. Péru, J. Genevey, F. Ibrahim, J. Lettry, and the ISOLDE Collaboration

Laser Induced Nuclear Reactions 229
    K. Ledingham, T. McCanny, P. Graham, X. Fang, R. Singhal, J. Magill,
    A. Creswell, D. Sanderson, R. Allott, D. Neely, P. Norreys, M. Santala,
    M. Zepf, I. Watts, E. Clark, K. Krushelnick, M. Tatarakis, B. Dangor,
    A. Machecek, and J. Wark

Spectral Studies Related to the 3.5 eV Isomeric State of Th-229 235
    J. P. Young, R. W. Shaw, and O. F. Webb

POSTER SESSION I: RIS AND CLUSTERS

Threshold Photoionization Spectroscopy of Li_nO and Li_nC Clusters 243
    P. Lievens, P. Thoen, S. Bouckaert, W. Bouwen, F. Vanhoutte, H. Weidele,
    and R. E. Silverans

POSTER SESSION II: ATOMIC RIS

Control of the Final State Photoionization Products by Laser-Induced Continuum Structure 249
    S. Cavalieri, R. Eramo, L. Fini, M. Materazzi, O. Faucher,
    and D. Charalambidis

Superelastic Collisions [e + Mg*] Following Resonant, 2-Photon Ionization of Mg Atoms 253
    S. A. Darveau and R. S. Berry

Studies on Autoionization States of Sm by 3-Step Resonance Photoionization 257
    H. Park, J. H. Yi, J. M. Han, Y. J. Rhee, and J. Lee

Monitoring of Gd Photoionization Process by Detection of Fluorescence Characteristics 261
    Y. Rhee, J. Yi, J. T. Kim, H. Park, J. Han, J. Lee, and M. G. Baik

POSTER SESSION III: RIS AND ULTRA-TRACE ANALYSIS

A RIS-TOF Instrument for the Measurement of Ultratrace Quantities of Uranium and Plutonium 269
    A. W. McMahon, J. D. Gilmour, M. B. Hernandez, and M. Rateitzak

Diode-Laser-Based Resonance Ionization Mass Spectrometry of Gadolinium 275
    K. Blaum, B. A. Bushaw, C. Geppert, P. Müller, W. Nörtershäuser, A. Schmitt, N. Trautmann, and K. Wendt

Trace Analysis of Plutonium by Resonance Ionization Mass Spectroscopy 279
    N. Erdmann, C. Grüning, N. Trautmann, A. Waldek, G. Huber, P. Kunz, M. Nunnemann, and G. Passler

POSTER SESSION IV: NEW LASER DEVELOPMENTS AND APPLICATIONS

A High Repetition Rate Solid State Laser System for Resonance Ionisation Mass Spectrometry of Actinides 285
    C. Grüning, N. Erdmann, G. Huber, P. Klopp, J. V. Kratz, P. Kunz, M. Nunnemann, G. Passler, O. Stetzer, A. Waldek, and K. Wendt

CW Laser-Initiated Resonance Ionization Mass Spectrometry of Gd Atoms ... 289
    D. Y. Jeong, Y. Rhee, J. Lee, and B. K. Rhee

Lasing Threshold Reduction and the Enhancement of Mode Selection in a Novel Littrow-Type Coupled Cavity 293
    D. K. Ko, S. H. Kim, B. H. Cha, and J. Lee

High-Efficiency Parametric Oscillation in Beta-Barium Borate with Pump Reflection 297
    J. Lee, S. W. Lee, D. K. Ko, S. H. Kim, J. M. Han, and B. H. Cha

Probing the Response Mechanism of the Thermionic Detector by Resonance Enhanced Ionization Spectroscopy 301
    A. W. McMahon and P. A. Schofield

On-Line Monitoring of Trace Compounds in the Flue Gas of an Incineration Pilot Plant: Formation of Polycyclic Aromatic Hydrocarbons 305
    H. J. Heger, R. Zimmermann, R. Dorfner, A. Kettrup, and U. Boesl

Laser Ionisation Mass Spectrometry (REMPI-TOFMS) for On-Line Analysis of Volatiles in Food Science: Coffee-Roasting and Headspace Experiments 309
    R. Dorfner, R. Zimmermann, C. Yeretzian, and A. Kettrup

POSTER SESSION V: PHOTOELECTRONS AND DISSOCIATION

Investigation of the Multiphoton Ionization of Ba Atoms with Photoelectron Spectroscopy 315
    A. Y. Elizarov

POSTER SESSION VI: MOLECULAR RIS

Higher Rydberg States of C_6H_6, C_6D_6, and C_6H_5F Studied by Two-Photon Resonance Ionization Spectroscopy 325
    S. Wang, R. J. Donovan, T. Ridley, and K. P. Lawley

POSTER SESSION VII: FEMTO-SECOND AND HIGH INTENSITY RIS

Ionization and Fragmentation of Small Molecules under psec and fsec Laser Excitation 331
    S. Couris, E. Koudoumas, and C. Fotakis

Dissociative Ionization and Angular Distributions of CS_2 and its Ions 341
    P. Graham, K. W. D. Ledingham, R. P. Singhal, D. J. Smith, S. L. Wang, T. McCanny, H. S. Kilic, A. J. Langley, P. F. Taday, and C. Kosmidis

POSTER SESSION VIII: SURFACE APPLICATIONS AND SPUTTERING

Desorption Dynamics Below and Above the Ablation Threshold of van der Waals Films 347
    M. Lassithiotaki, A. Koubenakis, J. Labrakis, and S. Georgiou

Electron Configuration of Dependence of Kinetic Energy Distributions of Ion-Beam Sputtered Ni Atoms Studied by Double Resonant Laser Ionization 353
    V. Philipsen, J. Bastiaansen, E. Vandeweert, P. Lievens, and R. E. Silverans

Development of a Biological Imaging Instrument 358
    N. P. Lockyer, S. C. C. Wong, and J. C. Vickerman

Author Index 365

PREFACE

The Ninth International Symposium on Resonance Ionization Spectroscopy: New Directions and Applications (RIS-98), was held at The University of Manchester Institute of Science and Technology, Manchester, UK, June 21-25, 1998. Following 17 years of tradition, this symposium reported on the basic science and technology of resonance ionization spectroscopy and its many varied applications. Attendance was markedly less than at previous meetings, reflecting cutbacks in sponsored research programs, particularly in the United States. Despite the smaller numbers, this carefully planned meeting maintained a balance between plenary, invited, contributed, and poster presentations, assuring a high quality program. Evidence of new ventures showed aggressive, well-staffed, and well-funded programs that find their roots in RIS. The proceedings of this well-received program are presented in this volume.

Reports on the progress in RIS development and related areas are a distinguishing feature of this symposium series. Work reported at RIS-98 demonstrated advances in RIS technology made possible by new, state-of-the-art advances in other areas, in particular laser development. Improvements in laser systems, such as femtosecond lasers and high power tunable diode lasers, provided fertile areas of research in the basic understanding of light and matter interactions, which are crucial to understanding RIS. In past symposia, basic science studies have shown limitations to the technology. For example, the measurement of isotopic ratios has been restricted by odd-even isotopic effects. However, this symposium revealed a clever method to overcome this problem.

RIS-98 followed the symposia custom of offering a short course on the principles and possible applications of RIS. The short course gave a brief introduction to the general theme of resonance ionization and highlighted the current state of the art from scientific and technological points of view. This short course was of interest to relative newcomers to the field in general, those who are possibly changing directions, and researchers who want to find out where RIS might be headed. Short course topics included: Laser Sources for RIS, RIS of Atoms, RIS of Molecules, and Applications of RIS. J. L. Collier, D. Charalambidis, H. H. Telle, and T. Krustev presented the topics. As in the past, the short course was very well attended with the participants coming a day early to participate.

Following the short course, the symposium continued with five days of presentations, including an afternoon outing with many informal discussions. Symposium topics were: RIS and Clusters, Atomic RIS, RIS and Ultra-trace Analysis, Applications, Photoelectrons and Photodissociation, Molecular RIS, Femtosecond and High Intensity RIS, Applications II and General Discussion Session, New Laser Developments and Applications, Surface Applications and Sputtering I and II, and RIS Applied to Nuclear and Particle Physics. The meeting concluded with a banquet recognizing the outstanding presentation by a graduate student and the outstanding poster presentation.

The RIS international advisory committee has long recognized that the future of this technology is with the graduate students who are working in the field. Graduate student participation is encouraged by making special arrangements for their attendance at the conference and by offering an award for the best work presented by a graduate student. The recipient of the 1998 award was Wilfried Nörtershäuser (Institut für Physik, Institut für Kernchemie, Universtität Mainz, Germany) for his work "Lineshapes and Optical Selectivity in Double-Resonance RIMS Measurements." This award, sponsored by the University of Tennessee Institute of Resonance Ionization Spectroscopy, acknowledges the most outstanding RIS research and development work presented by a graduate student, based on originality, innovativeness, analysis and interpretation of results, thoroughness of work, and the significance of the contribution to RIS.

The RIS-98 poster presentation was engaging and sparked stimulating and lively discussions. This forum for the exchange of ideas has become an integral part of the symposia, and to recognize outstanding work presented in a poster session, an excellence award for the best presentation is given. The 1998 poster award was presented to Yongjoo Rhee, Jonghoon Yi, Jin Tae Kim, Hyunmin Park, Jaemin Han, Jongmin Lee, and Moon-Gu Baik for their work entitled "Monitoring of Gd Photoionization Process by Detection of Fluorescence Characteristics." These researchers represent the Laboratory for Quantum Optics, Korea Atomic Energy Research Institute, and the Department of Physics, Kyungwon University. Their work is included in these proceedings, along with the work of Wilfried Nörtershäuser. These efforts are highly recommended for review as examples of the excellent work that is taking place in RIS research and development.

The host for this symposium was the University of Manchester Institute of Science and Technology (UMIST), with the Weston Conference Centre on the UMIST campus serving as the conference venue. Accommodations and meals were provided at the site, which allowed participants to give maximum attention to the details of the meeting. The success of this symposium was due to the hard work and advice of the RIS international advisory committee, the host's scientific committee, and the local organizing committee. Janet Adnams of the conference office, and Alex Henderson, who was responsible for web-page publicity, are commended for their excellent organizational work. The International Symposia on RIS and its Applications is coordinated through the Institute of Resonance Ionization Spectroscopy (IRIS) of The University of Tennessee. Catherine Longmire, Publications Coordinator for the Department of Physics and Astronomy and IRIS is appreciated for her work on all aspects of the conference planning and operations, including the final coordination of these proceedings.

John C. Vickerman
Ian Lyon
Nick P. Lockyer
James E. Parks

SPONSORS

Financial support from the following organizations is gratefully acknowledged:

Coherent Inc.

Edinburgh Instruments Ltd.

Elliot Scientific Ltd.

KORE Technology Ltd.

V A T Vacuum Products Ltd.

HOSTS

The University of Manchester Institute of Science and Technology hosted RIS-98, which was held at the Weston Conference Centre, UMIST, P.O. Box 88, Sackville Street, Manchester M60 1QD.

The International Symposium on Resonance Ionization Spectroscopy: Applications and New Directions is coordinated through the Institute of Resonance Ionization Spectroscopy of The University of Tennessee.

CONFERENCE ORGANIZERS

John C. Vickerman
Ian Lyon

CO-CHAIRMAN

James E. Parks

LOCAL ORGANIZING COMMITTEE

Nick P. Lockyer (Chair)
Janet Adnams (Conference Office)
Alex Henderson
Ian Lyon
John C. Vickerman

SCIENTIFIC COMMITTEE

John C. Vickerman (Chair)	Manchester	Ian Lyon	Manchester
Robert Donovan	Edinburgh	Klaus Müller-Dethlefs	York
Costas Fotakis	Crete	Helmut Telle	Swansea
Ken Ledingham	Glasgow		

INTERNATIONAL ADVISORY COMMITTEE

E. Arimondo	Italy	V. S. Letokhov	Russia
O. Axner	Sweden	N. Mikami	Japan
P. Benetti	Italy	C. M. Miller	USA
B. Bushaw	USA	J. C. Miller	USA
P. Camus	France	V. I. Mishin	Russia
J. E. Crawford	Canada	N. Omenetto	Italy
J. Grotemeyer	Germany	J. E. Parks	USA
W. Hogervorst	The Netherlands	J. C. Vickerman	UK
G. Hurst	USA	K. Wendt	Germany
K. Ledingham	Scotland	N. Winograd	USA
B. Lehmann	Switzerland	X. Xu	P. R. China

SESSION I
RIS AND CLUSTERS

Infrared Resonance Enhanced Multiphoton Ionization of Fullerenes

Gert von Helden*, Iwan Holleman*, André van Roij*, Guido M.H. Knippels[†], A.F.G. van der Meer[†], and Gerard Meijer*

*Department of Molecular and Laser Physics, University of Nijmegen, Toernooiveld 1,
NL-6525 ED Nijmegen, The Netherlands
[†] FOM-Institute for Plasma Physics Rijnhuizen, Edisonbaan 14,
NL-3430 BE Nieuwegein, The Netherlands

Abstract. Gas-phase fullerenes are resonantly heated by a train of high power sub-picosecond pulses from a Free Electron Laser (FEL) to internal energies at which they efficiently undergo delayed ionization with little or no fragmentation [1]. When the laser is tuned from 6–20 μm while the amount of laser produced parent ions is recorded, resonant absorption of 200–600 IR photons is observed and the resulting wavelength spectra have a close resemblance to the IR absorption spectrum of C_{60}. InfraRed Resonance Enhanced Multi Photon Ionization (IR-REMPI) with a FEL enables extremely sensitive IR spectroscopy with mass selective detection of gas-phase fullerenes. To gain insight into the ionization mechanism, the time evolution of the production of the ions following IR excitation is recorded. Ion creation is unexpectedly slow, occurring at times beyond 0.1 ms after laser excitation. The results indicate that reaching the first electronically excited state is a high hurdle on the way from hot molecules to ions.

INTRODUCTION

For a long time, the microscopic equivalent of the thermal emission of electrons from a heated surface, the thermionic emission of electrons from neutral hot molecules, was not observed. For thermionic emission to be possible a large amount of internal energy, at least more than the energy needed for ionization, has to reside in a molecule. Such super-heated molecules will often loose their energy preferentially via fragmentation and photon emission, the microscopic equivalents of the macroscopic cooling processes of evaporation and black-body emission. Thermionic emission can become a significant cooling channel when the energy needed for ionization is comparable to or smaller than the dissociation energy, as is the case for strongly bound molecules like the fullerenes [2] as well as for clusters of transition metal atoms [3]. To produce highly excited molecules, high power pulsed laser radiation in the visible (VIS) and ultraviolet (UV) region of the spectrum is commonly used. Although electronically excited states are initially prepared in this case, it is assumed that due to rapid thermalization and electronic-to-vibrational energy transfer, vibrationally excited, i.e. heated, molecules are created which can subsequently evaporate off an electron [4]. A more direct insight into the processes involved can be obtained when exciting the vibrational modes directly, using intense infrared (IR) light [5]. Over the last years Free Electron Lasers (FEL) have become available to users and their performance characteristics make them the ideal light sources for studies of this kind. Here we present results from experiments where gas phase C_{60} is being resonantly excited by IR light up to internal energies where autoionization becomes efficient [1].

EXPERIMENTAL

The experiments have been performed at the 'Free Electron Laser for Infrared eXperiments' (FELIX) user-facility in Nieuwegein, The Netherlands [6]. This laser produces pulsed IR radiation that is continuously tunable over the 100–2000 cm^{-1} range. The light output consists of macropulses of about 5 μs duration containing up to 100 mJ of energy. Each macropulse consists of a train of micropulses which are 1 ns apart. The macropulse repetition rate is 10 Hz. A scheme of the experimental setup is shown in Fig. 1. An effusive molecular beam of C_{60} is generated by evaporating C_{60} from a quartz oven. The FELIX beam enters the vacuum chamber and is focused onto the molecular beam in the region between the extraction plates of a Time of Flight (TOF) mass spectrometer. With an oven temperature of 875 K, the density of C_{60} in the interaction region can be estimated to be $(8\pm2) \times 10^9$ molecules/cm^3. After a delay of typically 30 μs after the FELIX pulse, ions can be pulse extracted and detected in a TOF mass spectrum on a MCP detector. To record IR resonance enhanced multi photon ionization (IR-REMPI) spectra of C_{60}, the total number of C_{60}^+ ions is then recorded as a function of IR laser wavelength.

To record the time-evolution of the production of ions, the total ion signal is measured as a function of time with the voltage on the TOF electrodes being constantly on (electric field of 600 V/cm in the extraction region) and the IR laser being fixed in frequency on a resonance. From the measured arrival time distribution of the ions at the detector, the flight time is subtracted to obtain the 'Time-Of-Birth' (TOB) distribution. These TOB distributions are the increment in total ion signal as a function of time, i.e. they correspond to the *rate* of ion production (multiplied by the time-constant of the detection system).

RESULTS AND DISCUSSION

Fig.1 shows the experimental setup and a typical mass spectrum obtained when FELIX is set to 19.2 μm. A strong peak of C_{60}^+ is observed. Surprisingly, the amount of fragmentation (C_{58}^+) is very small. It should be noted that the observation of fullerene ions, when exciting at this wavelength, results from the absorption of more than 500 IR photons by a single molecule! By monitoring the amount of

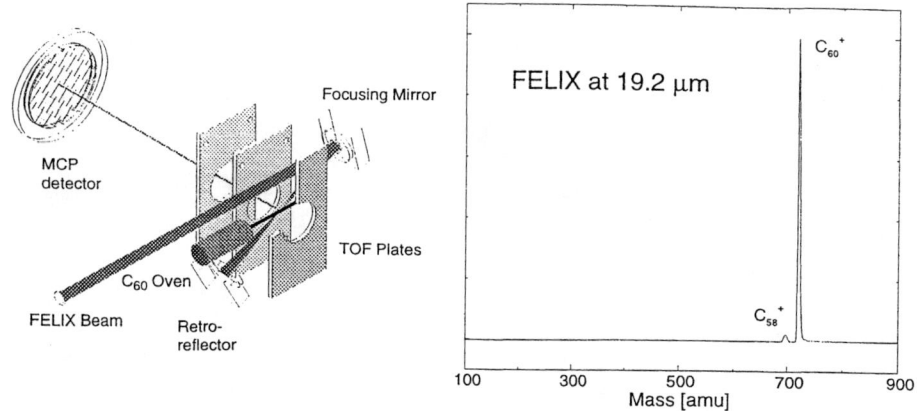

FIGURE 1. Experimental setup and TOF mass spectrum obtained when exciting at 19.2 μm

C_{60}^+ produced as a function of the FELIX wavelength, the IR REMPI spectrum, shown in the upper part of Fig.2, is obtained. Four peaks can readily be identified. In the lower part of the figure, the IR absorption spectrum of a thin film of solid C_{60} is shown. Again four lines can be seen and they correspond to the four well

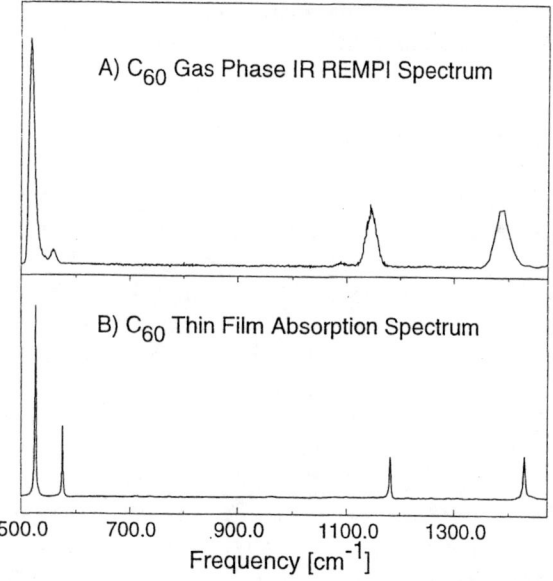

FIGURE 2. Shown is **a)** the IR-REMPI spectrum of gas-phase C_{60} effusing from the oven. Shown in **b)** is the thin film absorption spectrum of solid C_{60}, showing the four well known IR absorption lines.

known IR allowed F_{1u} fundamental modes of C_{60}. The two spectra show a clear correspondence in both, their line positions as well as the relative intensities of the lines. Nonetheless, it can be seen that the resonances in the IR-REMPI spectrum are systematically shifted to slightly lower frequencies than in the thin film absorption spectrum and that the lines in the IR-REMPI spectrum are slightly asymmetric. Further, the strongest lines in the IR REMPI spectrum are those which show the least frequency shift with respect to the corresponding lines in the thin film absorption spectrum.

The observation that C_{60} can be resonantly ionized by only IR photons seems astonishing. At least several hundred IR photons need to be absorbed by a single C_{60} molecule before it will ionize in the experimental time window and this might therefore appear to be a rather unlikely process. One also might expect that, after absorption of a few IR photons, the molecule runs out of resonance with the IR laser due to the intrinsic anharmonicities of the vibrational modes. The observation that so many IR photons can be absorbed by the molecules can be explained by fast internal vibrational redistribution (IVR) in combination with the pulse structure of FELIX. The molecule can therefore absorb one IR photon in a FELIX micropulse and, by the time the next micropulse arrives, the energy of the previous photon is completely randomized in the molecule. With 174 internal degrees of freedom and with an averaged vibrational frequency of 950 cm^{-1}, on average less than two quanta need to be put in each vibrational mode of C_{60} to reach the required energy of 35-40 eV. By sequential pumping of IR photons into the molecule, accompanied by thermalization via fast IVR, the C_{60} molecule can then be resonantly heated up to energies at which ionization becomes efficient.

To gain insight into the ionization mechanism, the time evolution of the ion production can be mesasured by monitoring the ion current as a function of time when keeping the high voltage on the TOF electrodes constantly on. Such obtained Time of Birth (TOB) distributions are shown in Fig 3, recorded with FELIX being set to 19.2 μm (520 cm^{-1}). The envelope of the FELIX pulse is shown in the lower part of the Figure. At long times, the chances for an ion to be detected diminish, since the neutral precursors escape from the detection region between the TOF electrodes. Using the experimental geometry and a Maxwellian C_{60} velocity distribution, the experimental detection efficiency curve shown in the lower part of the Figure is calculated. Several important points can immediately be recognized from Figure 2. First, at all excitation fluences, the TOB distributions extend to very long times. Taking the experimental detection efficiency into account, it is evident that ions are produced at 150 μs and beyond. This corresponds to a (1/e) autoionization timescale that is significantly larger than previously observed. Second, and even more surprising, the two TOB distributions taken at low fluence are clearly peaked after the end of the FELIX pulse and appear to consist of at least two components. When increasing the laser fluence, the fastest of these components gains dramatically in intensity relative to the slow component and both components are seen to have their maxima shifted to shorter times.

The observation that the TOB distributions are peaked after the FELIX pulse

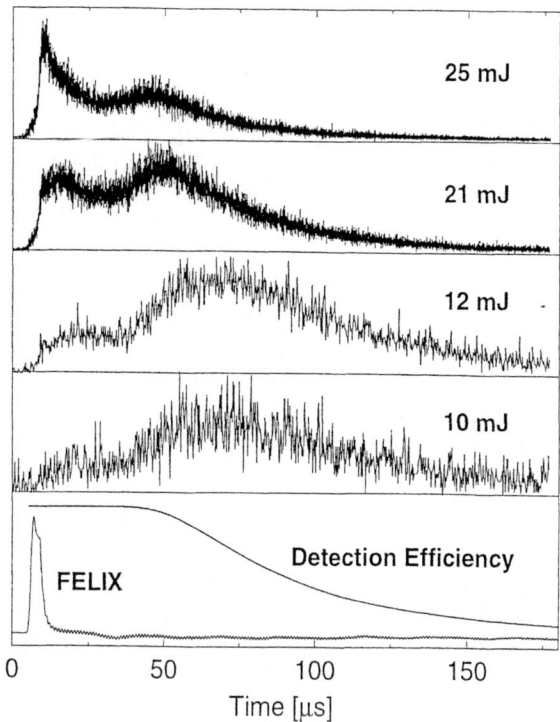

FIGURE 3. Time-Of-Birth (TOB) distributions of C_{60}^+ ions at three different macropulse energies. The envelope of the FELIX micropulses is shown in the lower part together with the time dependence of the experimental detection efficiency.

cannot be explained by a model that assumes a population of excited molecules that undergoes a first-order reaction. In such a model, the rate of ion production is directly proportional to the excited state population and an exponentially decaying function results. The simplest unimolecular model that can give rise to a distribution that peaks at a later time must include an intermediate step which then undergoes autoionization. In such a model, the TOB distibution can start with no ions being produces at t=0 and can then peak at a later time. Further evidence for such a mechanism comes from double resonance experiments, where C_{60} is first being electronically preexcited to a long lived triplet state and then being ionized by FELIX [7]. In those experiments, the TOB distributions are observed to decay almost an order of magnitude faster than the TOB distribution originating from the ground state surface of C_{60}. This indicates that electronic excitation to intermediate states is important in the autoionization process and that pre-excited triplet C_{60} has already taken the first, and perhaps highest hurdle towards the ionizing state. The data presented here show that the autoionization process of C_{60} is more complicated than previously thought. The observations are not consistent with the picture of a direct coupling of highly vibrationally excited ground-state C_{60} to the ion. In such a scenario at least 7.6 eV of vibrational energy, corresponding to a minimum of forty vibrational quanta, has to be exchanged to electronic energy in a single step, and the molecule appears to prefer a different route. It is observed that states that are intermediate between the vibrationally excited ground state levels and the electronic states of the ion play a crucial role in the thermal ionization process [7]. Vibrationally and electronically excited C_{60} is seen to undergo autoionization on a much faster time-scale than C_{60} that is only vibrationally excited. For C_{60}, the first electronically excited state is a triplet state at 1.5 eV. The first excited singlet state is around 2.0 eV above the ground state. From there on, the next higher singlet and triplet states follow in close succession, and the exchange of vibrational energy with electronic energy can occur efficiently, by exchange of only one or two quanta of vibrational energy, in a step-wise manner. A bottleneck in phase space will be the transition from the electronic ground state to the first electronically excited state.

REFERENCES

1. G. von Helden, I. Holleman, G.M.H. Knippels, A.F.G. van der Meer, G. Meijer, *Phys.Rev.Lett.* **79**, 5234 (1997).
2. E.E.B. Campbell, G. Ulmer, I.V. Hertel, Phys. Rev. Lett. **67**, 1986 (1991); P. Wurz, K.R. Lykke, J. Chem. Phys. **95**, 7008 (1991).
3. T. Leisner, K. Athanassenas, O. Echt, O. Kandler, D. Kreisle, E. Recknagel, Z. Physik D **20**, 127 (1991).
4. C.E. Klots, R.N. Compton, Surf. Rev. Lett. **3**, 535 (1996).
5. V.N. Bagratashvili, M.V. Kuz'min, V.S. Letokhov, A.N. Shibanov, JETP Lett. **37**, 112 (1983); M. Hippler, M. Quack, R. Schwarz, G. Seyfang, S. Matt, T. Märk, Chem. Phys. Lett. **278**, 111 (1997).
6. D. Oepts, A.F.G. van der Meer, P.W. van Amersfoort, *Infrared Phys.Technol.* **36**, 297 (1995);
G.M.H. Knippels, R.F.X.A.M. Mols, A.F.G. van der Meer, D. Oepts, P.W. van Amersfoort, *Phys.Rev.Lett.* **75**, 1755 (1995).
7. G. von Helden, I. Holleman, A.J.A. van Roij, G.M.H. Knippels, A.F.G. van der Meer, G. Meijer, *submitted*

Growth of CdSe clusters by laser ablation and mass spectroscopy of ion flow

A. Mazeikis, J.Miskinis, A.Petravicius, M.Robino*, J.Vaitkus, A.Zindulis

Vilnius University, Semiconductor Physics Department
and
Institute of Materials Science and Applied Research, Sauletekio 3-III, 2040 Vilnius, Lithuania
*Institut de Physique et Chemie des Matérioux de Strasbourg, CNRS, F-67087 Strasbourg, France

Abstract. Laser evaporation of CdSe has been analyzed by ion flow mass spectrometry and ultrafine spectrum structure was found. It was proposed the different strength atom bonds and plasma effects were responsible for additional mass spectrum components. The CdSe cluster growth on different substrates was investigated and a clear evidence of substrate structure influence on a cluster distribution was found. The two step mechanism of cluster growth regulation was demonstrated.

INTRODUCTION

Laser evaporation has become popular due to easy regulation of atomic layer formation (the separation of atom deposition and migration processes in material growth) and simplicity of target change. As it is described in different papers [1-3] a quality of laser pulse evaporated layers is very high and comparable with ones produced by MBE growth. The possibilities of regulation of the growth process has become attractive for investigation of different compound cluster growth and attempts of incorporation of clusters into epitaxial layer of different materials [4,5]. But properties of layers depend on the laser excited plasma plume which needs more detailed investigation.

This work presents the data that allow improving the possibilities to control the cluster growth by use of different substrate and test materials. The main attention is paid to CdSe material growth and the peculiarities of laser induced plasma spread were found.

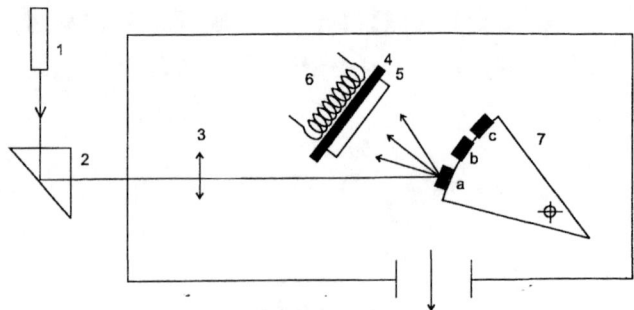

FIGURE 1. Experimental setup for laser deposition: 1 – YAG:Nd laser, 2 – prism, 3 – lens, 4 – substrate, 5 – deposited layer, 6 – heater, 7 – target holder (a,b,c – targets), arrows – the connection to a pump and control chamber.

EXPERIMENT SETUP

The deposition of layers was performed by the laser ablation method. The low frequency Nd:YAG laser second harmonics ($h\nu = 2.33$ eV, pulse duration – 20 ns, power 20 MW/cm^2) was used [4]. The experimental setup is presented in Fig.1.

Two modifications of the evaporation chamber were used: one was placed in sample preparation chamber in Riber LAS 3000 equipment and another was constructed in a high vacuum system with a possibility to change ratio of components in the surface layer by annealing in low pressure Se ambient. The target holder and sample holder with a heater (temperature was controled with thermo-couple) were constructed. The distance between the target and the substrate was 2 cm and the observed front of plume was parallel to the substrate surface. A laser beam was focused on a target in 0.05 mm^2 spot. The single crystals of ZnSe, CdSe and PbS were used as the source for cluster and layer evaporation. During deposition the substrate temperature was between 500 and 560 K and the ambient pressure was 5 μPa. The laser pulse energy was chosen to ensure a slow (less than 0.1 monolayer per pulse) growth of layer to reduce the probability of microcrystallite formation, to increase a migration of deposited atoms to the atomic step edge on the substrate. The experimental number of monolayers has been evaluated from the growth rate of layer thickness for thick layers (their thickness was measured by optical interferometer).

A few types of substrate were used: a) NaCl substrate with terrace-step structure was made by cleavage and the direction of the knife was misoriented by a few degrees from the (100) plane; b) Si substrate was a Si (111) wafer misoriented by 4° towards (110) direction (the wafer was oxidised at 1320 K temperature in wet oxygen, at the end of the procedure in dry oxygen and an area for layer deposition was etched in alkaline solution); c) ZnSe crystal with different orientation blocks.

Thin layers of CdSe on different substrates were deposited. Deposition rate was less than 1 monolayer per pulse. The ion flow was controlled with ion energy

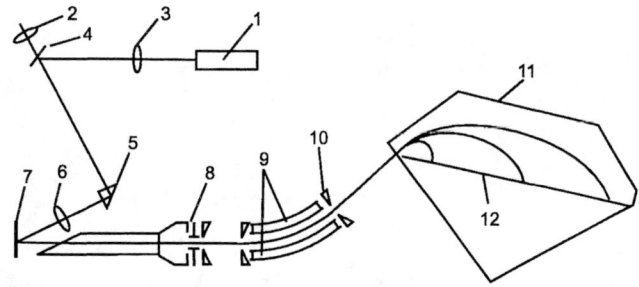

FIGURE 2. Scheme of double focussing mass spectrometer: 1 – laser, 2 – ocular, 3 - 6 – beam focussing elements, 7 – sample, 8 – spherical diode and ion beam forming system, 9 – cylindrical electrostatic analyser (focussing according energy), 10 – block of slits, 11 – magnetic analyser, 12 – ion sensitive plate

mass analyzer EMAL-2, the Mattauch-Herzog system double focussing instrument (Fig.2). This scheme is presented to explain a possibility of ultrafine mass spectrum structure.

An ion mass resolution was enough to distinguish ion mass spectra components separated by a few percents of proton mass). The main attention in this work was paid to Cd and Se ion flow excited by Nd:YAG laser second harmonics.

The CdSe single crystals were grown by Reynolds-Green methods, thin layers were grown by vacuum evaporation and their activation was performed in air in ambient of $CdCl_2$. For investigation of layer thickness effect amorphous Se-Te layers terrace-step shape were formed.

RESULTS AND ANALYSIS

a) CdSe cluster growth.

The initial stages of CdSe growth on ZnSe were investigated by atomic force microscope (AFM). If the ZnSe sample has a vicinal surface the CdSe crystallites revealed the terrace-steps on the surface (Fig.3a).

The two step cluster growth procedure was realised if the Si crystal was used as a substrate and the terraces were formed on surface. The ZnSe crystallites were grown on terraces and the terrace width determined the magnitude of crystallites. If CdSe was deposited on ZnSe crystallites the CdSe nanocrystals grew up. Fig.3b,c illustrate this situation. There are seen the probably coalescent ZnSe crystallites and CdSe column type crystallites. The ZnSe nanocrystallites were grown on Si(111) and Si(100) substrates and different net of nanocrystallites was formed on different substrate. The preliminary results were presented earlier [6]. The distribution of CdSe crystallites demonstrated a rather big migration of atoms on substrate which contains the coalesced crystallites of ZnSe (Fig.3b). The similar CdSe clusters on ZnSe were grown by MBE [7].

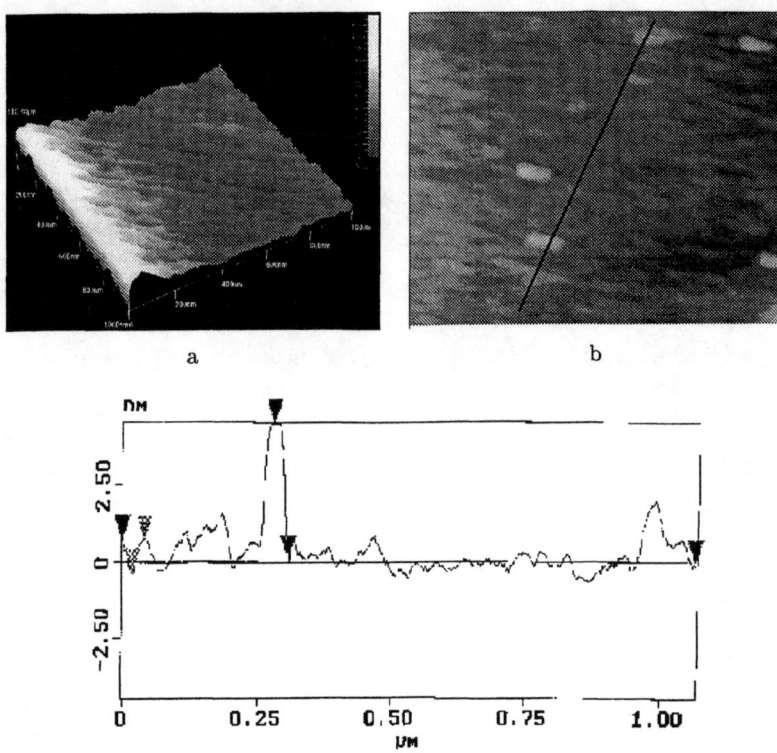

FIGURE 3. CdSe clusters grown on different substrates: a – on a vicinal ZnSe crystal; b – on coalescented ZnSe clusters, grown on a vicinal Si crystal; c – profile of surface in direction of line on previous picture

The results illustrate the possibility to create the conditions for self-organised cluster growth but for analysis of the composition of clusters it is necessary to know the composition of ion flow to the substrate.

b) Cd and Se ion flow.

The regime of laser evaporation was controlled by investigation of the components ion flow. The Cd and Se isotope mass spectra fine structure was investigated. (Preliminary results are presented in [8]). It was found that there is an additional structure in the spectra and the different types of target (CdSe single crystals, polycrystalline layers and different element mixtures) were used to investigate the origin of this ultrafine mass spectra structure.

Se spectra normally were simple if a single crystal were used as a target but if the polycrystalline target were used normally the doublet or triplet spectral lines were observed (Fig.4a). These spectra were measured in a part of sample with polycrystalline blocks. If the measurement were done in the high quality

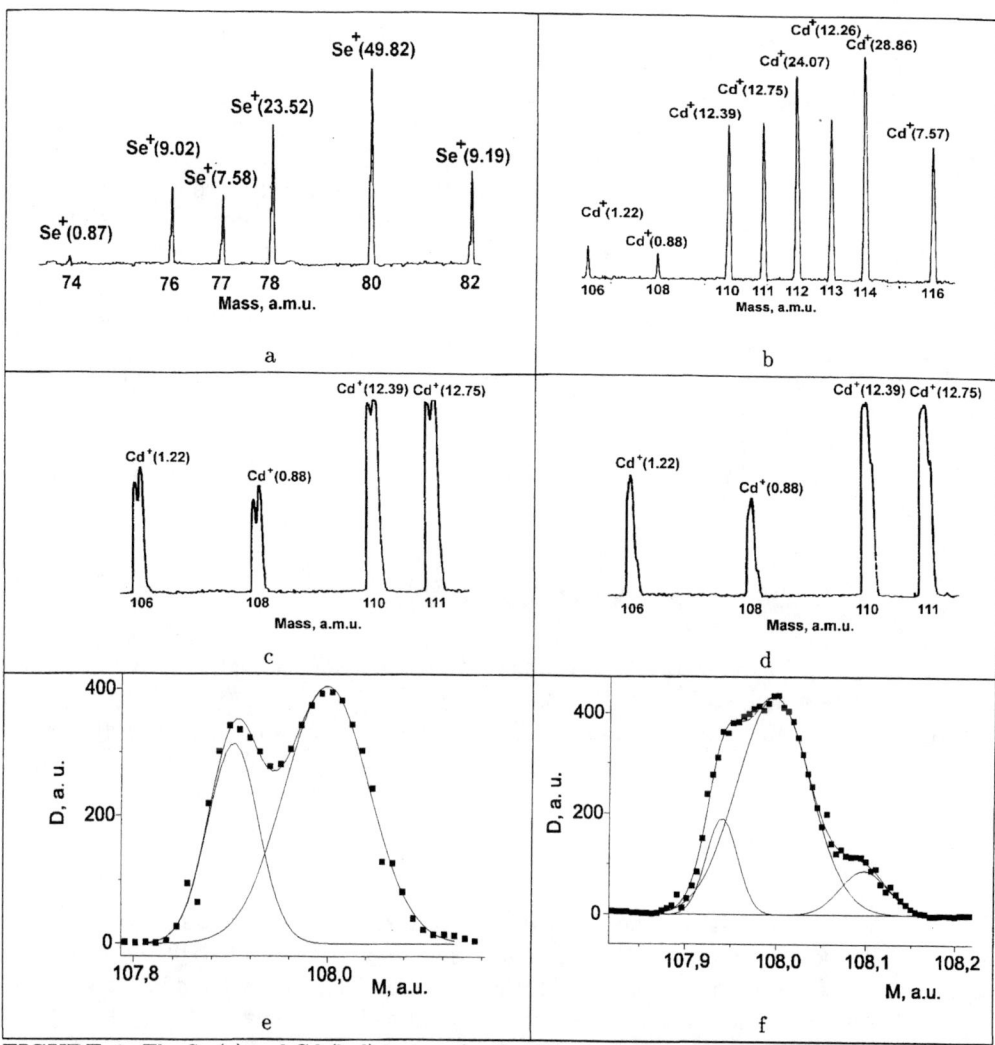

FIGURE 4. The Se (a) and Cd (b-d) isotope family if the ions were evaporated from the CdSe target. (a, b) – crystalline and (c, d, e, f) polycrystalline before and after annealing, respectively

region a lower mass component was absent. The separation of components was approximately 8% of mass unit. For a control of spectra structure nature the different thickness Se-Te-P structure on Al substrate was investigated. The Se mass spectrum was simple and widened if the thin layer was analysed and an additional component appeared in case of a thick layer.

The Cd ion spectra was also single line or doublet and in a few cases triplet type. The additional components were observed in the lower and higher mass regions. The separation of components was approximately 6% of mass unit. A few samples are shown in figure 4b-d.

The origin of observed ultrafine mass spectra structure could be related with two objections: a) the main and higher mass components are explained as a result of different binding of atom in the crystal lattice or existence of another compound binding the same ion (here: CdSe and possible $CdCl_2$ [9]); b) a lower mass component appears due to the reduction of double ionized ion charge [10] and loss of energy in the plasma plume in its initial stage or due to the plasma and thermal spreading of ions from a plume.

CONCLUSIONS

Laser evaporation on terrace step surface permitted nanocluster growth and their dimensions were defined by substrate properties.

Ion flow mass spectrometry allows the control of ions of similar energies to be transmitted.

Acknowledgements

Authors express gratitude to Lithuanian Science and Studies Foundation for support of research and to Dr.V.Jasutis for help in performance of RHEED measurements, one of authors (A.M.) – to Lithuanian Foreign Studies Foundation and Open Society Fund-Lithuania for support of travel to Conference.

REFERENCES

1. Deiss J. L., Chergui A., Koutti L., Loison J.L., Robino M., Grun J. B., *Applied Surface Science*, **86**, pp.149-153 (1995).
2. Mertin M., Offenberg D., An C.W., Wesner D.A., Kreutz E.W., *Applied Surface Science*, **96-98**, pp.842-848 (1996).
3. Chergui A., Deiss J. L., Grun J. B., Loison J. L., Robino M., Beserman R., *Applied Surface Science*, **96-98**, pp.874-880 (1996).
4. Vaitkus J., Kazlauskiene V., Miskinis J., Sinius J., *Mat.Res.Bull.*, **33**, No. 5, pp.711-716 (1998).

5. Vaitkus J., Barauskaite L., Kazlauskiene V., Miskinis J., Sinius J., *2nd International Conference on Photo-excited Processes and Applications*, Abstract Book, pp. 98, CTEH, Tel-Aviv.
6. Vaitkus J., Mazeikis A., Miskinis J., Kazlauskiene V., Robino M., Sinius J., Zindulis A., *Abstracts of IV Baltic Symposium on Atomic Layer Epitaxy*, p.24-25, IP, Tartu (1997).
7. Ko H. C., Park D. C., Kawakami Y., Fujita Sz., Fujita Sh., *Appl.Phys.lett.*, **70**, pp.3278-3281 (1997).
8. Petravicius A., Zindulis A., *Lithuanian J.Physics*, **36**, pp.247-249 (1996).
9. Kutra I., Zindulis A., Sakalas A., *Sov. Phys. - Collections (Alerton Press)*, **3**, pp.105-111 (1980).
10. Bykovskii Yu.A., Vasij'ev N.M., Degtyarenko N.N., Elesin V.F., Laptev I.D., Nevolin V.N., *JETP Lett.*, **15**, pp.217-218 (1972).

SESSION II
ATOMIC RIS

Experimental Studies of Quantum Control in Atoms

D. Charalambidis[‡], N. E. Karapanagioti[*], D. Xenakis and C. Fotakis[‡]

Foundation for Research and Technology–Hellas, Institute of Electronic Structure and Laser, PO Box 1527, GR-711 10 Heraklion, Greece.
[‡]*Also: Physics Department, University of Crete, GR-713 10 Heraklion, Greece.*
[*]*Present Address: Max-Planck-Institut für Quantenoptik, Hans-Kopfermannstr. 1, D-857 48 Garching, Germany*

O. Faucher and E. Hertz

Laboratoire de Physique, Université de Bourgogne, Faculté des Sciences Mirande, BP 138, FR-21004 Dijon, France.

S. Cavalieri, R. Eramo, L. Fini and M. Materazzi

Dipartimento di Fisica and European Laboratory for Non-Linear Spectroscopy (LENS), Universitá di Firenze, Largo E. Fermi 2, I-50125 Firenze

Abstract. We present selected examples of recent experimental results in coherent quantum control of processes and products of laser atom interactions. These include control of ionization and autoionization rates, of light induced atomic structure, ionization branching ratios and frequency up-conversion. Quantum interference of atomic transitions, induced by a bichromatic electromagnetic field, is underlying all the results of this work. Since the quantum interference is induced by the field, its parameters govern control of products and processes. Thus control is demonstrated through the wavelength, intensity and phase of the field.

INTRODUCTION

Interaction of atoms with a bichromatic laser field gives rise to excitations of atoms via multiple indistinguishable pathways. Since the time dependent Schrödinger equation describing the temporal evolution of the system is an amplitude equation, the existence of multiple coherent excitation channels results into intra-atomic quantum interferences that govern excitation probabilities and populations. Constructive or destructive interferences can be classically understood as arising from different dipole moments of the atomic system, induced by the different field combinations setting up the excitation channels. In the simplest case of two interfering excitation channels the two dipole moments may have any amplitude and phase relation and thus result into any degree of constructive or destructive interference. In some cases it proves more convenient to adopt an alternative equivalent quantum mechanical picture to that of interfering amplitudes. This is to consider one of the E/M fields as dressing the atom while the second field is probing the combined system 'field plus atom' which now exhibits new structure and excitation dynamics.

In such excitation schemes involving quantum interference, the characteristics of the process are governed and thus controlled by the interference parameters. The latter depend on and thus may be controlled in the laboratory by the field parameters, wavelength, polarization, intensity and phase. What can be controlled are of course excitation and decay rates but also quantities that traditionally are considered to be determined by the atomic parameters, such as autoionization profiles and ionization branching ratios.

In the present work we review a variety of experiments demonstrating coherent quantum control of ionization including autoionization rates and profiles as well as control of ionization branching ratios to different continua. Although all schemes investigated are based on quantum interference and thus control is coherent, not all of them are phase-dependent as in some the phase is eliminated e.g. through successive absorption and emission of a photon of the same field (Raman type processes). Thus we will classify the experiments into two categories: Phase insensitive and phase sensitive.

PHASE INSENSITIVE QUANTUM COHERENT CONTROL THROUGH LASER INDUCED CONTINUUM STRUCTURE

The basic scheme of laser induced continuum structure (LICS) involves two discrete states $|1\rangle$ and $|2\rangle$ (of which only $|1\rangle$ is initially populated) that are coupled via two electromagnetic fields ω_1, ω_2 to a continuum state $|c\rangle$ (Fig. 1a). ω_2, which will be referred to as the coupling laser, is usually strong in order to induce the continuum structure (embedding one of the discrete states in the continuum), while ω_1 can be a weak probe of the induced structure. The result of the interaction is a quantum mechanical interference which is usually probed through ionization or polarization rotation as a function of one of the laser wavelengths. In the case where the coupling of state $|1\rangle$ with the continuum involves more than one photons the effect can be also probed through the non-linear scattering of this field. LICS can be described in different equivalent ways, which will be briefly outlined further on.

Firstly, it can be described as an interference of different excitation channels of the continuum of the bare atom (Fig. 1a). The two fields also couple the two discrete states to each other through two-photon Raman processes, thus opening different pathways to the continuum. State $|1\rangle$ can ionize directly absorbing one photon of frequency ω_1, or via the state $|2\rangle$ due to the two Raman couplings. The process involves two discrete states coupled to each other and to the same continuum satisfying the energy balance condition $E_{|1\rangle} + \hbar\omega_1 = E_{|2\rangle} + \hbar\omega_2$ and as such is closely related to an

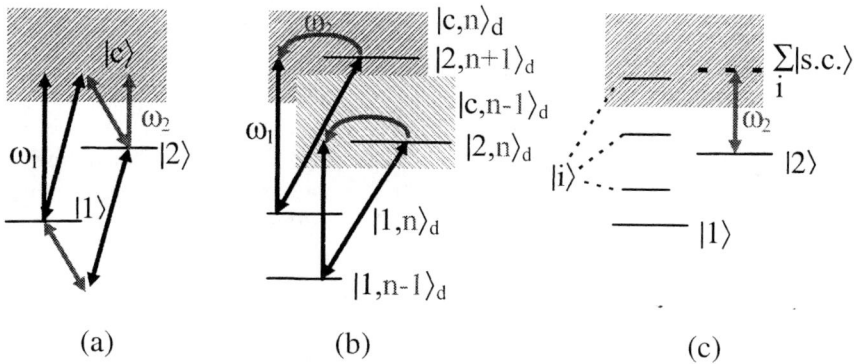

Figure 1. Three equivalent represenations of LICS.

autoionization process, even if different types of couplings underlie the two processes. The full analogy between LICS and autoionization has been shown by Dai et al. (1). Consequently the interference process manifests itself as an asymmetric autoionizing-like resonance, the shape of which is determined, under low intensity and single rate approximation conditions by a q parameter equivalent to the Fano parameter in autoionization:

$$q = \frac{\text{Re}\,\Omega_{12}}{\text{Im}\,\Omega_{12}} = \frac{M_{12}}{\mu_{1c}\mu_{2c}}, \quad M_{12} = \sum_{l} \frac{\mu_{1l}\mu_{l2}}{-\omega_2 + \omega_{1l}} + P\!\!\int \frac{\mu_{1l}\mu_{l2}}{\omega_2 + \omega_{2l}}d\omega_c \quad (1)$$

where M_{12} is the two photon transition moment between states $|1\rangle$ and $|2\rangle$, μ_{1c}, μ_{2c} are the bound-free transition moments, Ω_{12} is the two photon Rabi frequency and $\sum\!\!\!\!\int_l$ implies summation over the discrete and integration over the continuum part of the spectrum.

Due to incoherent channels involved, such as ionization of the system from state $|1\rangle$ via multiphoton absorption of the field ω_2 or ionization of state $|2\rangle$ via ω_1, a practically constant ionization background will contribute to the ionization spectrum. Thus the ionization probability per unit time will be:

$$\partial P_{ion}/\partial t \sim [1 + kf(q,\delta)] \quad (2)$$

where P_{ion} is the ionization probability, k is a laser intensity dependent contrast parameter involving the widths of states $|1\rangle$, $|2\rangle$ and the laser bandwidths and δ is essentially the detuning $(E_1+\hbar\omega_1) - (E_2+\hbar\omega_2)$. If LICS is probed through ionization the degree of the asymmetry is given by the q parameter and the observability of the structure depends on the contrast parameter k. At higher laser intensities and conditions in which the single rate approximation breaks down, the shape of the LICS resonance depends on the laser intensity.

An equivalent way of describing the same effect is utilizing the dressed atom picture (Fig 1b). The strong filed ω_2 dresses the system. The atom plus field, after introduction of the interaction with the field, exhibits a ladder of new eigenstates $|i, n\pm m\rangle_d$, n being the number of photons of the field ω_2 and m and integer. The $|2, n\pm m+1\rangle_d$ discrete states are now embedded in the $|c, n\pm m\rangle_d$ continua in which they can decay radiatively. ω_1 couples the $|1, n\pm m\rangle_d$ states with the $|2, n\pm m+1\rangle_d$ discrete and $|c, n\pm m\rangle_d$ continuum state, thus resulting in an autoionization like process.

Adopting a.c. Stark splitting terminology (Fig. 1c), it is worth noting that the dressed atom-photon states that are embedded in the continuum can be understood as the sum of the ac Stark split components $|s.c.\rangle$ resulting from the coupling of state $|2\rangle$ with all allowed states $|i\rangle$ of the system. This picture may be more convenient in describing LICS in structured continua, e.g. in the vicinity of autoionizing states. In the particular case in which one of these states is near resonant, its split component dominates and the sum may reduce to one term. If the detuning is less than the width of the states involved, the a.c. Stark splitting of the dressed autoionizing state may give rise to a window resonance in the spectrum due to the destructive interference of the two Autler-Townes components. In other words, the coherent superposition of the two states prepared by the coupling laser produces a dipole moment that has the same amplitude and is out of phase with the dipole moment induced by the probing laser.

As far as field phase is concerned, LICS is immune to the relative phases of the two electromagnetic fields. As the process involves absorption and emission of the frequency ω_2 the phase of the corresponding wave vanishes in the interaction.

Some years ago we have reported the first observation of LICS in the smooth (unstructured) continuum of the Na atom (2). The rate of ionization has been controlled through the detuning δ, resulting into an asymmetric ionization structure that manifested the destructive and constructive interference at different detunings. Laser intensity effects have also been studied in these works. The earlier experiment has been verified in an advanced version employing energy resolved photoelectron spectroscopy that has shown a clean asymmetric Fano type resonance (3).

A vast variety of experiments has followed these first successful observations. LICS in three photon excited structured continua has resulted into several interesting observations in atomic Ca. In these schemes LICS is in the vicinity of an autoionizing state (AIS). The virtual state of the Raman process that goes through the continuum is now lying near or is replaced by a third discrete state $|3\rangle$, the AIS. This third state plays a dominating role and controls the modification characteristics. Experiments have been performed in a Ca effusive beam as well as in Ca vapor in a heat pipe in a three photon ionization scheme (4,5). Ionization rates have been demonstrated to be controlled through detuning, laser intensity (as has been interpreted due to a.c. Stark shifting of the states and thus change of the detuning) and laser polarization (5). The three photon excitation of the continuum (or AIS) allowed the study of the effect also in third harmonic generation (THG) (6). THG could be controlled through the detuning and the dressing laser intensity. Enhancement or supression of the harmonic by a factor of up to three could be observed. These experiments have also established the relation of LICS with a.c. Stark splitting (7), which due to the dominance of the near resonant AIS provided a convenient way for the interpretation of the results.

Very closely related to the experiments employing Λ-type coupling schemes of LICS are investigations on the resonant mixing of two Mg AIS ($3p^2$ ^1S and $3p3d$ ^1P) probed by a second electromagnetic field from the ground state of the atom. The experiments have been performed in a Mg effusive beam and have shown strong suppression of autoionization due to coherent population trapping in the ground state (8). Furthermore, by detuning far from resonant coupling, a clear 'above threshold' LICS has been observed caused by the 're-embedding' of the 3p3d ^{1}P AIS of Mg at a position in the continuum lower by the photon energy of the coupling laser. High resolution photoelectron spectra have verified (9) that the observed structure in the total ionization spectrum is due to an 'above threshold' LICS and not to the autoionization decay of the resonant 3p3d ^{1}P state. This result has been verified by theoretical calculations by Bachau et al. (10), who have also demonstrated stabilization of ionization in the case of the resonant coupling due to the coherent population trapping. (9)

When the atomic system is radiatively decaying in more that one continua, structure can be induced in all the continua by dressing them with a strong laser field, the latter establishing a coupling between the continua and an atomic bound state. This structure can then be probed through a second field from e.g. the ground state. For the different continua there are different interfering channels causing ionization and thus LICS can be observed separately for each continuum by employing energy resolved photoelectron spectroscopy. Since the coupling of the ground and the excited bound state to each one of the continua is in general different, the LICS profiles are expected not to be the same for each continuum. Thus the ratio of the ionization products, namely electrons resulting from the decay into the different continua is expected to be modified through the LICS process and variable in the vicinity of the induced structure.

Recently (11) we have demostrated control of ionization product ratios in atomic Xe. Since the experiment is described in another contribution of this volume we only shortly summarize the results of this study. Ionization of the Xe ground state is in the two electronic continua, that correspond to the two fine structure levels of the ground state of the Xe$^+$ ion. Both continua are dressed through a second laser that couples the $5p^5 10p[\frac{1}{2}]_0$ state with each of them (see figure 2a). The dressed $^2P_{1/2}$ continuum, probed from the atomic ground state through three photon absorption, exhibit a clear induced structure (window resonance), while the $^2P_{3/2}$ continuum possess no observable structure under the present experimental conditions, as manifested through

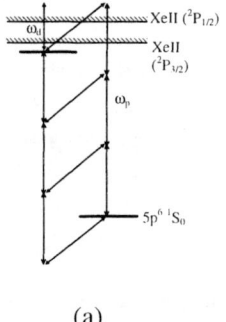

(a)

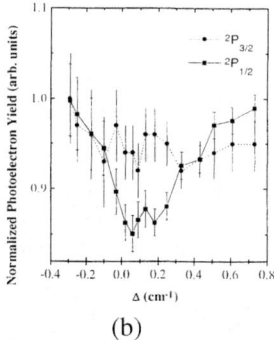

(b)

Figure 2. Control of ionization branching. (a) The coupling scheme, (b) normalized photoelectron yields versus detuning $\Delta = E_{5p^6} + 3\hbar\omega_p - E_{5p^5 10p} - \hbar\omega_d$

high resolution photoelectron spectra. The normalized photoelectron yields for the decay into the two continua versus detuning $\Delta = E_{5p^6} + 3\hbar\omega_p - E_{5p^510p} - \hbar\omega_d$ are shown in figure 2b where E_{5p^6} is the ground state energy and E_{5p^510p} is the energy of the $5p^510p[½]_0$ state. Hence the modified ionization ratio is controlled by the wavelength of the dressing field. In a similar coupling scheme control of the photofragmentation product ratio has been also demonstrated by LICS in dissociative continua (12). Furthermore control of ionization branching ratios has been achieved in atomic Ba through a specific case of interference of two (one plus one) photon ionization channels, each of them being near resonant with two different excited states of the system. (13)

PHASE SENSITIVE QUANTUM CONTROL

Interaction of an atomic system with a laser field and one of its harmonics may lead to the excitation of a bound or continuum state via two different channels with different degree of non-linearity, one of which involves a harmonic photon. Since the two fields are coherent with correlated phases, the excitation rate will be proportional to the square of the modulus of the sum of the two excitation amplitudes. The interference cross term and thus the excitation rate will modulate with the relative phase of the two fields.

We have recently demonstrated (14) control of the ionization rate in a four photon resonant (with the 5p[5/2]$_2$ state) five photon ionization of Kr. The 5p[5/2]$_2$ state is excited via (i) four photon absorption of the fundamental laser frequency ($4\omega_1+\omega_1$ ionization) and (ii) one third harmonic and one fundamental frequency photon absorption ($\omega_3+\omega_1+\omega_1$ ionization), a particular aspect of the scheme being that interference occurs at a virtual level.

The excitation probability of the 5p[5/2]$_2$ state is:

$$W \propto [\mu^{(4)} E_{10}^4]^2 + [\mu^{(2)} E_{10} E_{30}]^2 + 2\mu^{(4)} \mu^{(2)} E_{10}^5 E_{30} \cos(\vartheta_3 - 3\vartheta_1) \qquad (3)$$

where $\mu^{(n)}$ is the effective n-photon (4 visible, and 1 visible + 1 VUV photon) electric dipole moment of the transition between the ground and the 5p[5/2]$_2$ state, E_{l0} electric field amplitude of the fundamental ($l=1$) and the third harmonic ($l=3$) and $\Delta\vartheta = \vartheta_3 - 3\vartheta_1$ where ϑ_1, ϑ_3 the phase of the fundamental and third harmonic waves. The excitation probability oscillates, as in the case of three- photon excitation, as $\cos(\Delta\vartheta)$. This is expected since the interference in the present scheme occurs at the three photon level i.e. in the excitation of a virtual state and the absorption of the fourth fundamental photon does not affect the interference process as it adds the same phase and hence no phase difference in the two interfering channels and correspondingly for the harmonic.

By varying the relative phase of the two electromagnetic fields in a gas phase shifter a large modulation has been observed which is due to the chosen non-linear excitation scheme. The modulation depth, defined as $\dfrac{I_{max} - I_{min}}{1/2(I_{max} + I_{min})}$, where I_{max} is

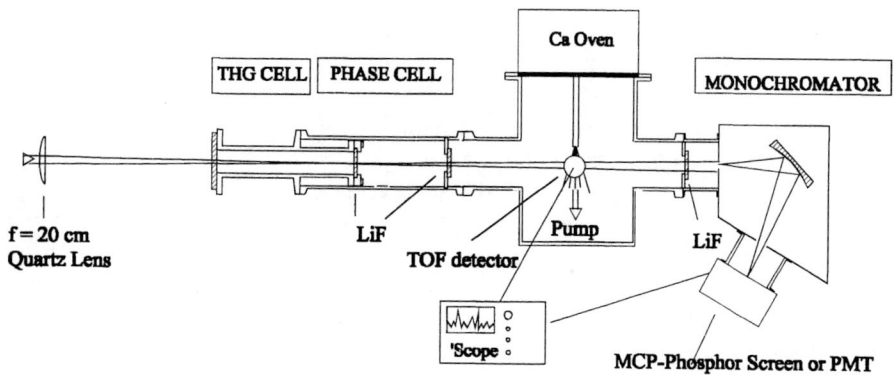

Figure 3. Experimental Set-up in the phase control experiment of autoionization in Ca

the maximum and I_{min} the minimum ion signal of the modulation, has a maximum value close to 1.0 in the present experiment. In a similar scheme phase control of harmonic generation has been demonstrated (15). Non-resonant phase control of ionization has also been demonstrated from an excited Na state (16).

The demonstrated interference at a virtual level is of importance for schemes that involve excitation of highly excited bound or continuum states, such as excitation of multiple continua aiming at the phase control of ionization branching ratios. Currently such an experiment is in progress in the above scheme by employing energy resolved photoelectron spectroscopy. It should be noted that phase control of molecular dissociation branching ratios has been achieved (17).

Recently we have been able to demonstrate field phase dependent autoionization in atomic Ca. Excitation is in the region of the $4p7s[½]^0{}_1$ doubly excited state autoionizing state and occurs from the atomic ground state through a three photon channel ($3\hbar\omega$) and a single photon channel ($\hbar\omega_3$), ω_3 being the third harmonic of ω (Fig. 4a).

Due to the given excitation cross sections of the scheme, the energies of the fundamental and third harmonic, as well as the atomic number density available, a non

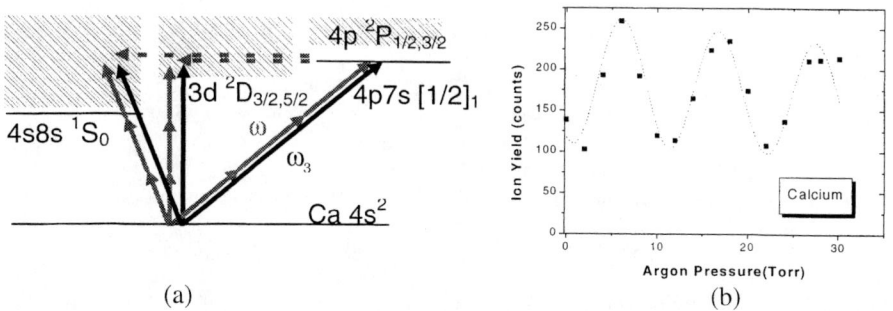

Figure 4. Phase control of autoionization rate. (a) Coupling scheme, (b) Ion yield vs relative phase

conventional (14, 18) phase control experimental set-up (shown in fig. 3) had to be used. This employed an unfocused beam ionization geometry (by the two diverging fields) and an ion counting technique for the data acquisition procedure. The autoionization rate exhibits a sinusoidal modulation as a function of the relative phase of the two excitation fields, shown in figure 4b.

This result is the initial step in the achievement of control of an autoionizing lineshape, which has been theoretically predicted (19) but has not been experimentally demonstrated so far. Furthermore, it demonstrates the feasibility of phase control in unfocused geometries with their corresponding advantages for applications of phase control in a large interaction volume and hence large number of species, e.g. for the control of chemical reactions.

ACKNOWLEDGMENTS

This work has been carried out in the Ultraviolet Laser Facility and the LENS facility. (TMR Program, contract number ERBFMGECT950017).

REFERENCES

1. Bo-nian Dai, and P. Lambropoulos, *Phys. Rev. A* **36**, 5205(1987)
2. Y.L. Shao, D. Charalambidis, C. Fotakis, J. Zhang, and P. Lambropoulos *Phys. Rev. Lett.* **67**, 3669 (1991); S. Cavalieri, F.S. Pavone, and C. Matera, *Phys. Rev. Lett.* **67**, 3673 (1991); S. Cavalieri, C. Matera, F.S. Pavone, J. Zhang, P. Lambropoulos, and T. Nakajima, *Phys. Rev. A* **47**, 4219 (1993)
3. S. Cavalieri, R., Eramo, and L. Fini, *J. Phys. B* **28**, 1793 (1995)
4. O.Faucher, D. Charalambidis, C. Fotakis, J. Zhang,, and P. Lambropoulos, *Phys. Rev. Lett.* **70**, 3004 (1993)
5. O.Faucher, Y.L.Shao, D.Charalambidis, and C. Fotakis, *Phys. Rev. A* **50**, 641(1994)
6. O.Faucher, Y.L.Shao, and D. Charalambidis, *J. Phys. B* **26**, L309(1993)
7. Y.L. Shao, O. Faucher, J. Zhang and D. Charalambidis, *J. Phys.* B**28**, 755 (1995)
8. N.E.Karapanagioti, O.Faucher, Y.L.Shao, D.Charalambidis, H.Bachau, and E. Cormier, *Phys. Rev. Lett.* **74**, 2431 (1995)
9. N.E.Karapanagioti, D.Charalambidis, C.J.G.J.Uiterwaal, C.Fotakis, H.Bachau, I.Sánchez, and E.Cormier, *Phys. Rev. A*.**53**, 2587 (1995)
10. H.Bachau, and I.Sánchez, *Z. Phys. D* **38**, 19 (1996)
11. S. Cavalieri, R. Eramo, L. Fini, M. Materazzi, O.Faucher and D. Charalambidis, *Phys. Rev. A***57**, 2915 (1998)
12. A. Shnitman, I. Sofer, I. Golub, A. Yogev M. Shapiro, Z. Chen, and P. Brumer, *Phys. Rev. Lett.* **76**, 2886 (1996)
13. Feng Wang, Ce Chen, and D.S. Elliott *Phys. Rev. Lett.* **77**, 2416 (1996)
14. N.E. Karapanagioti, D. Xenakis, D. Charalambidis and C. Fotakis *J. Phys. B* **29**, 3599 (1996)
15. D. Xenakis., N. E. Karapanagioti, C. Fotakis and D. Charalambidis, *Optics Commun.* **152**, 83 (1998)
16. S. Cavalieri, R. Eramo, L. Fini, *Phys. Rev. A* **55**, 2941 (1997)
17. Langchi Zhu, V. Kleiman, Xiaonong Li, Shao-Ping Lu, K. Trentelman, Yongjin Xie and R. J. Gordon, *Science* **270**, 77 (1995)
18. Chen Ce, Yi-Yian Yin and D. S. Elliott, *Phys. Rev. Lett.* **64**, 507 (1990); Chen Ce and D. S. Elliott, *Phys. Rev. Lett.* **65**, 1737 (1990)
19. T. Nakajima and P. Lambropoulos, *Phys. Rev. Lett.* **70** 1081 (1993); T. Nakajima and P. Lambropoulos, *Phys. Rev .A* **50** 595 (1994)

Photoionisation of laser-cooled cesium 6 $^2P_{3/2}$ atoms

F. Fuso, O. Maragò[1], D. Ciampini, E. Arimondo,
C. Gabbanini* and S.T. Manson[†]

*Istituto Nazionale Fisica della Materia, Dipartimento di Fisica, Università di Pisa
Piazza Torricelli 2, I-56126 Pisa, Italy*
* *Istituto Fisica Atomica e Molecolare del CNR, Via del Giardino 7, I-56127 Pisa, Italy*
[†] *Department of Physics and Astronomy, Georgia State University, Atlanta, GA 30303, USA*

Abstract. We have investigated photoionisation of laser-cooled cesium atoms excited to the 6 $^2P_{3/2}$ state by irradiating a cesium magneto-optical trap with a beam from either an Ar^+ laser or a frequency doubled diode laser. From the analysis of time resolved and steady state fluorescence, we have derived the ionisation cross section on the basis of a simple model of the trapping dynamics in the presence of an ionising beam.

INTRODUCTION

In the last few years, laser-cooling has been established as a powerful technique for the preparation of cold and dense atomic samples, which, thanks to their low temperature, exhibit interesting spectroscopic features. So far, a large number of spectroscopic experiments on laser-cooled samples have been devoted to the investigation of diatomic molecular systems, mainly through photoassociation spectroscopy [1]. In this paper we report on photoionisation of laser-cooled cesium atoms excited to the 6 $^2P_{3/2}$ state following interaction with an ionising laser beam. The trapping beams are used to populate the excited state in a controlled way. By comparing the fluorescence emission from the MOT with or without the ionising beam we could derive the ionisation cross section for cesium 6 $^2P_{3/2}$ atoms. Measurements were carried out by using different wavelengths of the ionising beam, in the range λ_p = 501–422 nm, corresponding to an electron excess energy $\Delta E \sim$ 47–511 meV.

Photoionisation of laser-cooled atoms was first introduced by Dinneen et al. [2], who used a rubidium magneto-optical trap (MOT) loaded from an atomic beam. More recently, Gabbanini et al. [3] exploited the same method with a rubidium MOT loaded from the background, a system similar to that used in the present experiment. Ionisation of excited atoms in a MOT leads to strong modifications in the trapping dynamics and in the steady-state density of trapped atoms. Both quantities could be easily monitored in our experiment and provided two independent methods to derive the ionisation cross-section on the basis of a simple model of trapping dynamics. Results were compared to theoretical and experimental data available in the literature

---

[1] Now at: Clarendon Laboratory, Oxford University, Oxford, Great Britain

for thermal samples. The agreement is satisfactory, though our model could not fully explain some experimental findings, possibly arising from the peculiar behaviour of ionisation and recombination processes in cold samples.

EXPERIMENTAL

The cesium MOT employed in the present experiment, already described in detail elsewhere [4], exploits three pairs of counterpropagating σ^+/σ^- beams and a quadrupolar magnetic field produced by a pair of anti-Helmoltz coils. The trapping beams were tuned on the red side (detuning $\Delta\nu_T$) of the $6\ ^2S_{1/2}$ (F = 4) → $6\ ^2P_{3/2}$ (F' = 5) transition around 852.1 nm. In addition to the trapping beams, repumping radiation tuned on the $6\ ^2S_{1/2}$ (F = 3) → $6\ ^2P_{3/2}$ (F' = 4) was used to prevent loss of atoms due to off-resonant excitation of the F' = 4 state, this state lying close to the F' = 5 ($\Delta\nu_{hf} \sim$ 251 MHz). All radiation needed for MOT operation was produced by a diode laser system working in a master-slave configuration. Master lasers for both trapping and repumping were coupled to external cavities in order to reduce the linewidth and to achieve a continuous tunability in the range of interest. They were locked to the desired hyperfine cesium transition by Zeeman-lock technique, and precise tuning was accomplished by acousto-optic modulators (AOMs) which were also used to switch on/off the beams. Cesium vapour was contained in a pyrex UHV cell connected to a liquid cesium reservoir, kept at room temperature, and to a 6 l/s ion pump. The background pressure in the cell was $< 10^{-9}$ mbar. Typical operating conditions for the MOT were the following: total intensity of the laser beams $I_T \sim 170$ mW/cm^2 and $I_R \sim$ 30 mW/cm^2 for the trapping and repumping radiation, respectively, magnetic field gradient ~0.09 T/m. A large detuning was chosen for the trapping beams ($\Delta\nu_T = -3\Gamma$, Γ being the natural linewidth of the transition) in order to reduce the MOT density and to prevent undesirable effects due to collisions between trapped atoms.

Diagnostics of the MOT operation was accomplished by using a CCD camera to monitor the shape of the atomic cloud, and a calibrated photodiode to measure the fluorescence emission flux from the MOT. By carefully choosing the intensity for each trapping beams to balance the magnetic field gradient asymmetries, an almost spherical shape was realised, with a typical radius ~0.3 mm. The number of trapped atoms was ~10^7, corresponding to a density around 8×10^{10} cm^{-3}. The temperature was not measured during the present experiment, but a comparison with previous analysis of the same system suggests a typical temperature of a few μK.

Photoionisation of laser-cooled atoms was accomplished by shining an additional laser beam on the MOT. As the photoionising source, different lines from an Ar$^+$ laser (maximum power 100 mW in single-line operation) were used. In addition, a frequency-doubled diode laser was used, operating at 422 nm with a maximum power ~10 mW. Figure 1 shows a simplified sketch of the cesium energy levels involved in the experiment, along with a list of the photoionising wavelengths, λ_p, used in the experiment with the corresponding electron excess energy, ΔE. The photoionising beam was mildly focused at the centre of the MOT, and its power and dimension were carefully measured. Since the waist of the ionising beam, 3.3 mm, was much larger than the MOT radius, the intensity of the ionising beam, I_p, was considered uniform all over the MOT and corresponding to the peak intensity within the Gaussian distribution. Measurements were performed at various power of the ionising beam, the corresponding maximum I_p value during the experiment being ≤ 1 W/cm^2 for the strongest Ar$^+$ laser lines.

During the experiment, the MOT was continuously switched on/off by sending a pulse train to the AOMs controlling trapping and repumping beams. Trapping dynamics with or without the ionising beam was monitored by recording the signal from the calibrated photodiode with a digital oscilloscope triggered by the same pulse train. Measurements were carried out by averaging over ~100 switching cycles.

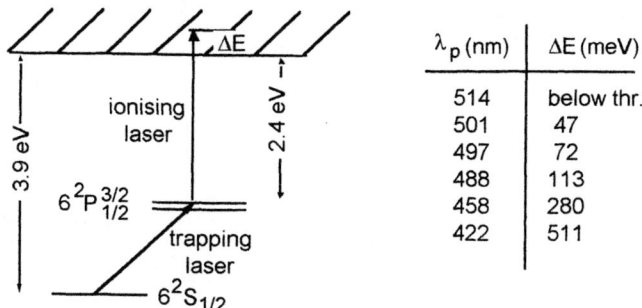

FIGURE 1. Sketch of the cesium energy level diagram and list of the photoionising laser wavelengths λ_p used in the experiment along with the corresponding electron excess energies ΔE.

RESULTS AND DISCUSSION

The typical trap dynamics without photoionising laser, monitored through the time behaviour of the fluorescence emission from the MOT, is reported in Fig. 2(a). According to literature [5], the rate equation governing the number of trapped atoms is:

$$\frac{dN}{dt} = L - \gamma N - \gamma'(N)N, \qquad (1)$$

where L is the loading rate, γN and $\gamma'(N)N$ are the loss rates due to collisions with atoms belonging to the background and to the MOT, respectively. Due to the low density of our MOT, collisions with background are the dominant loss channel, and we can neglect the dependence on N of γ', obtaining:

$$\frac{dN}{dt} = L - RN, \qquad (2)$$

where R is the total loss term. Solution of Eq. 2 gives an exponential growth $N(t) = N_0 [1 - \exp(-Rt)]$, with a steady-state value $N_0 = L/R$.

The experimental data acquired without the ionising beam are well described by Eq. 2. A similar behaviour was obtained by irradiating the MOT with the Ar^+ laser line below threshold (λ_p = 514 nm). This finding confirms, within the sensitivity of our experiment, that the results discussed in the following are due to single photon ionisation of cesium 6 $^2P_{3/2}$ atoms, ruling out, in agreement with the low intensity of the ionising beam, multiphoton non-resonant ionisation of both ground state and excited atoms. In addition, by observing the steady-state fluorescence flux and the CCD image of the MOT, it was checked that the presence of the ionising beam did not produce any additional observable modification to the MOT operation.

On the contrary, irradiation of the MOT with laser lines above ionisation threshold led to strong modifications in both the time behaviour and the steady-state fluorescence from the MOT, as shown in Fig. 2(b). Photoionisation is expected to produce two main effects in the MOT operation, namely the opening of a new loss channel due to ionisation of trapped atoms, and a decrease of the loading rate due to ionisation of background atoms.

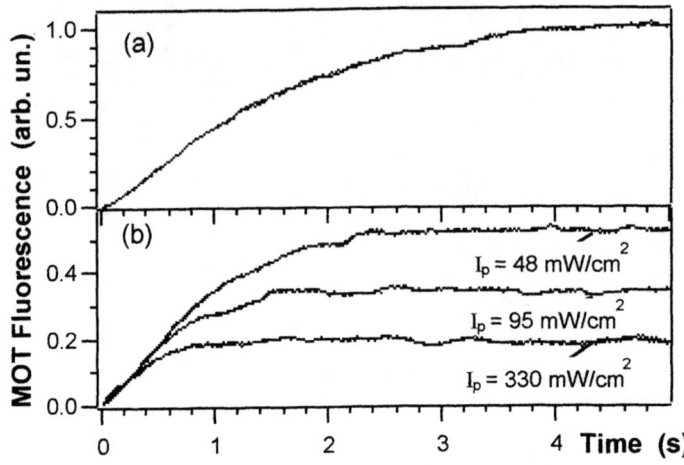

FIGURE 2. Time behaviour of MOT fluorescence without (a) and with (b) photoionising laser beam ($\lambda_p =$ 497 nm) at different intensities I_p.

The additional loss rate for photoionisation, R_p, can be expressed, assuming a single photon unsaturated process, as:

$$R_p = p_e \frac{\sigma_p I_p \lambda_p}{hc}, \qquad (3)$$

where p_e is the fraction of excited atoms in the MOT, σ_p is the ionisation cross section, I_p and λ_p are parameters (intensity and wavelength) of the ionising beam. The value of p_e can be evaluated theoretically [6] as a function of $\Delta\nu_T$ and of the Rabi frequency of the trapping laser for the operating conditions of our MOT, giving a value $p_e = 0.3$ with a relative uncertainty around 10 %.

The effects on the loading rate, which are expected to be less important, can be described according to a simplified model of the trapping mechanism, presented in more detail in [7], that accounts for ionisation of excited atoms in the background acting during the typical time, τ_c, during which atoms are captured from the background. According to this model, the loading rate L_p in the presence of the photoionising beam depends exponentially on the ionisation rate R_p, and on the capture time τ_c:

$$L_p = L \exp(-R_p \tau_c). \qquad (4)$$

By including the effects of Eqs. 3, 4, the following time behaviour is found:

$$N(t) = N_s[1 - \exp(-(R + R_p)t)], \qquad (5)$$

with a steady-state value:

$$N_s = N_0 \frac{\exp(-R_p \tau_c)}{1 + R_p/R}. \qquad (6)$$

Both the time behaviour N(t) and the steady-state value N_s (or, better, the ratio N_s/N_o of the steady-state values with or without the ionising beam) could be easily determined experimentally from MOT fluorescence. The value of R_p as a function of the ionising laser intensity I_p and wavelength λ_p was then determined through two different methods, the first one based on a best fit procedure of the time resolved data to Eq. 5, and the second one based on a best fit procedure of steady-state values to Eq. 6. It must be noted that, in the latter case, τ_c is a fitting parameter. From the determination of R_p at different photoionising laser intensities I_p, the cross section σ_p could be derived exploiting the linear dependence of Eq. 3.

Good agreement between experimental results and the best fit was found for both methods in all sets of data [7]. The values of the cross section σ_p obtained according to both methods are shown in Fig. 3(a) as a function of the electron excess energy ΔE, the main source of error being the uncertainty in the evaluation of the parameter p_e entering Eq. 3. Previous experimental results by Nygaard et al. [8], who investigated ionisation of a cesium beam by a discharge lamp, are reported in the same graph. The agreement with those data is rather poor, our results being consistently smaller (by approximately 30 %). For comparison, theoretical evaluations of the cross section in the energy range of our interest are also reported obtained through different techniques, namely semiempirical methods by Weisheit [9] and Norcross [10], and *ab-initio* calculations based on Hartree-Fock [11] and Hartree-Slater [12] methods. Our results are well reproduced by the Hartree-Slater calculation, the agreement being within the experimental uncertainties except for the point corresponding to the largest ΔE, probably due to the unsatisfactory stability in the intensity of the frequency doubled diode laser source employed for that point.

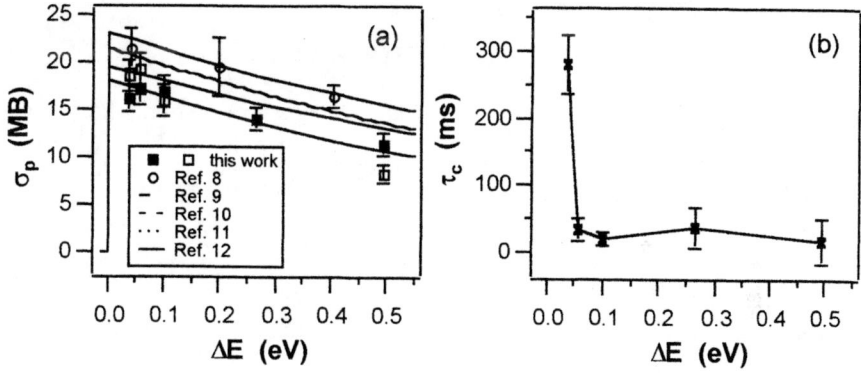

FIGURE 3. Ionisation cross section σ_p (a) and τ_c values (b) as a function of the electron excess energy ΔE obtained as described in the text. Filled and open squares in panel (a) refer to data obtained from MOT fluorescence dynamics and steady-state measurements, respectively.

As already pointed out, our experimental results allowed us also to estimate the time τ_c appearing in Eq. 6. Figure 3(b) reports τ_c as a function of ΔE. According to our model, τ_c should depend only on the trap dynamics, being related to the time needed for atoms in the background to be captured. In the typical conditions of our experiment, τ_c is expected to be in the range 20–50 ms [13], but, in any case, it should be independent of ΔE. On the contrary, from our measurements it turns out to be strongly energy dependent, reaching the rather large value $\tau_c > 250$ ms for the minimum ΔE explored in the present work. In other words, it appears that an energy-dependent process plays some role in our ionisation experiment. Such unexpected findings might be related to the peculiar behaviour of recombination processes occurring between cold ions and (relatively) cold electrons. Further work will be carried out to shed more light on this phenomenon from both theoretical and experimental points of view, by implementing a more sophisticated model and by investigating the behaviour of the system as a function of the geometry of the ionising laser beam.

CONCLUSIONS

We have carried out a photoionisation experiment on laser-cooled cesium atoms. Despite the simplicity of the excitation and ionisation scheme and of the detection technique, where neither absolute calibration procedures nor evaluation of the collection efficiency were required, we obtained accurate measurements of the cross section in the excess energy range explored in the experiment. Two independent methods were used to find cross section values, with good self consistency. This fact, along with satisfactory agreement with theoretical predictions, demonstrate the validity of our method in providing quantitative measurements of ionisation cross sections. On the other hand, our study indicates that recombination processes involving cold ions are an important issue, deserving future more detailed theoretical and experimental investigations.

ACKNOWLEDGMENTS

N. Beverini and C. Notini are gratefully acknowledged for providing the frequency doubled diode laser source used in the experiment. The investigation was supported by EC TMR Grant No. ERB4061PL950044, by Progetto Integrato IFAM/Università di Pisa, by INFM BEC-PRA Project, and by NSF Grant No. PHYS9722501.

REFERENCES

1. See, for instance, A. Fioretti et al., *Phys. Rev. Lett.* **80**, 4402 (1998); A. Fioretti et al., this volume.
2. T.P. Dinneen et al., *Opt. Lett.* **17**, 1706 (1992).
3. C. Gabbanini et al., *Opt. Commun.* **141**, 25 (1997).
4. A. Fioretti et al., *Phys. Rev. A* **55**, R3999 (1997).
5. P. Feng et al., *Phys. Rev. A* **47**, R3495 (1993).
6. P.D. Lett et al., *J. Opt. Soc. Am. B* **6**, 2084 (1989); C.G. Townsend et al., *Phys. Rev. A* **52**, 1423 (1995).
7. O. Maragò et al., *Phys. Rev. A* **57**, R4110 (1998).
8. K.J. Nygaard et al., *Phys. Rev. A* **12**, 1440 (1975).
9. J.C. Weisheit, *J. Quant. Spectrosc. Radiat. Transf.* **12**, 1241 (1972).
10. D.W. Norcross, *Phys. Rev. A* **7**, 606 (1973).
11. A.Z. Msezane, *J. Phys. B* **16**, L489 (1983).
12. J. Lahiri and S.T. Manson, *Phys. Rev. A* **33**, 3151 (1986).
13. A. Steane and C. Foot, *Europhys. Lett.* **14**, 231 (1991).

High–Lying Bound Rydberg States of Excited Hg(6s6p $^3P_1$) Atoms from Two–Color Resonance Ionization Mass Spectroscopy

Peter Bisling, Jan Dederichs, Bernd Neidhart, and Claus Weitkamp

GKSS–Forschungszentrum Geesthacht GmbH, Postfach 11 60, D–21494 Geesthacht, Germany

Abstract. Mercury isotopes are investigated with two–color resonance ionization mass spectroscopy (RIMS). Isotope shifts, hyperfine structure splittings, and the lifetime of the intermediate 6s6p $^3P_1$ state are determined by RIMS. Ion yields at the threshold region in various static electric fields are measured in order to determine an extrapolated ionization energy value at zero field strength. New energy values for high–lying bound 6s nd ^3D (21 < n < 70) Rydberg states of Hg isotopes are found under field–free conditions and with delayed, pulsed electric field ionization. To the author' knowledge this is the first time that isotopic effects on the ionization energy are deduced from the convergence limit of the Rydberg series.

INTRODUCTION

A performance criterion for analytical methods is the detection of mercury at low concentration levels. One of these methods is resonance ionization spectrometry (1). Spectroscopic data for multiphoton ionization schemes involving, in particular, Rydberg states are useful for further improvements in this field. Resonance ionization mass spectrometric (RIMS) experiments with Hg are described using two colors to excite the 6s6p $^3P_1$ intermediate and high–lying bound Rydberg states. Data on numerous new levels as well as additional values of atomic properties are investigated. Although they are known from different spectroscopic methods, many of them have not yet been obtained by RIMS. The principal aim of these investigations is to evaluate the experimental parameters for their determination, because they have fundamental significance for various detection schemes and may help to optimize the selectivity and sensitivity of RIMS for analytical applications.

EXPERIMENTAL

Two pulsed, tunable, frequency–doubled dye lasers provide the radiation for selective intermediate and Rydberg excitation. The excitation spectra of the 6s6p $^3P_1$ state around 254 nm are recorded at a bandwidth of 0.04 cm^{-1} of the fundamental and with a fixed second wavelength of 219 nm. The ionization–threshold and Rydberg–series spectra are measured with the first laser fixed at 254 nm while the second is scanned in the spectral region near the threshold. The bandwidth of both laser fundamentals is 0.2 cm^{-1}. For exact field strengths of the static electric field at the ion source of the time–of–flight mass spectrometer, equipotential surfaces are calculated numerically for the given boundary conditions. Thus the influence of fringing fields on the spot of ionization is taken into account. Rydberg states are ionized by pulsed electrical fields applied at the ion source 1 µs after the laser excitation. A detailed description of the setup is given in Ref. (2).

RESULTS

Ten components of the seven stable Hg isotopes are resolved in the excitation spectra of the $6s6p\ ^3P_1$ level. Fig. 1a shows the nine most significant components. The centers of gravity (CoG) of their energies reflect the nuclear mass and volume effects of the isotopes, which also manifest in a small odd–even staggering (Fig. 1b). Hyperfine structure (HFS) splittings due to the nuclear spin (I) allow to calculate the magnetic dipole and electric quadrupole interaction constants A and B using the Casimir formula (3). The results are $A = 0.492(6)$ cm^{-1} for ^{199}Hg ($I = {}^1\!/_2$) and $A = -0.182(2)$ cm^{-1} and $B = -0.0089(5)$ cm^{-1} for ^{201}Hg ($I = {}^3\!/_2$). These values agree well with literature data as summarized e. g. in Ref. (4). An anomalous response of the line strengths of the odd–mass isotopes relative to those of the even–mass isotopes gives reason to the assumption that the application of a simple sum rule for the intensities of HFS components does not always reproduce natural odd– to even–mass isotope abundance ratios (5). A variable delay between the two laser sources yields 120(10) ns for the lifetime of the $6s6p\ ^3P_1$ state.

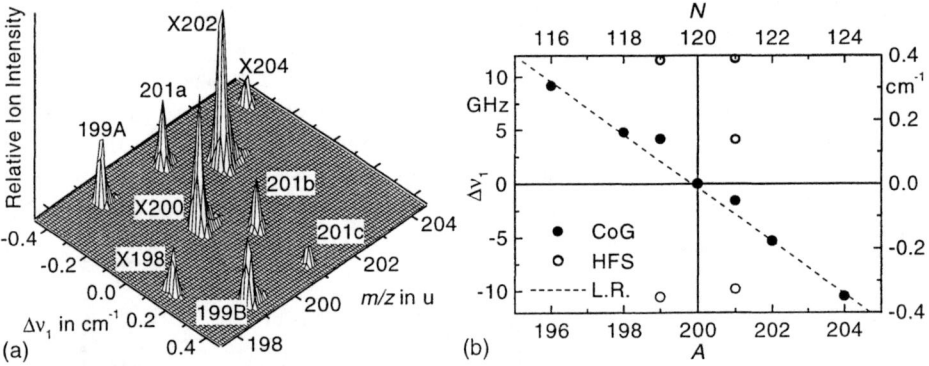

FIGURE 1. (a) Three–dimensional plot of the intensity of ions obtained via intermediate excitation at $\lambda_{air,\,1} = 253.652$ nm versus shift Δv_1 relative to ^{200}Hg and mass number. (b) Relative positions of IS, HFS splitting, and CoGs versus mass and neutron numbers A and N showing odd–even staggering. Dashed line represents even–isotope linear regression (L.R.).

In order to study threshold ion yields, appearance energies of Hg$^+$ are measured in the presence of various static electric fields E at the ion source (Fig 2a). A great number of autoionizing Rydberg series from multiple Stark states with growing resonance enhancements is observed at increasing field strengths. According to the saddle–point model for the ionization, the extrapolation of the square root of E to zero E gives a first ionization energy $IE = 84184(1)$ cm^{-1} (Fig. 2b).

Previous work on Rydberg series under field–free conditions (6) is continued here (Fig. 3a) using delayed pulsed field ionization. Energy values of the 6s nd ^3D series from the NBS tables (7) match with the assignment of the observed series. The energy scale is calibrated to the reference values at overlapping nd terms (21 < n < 30) and to well–known bound states, namely $6s7s\ ^3S_1$ at $\lambda_{air,\,2} = 435.831$ nm, using the fundamental wavelength of the second laser for double–resonance three–photon ionization. A more precise value of IE is determined from the convergence of the ^{200}Hg 6s nd ^3D (21 < n < 70) levels at the observed energy positions $E_{nd} = \nu_1 + \nu_2$ according to the Rydberg–Ritz series formula (8). The series proves to be at sufficiently high energy for Ritz corrections to be insignificant. With the assumption that perturbations are absent and that the quantum defect is a slowly varying function of energy, $IE = 84184.2(2)$ cm^{-1} is obtained for the ^2S-limit (Fig. 3b), in excellent agreement with the most recently published value (9), namely 84184.15(7) cm^{-1}.

A closer view of the Rydberg spectra of the odd–mass isotopes in Fig. 3a clearly shows HFS effects on the peak positions and ionization efficiency minima where the even–mass isotopes have maxima, and also reveals systematic IS of related lines for the even–mass isotopes. The application of the described procedure for the determination of IE values of the significant even–mass isotopes yields a small relative shift of $\Delta IE = 0.05$ cm^{-1} for $IE(^{204}\text{Hg}) - IE(^{198}\text{Hg})$.

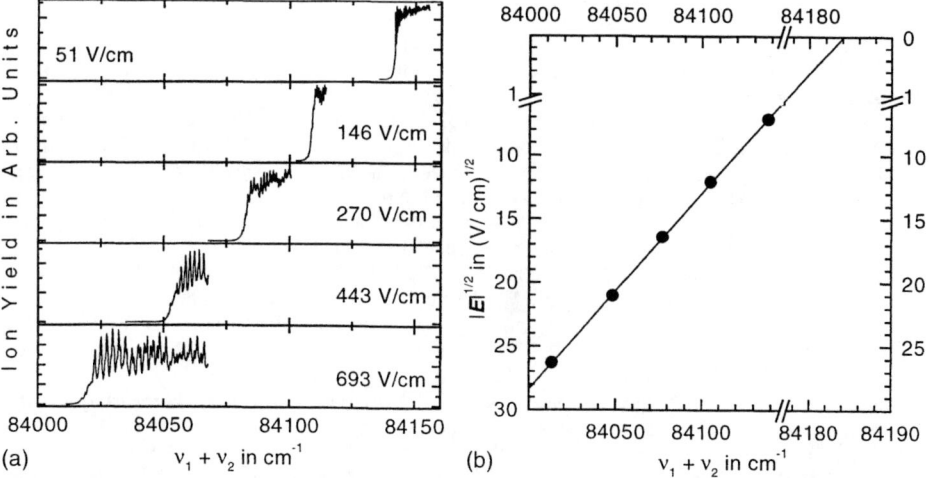

FIGURE 2. (a) Appearance energy measurements of ^{200}Hg$^+$ in various static electric fields E. (b) Determination of the ionization energy IE of ^{200}Hg. Extrapolation of the square root of E gives IE. Note the extended scales in the upper right–hand corner.

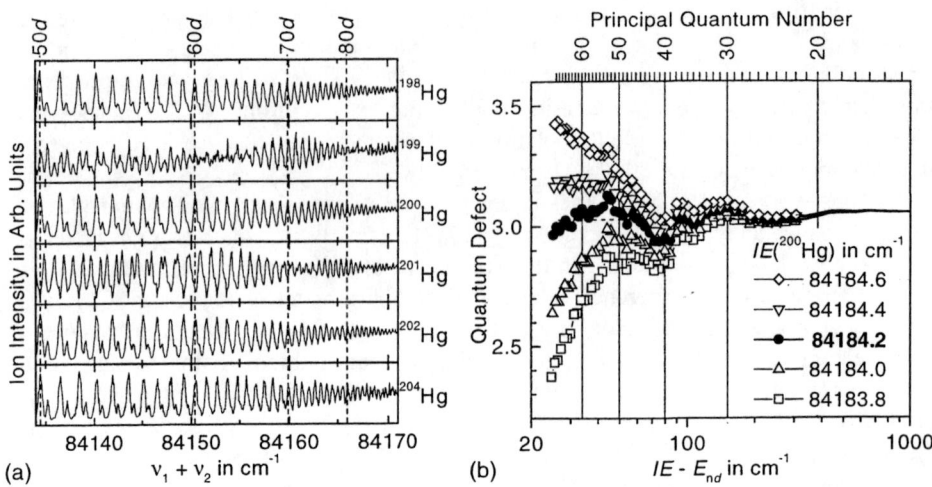

FIGURE 3. (a) Measurements of the $nd\ ^3D$ Rydberg series of mercury isotopes. (b) Dependence of the quantum defect on the principal quantum number (top axis) and on the term energy values $IE - E_{nd}$ (bottom axis). The solid circles are obtained with the best-fit values of IE.

Where available, previous data obtained with different spectroscopic methods agree convincingly with present results. In addition, valuable new information may be deduced from this work in order to optimize the performance of resonance ionization methods particularly in isotope-selective analytical applications.

REFERENCES

1. Clevenger, W. L., Matveev, O. I., Cabredo, S., Omenetto, N., Smith, B. W., and Winefordner, J. D., Anal. Chem. **69**, 2232–2237 (1997).
2. Bisling, P., Heger, H. J., Michaelis, W., Weitkamp, C., and Zobel, H., "New Perspectives in Laser Analytics: Resonance–Enhanced Multiphoton Ionization in a Paul Ion Trap Combined with a Time–of–Fight Mass Spectrometer," in *Resonance Ionization Spectroscopy 1994, AIP Conference Proceedings 329*, New York: AIP Press, 1995, pp. 511–514.
3. Casimir, H. B. G., *On the Interaction between atomic nuclei and electrons*, (Haarlem, The Netherlands: Teyler's Tweede, 1936) San Francisco, Calif.: Freeman, 1963.
4. Bonn, J., Huber, G., Kluge, H.–J., and Otten, E. W., Z. Physik A **276**, 203–217 (1976).
5. Bisling, P. Dederichs, J., Neidhart, B., and Weitkamp, C., "Two–Color Resonance Ionization Mass Spectrometry of Mercury: Isotope Shifts and Hyperfine Structure at 253.7 nm," in *XX. ICPEAC, Abstracts of Contributed Papers*, Vol. II, edited by F. Aumayr, G. Betz, and H. P. Winter, Vienna, Austria, 23–29 July 1997, p. TU011.
6. Bisling, P., Weitkamp, C., and Zobel, H., "RIS of Mercury for Analytical Applications," in *Resonance Ionization Spectroscopy 1996, AIP Conference Proceedings 388*, New York: AIP Press, 1997, pp. 283–286.
7. Moore, C. E. *Atomic Energy Levels, Volume III*, Nat. Stand. Ref. Data Ser. – Nat. Bur. Stand. **35**, Washington, D.C.: U.S. Government Printing Office, 1971, pp. 191–195.
8. Condon, E. U., and Shortley, G. H., *The Theory of Atomic Spectra*, New York: The Cambridge University Press, 1977, ch. 5, pp. 141–147.
9. Baig, A. M., Ali, R., and Bhatti, S. A., *J. Opt. Soc. Am. B* **14**, 731–736 (1997).

SESSION III
RIS AND ULTRA-TRACE ANALYSIS

Ultratrace Determination of Long-lived Radioactive Isotopes

K. Wendt

Institut für Physik, Johannes Gutenberg-Universität Mainz, D-55099 Mainz, Germany

Abstract: A review on the experimental approaches to ultratrace determination of long-lived radioactive isotopes and the broad range of interest of these studies is given. Activities cover a wide variety of fundamental and applied research in the fields of geochemical and environmental studies, cosmochemistry and astrophysics, nuclear and particle physics as well as applications in bio-medical and material sciences. The performances and advantages of Resonance Ionization Mass Spectrometry are worked out and compared to competitive techniques. The history of the high resolution measurements and the present development towards compact, reliable and simplified experimental systems is briefly outlined.

INTRODUCTION

The determination of traces and ultratraces of elemental and molecular species in environmental and artificial samples has been a principal task of analytical chemistry since ancient times. It has led to numerous fundamental discoveries and formed an important basis for the scientific understanding of the world around us. The techniques developed for this purpose must always be coordinated somewhere between physics and chemistry with classical chemical analytical techniques such as electrochemical methods or chromatography and more physics-based technologies like mass spectrometry, nuclear magnetic resonance or X-ray spectroscopy. Nevertheless, the widest spectrum of analytical tools is based on a variety of optical spectroscopic techniques, which all utilize the interaction of matter with light and which have been developed throughout many centuries. The methods started from the most simple spectral analysis by observing the colour change of a flame and presently apply rather sophisticated combinations of particle and laser beams, complex excitation processes, photon or particle detectors and advanced data acquisition systems.

One of the most refined of these optical techniques is Resonance Ionization Mass Spectrometry (RIMS), which nowadays is applied for the ultratrace determination of elemental species, coming close to the ultimate scientific goal of detecting „a few" or even one individual atom and suppressing a large surplus of contaminators (1).

One particular field of successful application of RIMS is the ultratrace analysis of long-lived radionuclides. Here, the detection of the radioactive decay within a reasonable measuring time by radiometric means is very inefficient and direct counting of the number of atoms of the species becomes favoured. Corresponding experimental techniques are mostly based on mass spectrometric methods to avoid background, but apply widely different designs and ionization methods to achieve the required overall efficiency and isobaric as well as isotopic selectivity. They include conventional mass spectrometry, inductively coupled plasma mass spectrometry (ICP-MS), accelerator mass spectrometry (AMS) and RIMS. Alternative concepts are neutron activation analysis (NAA) and proton induced X-ray spectroscopy (PIXE). Nevertheless both techniques are limited in the range of applicability and require access to either a nuclear reactor or a proton-accelerator, and will not be discussed here.

DETERMINATION OF LONG-LIVED RADIOACTIVE ISOTOPES

Long-lived radionuclides are omnipresent in our environment. Sources and origin of individual isotopes can generally be ascribed either to

- Natural Radioactivity, including
  Terrestrial (or Primordial) Radionuclides and
  Cosmogenic Radionuclides, or to

- Anthropogenic Radioactivity, with contributions arising from
  Nuclear Explosions,
  Releases from Peaceful Use of Nuclear Power,
  Medical Use of Radionuclides, and further
  Miscellaneous Sources of Man-made Low-Level Radioactivity.

The first class of about 50 terrestrial radionuclides have been formed either before or together with the formation of the earth about 4.65×10^9 years ago and thus have half-lives longer than this value. They range from ^{40}K up to ^{209}Bi with relative isotopic abundances of minimum 10^{-4} up to 100%. Hence, they are not ultratrace isotopes but accessible to standard analytical techniques and are normally even not quoted as being unstable nuclides.

The second class of terrestrial radionuclides are the members of the three actinide transformation series, which are presently still naturally occuring: the uranium series, starting at ^{238}U and ending at ^{206}Pb, the thorium series from ^{232}Th to ^{208}Pb, and the actinium series (named for its first discovered member ^{227}Ac) from ^{235}U to ^{207}Pb. The fourth series, the neptunium series, was headed by ^{241}Pu, but consisted entirely of radionuclides with half-lifes at least three orders of magnitude shorter than the age of the earth and thus has no measurable amount of survivors left.

In addition to primordial nuclides, cosmic rays generate a number of stable and unstable nuclides in the atmosphere, biosphere and lithosphere by a variety of nuclear reactions. The primary cosmic ray fluence has a maximum of about 1 cm^{-2} s^{-1} at the upper layer of the atmosphere, being composed primarily of high-energy protons ($\approx$ 90%) and He-nuclei ($\approx$ 10%). Production rates of long-lived radionuclides are as high as 2.5 atoms/cm^{-2} s^{-1} for ^{14}C down to about 10^{-6} atoms/cm^{-2} s^{-1} for heavier radionuclides like ^{81}Kr. Global inventories of these nuclides ranges from many tons (^{10}Be, ^{14}C, ^{26}Al, ^{36}Cl) and kg (^{3}H, ^{32}Si, ^{39}Ar, ^{81}Kr) down to g (^{7}Be, ^{22}Na, ^{32,33}P, ^{35}S), and relative isotopic abundances are in the range of 10^{-10} (for ^{10}Be) down to below 10^{-20} (for ^{33,32}P) (2).

The amount of anthropogenic radioisotopes released into the environment by the different scenarios listed above cannot be quantified easily. Even for the widely distributed fallout from nuclear explosions, this value undergoes strong local variations, with presently about a factor of four higher concentrations of all anthropogenic radionuclides on the northern hemisphere. For releases from nuclear power plants and during nuclear accidents strongly fluctuating local contaminations must be determined. Nevertheless for a few isotopes, the present-day global inventory can be estimated. E.g for both, ^{90}Sr and ^{137}Cs, which are known to be the most important fallout constituents because of their biological radiation hazard, a value of $\approx$5x10^{17} Bq ($\approx$5x10^{26} atoms), for the α-emitters ^{239,240}Pu a sum of 1.2x10^{16} Bq ($\approx$5x10^{27} atoms) is obtained (2). The amount of ^{3}H in nature increased from a value of 10^{27} atoms, arising dominantly from cosmogenic generation, by a factor of more than 100 to its maximum value in 1962 due to nuclear explosions. In surface water in the US the ^{3}H-concentration was raised by more than a factor of 400. Both numbers are now slowly decreasing with the half-life of 12.3 years of ^{3}H. Another radiotoxic nuclide of interest is ^{129}I. It is nowadays released from reprocessing plants in doses higher than during the nuclear weapon tests in the 1960s. In this way the relative abundance of ^{129}I/I has been raised in Europe by up to four orders of magnitude (3). The actual distribution and migration of all these long-lived species is a major task in the surveillance of radiotoxic hazard.

Apart from the determination of radiotoxic exposure, selective ultratrace determination of the various long-lived radionuclides mentioned above gives access to a variety of fundamental research problems. As shown in Figure 1, the natural abundances of many of the long-lived radionuclides of interest lie in the range of less than 10^{-12} down to about 10^{-17} of the total sample or in unfortunate cases even of the content of neighbouring isotopes. Nevertheless, they are well suited as probes and sensors in geochemical and environmental studies in the atmosphere (i.e. cosmic ray productions via ^{10}Be, ^{14}C, ^{26}Al, ^{39}Ar, origin and chemistry of trace gas by ^{14}C), the hydrosphere (circulation studies, dating of oceanic and groundwater via ^{14}C, ^{36}Cl, ^{39}Ar, ^{81}Kr), the cryosphere (dating via ^{14}C, detection of variations of cosmic ray fluence in ^{10}Be), and the lithosphere (geomorphological studies with ^{26}Al and ^{36}Cl, characterization of vulcanism via ^{3}He and ^{10}Be). Further applications cover cosmochemistry and astrophysics, nuclear and particle physics as well as bio-medical and material sciences (4,5). Extremely high selectivity and efficiency in the range of 10^{-3} is required for all these studies, as otherwise the technique has to deal with rather large samples. In case of a relative abundance of 10^{-12} the determination technique has to handle at least 10μg and for 10^{-17} even about 1g of sample material to give statistically significant results. The assumed efficiency is usually realistic for mass spectrometric determination techniques, while the requirements of rather complete suppression of background and the extremely high selectivities are only achieved with dedicated experimental techniques.

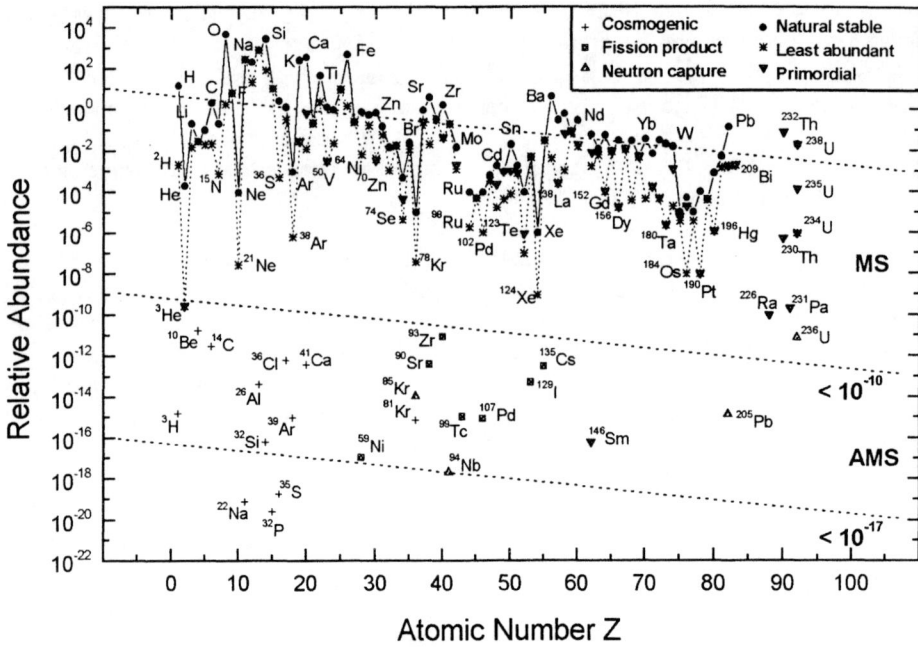

Figure 1. Relative abundances of the elements in the earth's crust, listing most and least abundant stable isotopes together with long-lived radionuclides from natural and anthropogenic sources.

Experimental Approaches: RIMS and Competitive Techniques

The isotopic abundance sensitivity of a conventional mass spectrometer, which is the reciprocal of the selectivity, is limited at a value of about 10^{-10} even under most favourable conditions (6). This limit is primarily due to gas kinetic and wall scattering, charge changing collisions, and background from particles with identical ratio of energy/charge and/or mass/charge (i.e. $^{28}Si^{2+} \Leftrightarrow ^{14}C^{+}$), which together are the dominant processes at beam energies around 10 to 100 keV. Furthermore isobaric interferences from neighbouring elements or molecular contaminations must be fully suppressed. Thus, conventional mass spectrometry is succesfully applied to selective ultratrace radioisotope determination either in the range of very light radionuclides, e.g. ^{3}He (7), or for very heavy isotopes, i.e. in the range of the actinides. In both of these extreme cases, as well as for a few other isotopes, e.g. ^{99}Tc, serious elemental isobaric interferences do not exist, molecular contributions can efficiently be avoided and isotopic selectivity requirements do not surpass 10^{9} after chemical extraction. For these cases detection limits of conventional mass spectrometry in the range of 10^{6} atoms are reported (8-10). Due to simple chemical pre-treatment, wide range of applicability and high ionization probability, ICP-MS has also attained a lot of attention in elemental trace analysis during the last years. Instrumental limitations are comparable to the numbers discussed above and results are reported on different radionuclides in the actinide region with similar detection limits of 10^{6} to 10^{7} atoms (11,12).

AMS overcomes many of the limitations of conventional mass spectrometric techniques by using high beam energy, which drastically reduces all scattering effects, by efficiently suppressing molecular interferences in a negative ion source and during the stripping process inside the tandem accelerator, and by practically eliminating all remaining atomic isobaric interferences by using elemental-selective detection. Thus selectivity can be raised up to extremely high values above 10^{15}. Nevertheless, problems arise in the formation of negative ions, which is possible only for a limited number of elements and involves sophisticated chemical preparation. Hence, routine AMS operation so far is limited to only six long-lived radioisotopes: ^{10}Be, ^{14}C, ^{26}Al, ^{36}Cl, ^{41}Ca and ^{129}I, while a few test measurements on isotopes like ^{90}Sr, ^{205}Pb, ^{236}U and ^{239}Pu are also reported (13). Especially the case of ^{14}C has made AMS famous, as it is the work horse for radiodating of various materials with thousands of samples analyzed in a few specialized AMS facilities world wide.

During the last two decades a number of RIMS-techniques for efficient and highly selective determination of radioisotopes have been developed. Incorporating the resonant optical excitation and ionization process into mass spectrometry provides striking features in respect to competing techniques:

- Isobaric suppression by many orders of magnitude is due to the uniqueness of each individual optical resonance line especially in resonant multi-step excitation.
- High isotopic selectivity of up to 10^9 is achieved with narrow-band lasers in coherent two- or three-step resonant excitation. This is due to the shift of resonance lines of individual isotopes caused by isotope shift and hyperfine structure. In case of insufficient natural shifts, the difference in Doppler-shift of individual isotopes after acceleration can be used for artificial increase.
- Good overall efficiency is due to the high optical cross sections for resonant excitation and corresponding saturation of the excitation step with existing lasers. Coherent effects can further suppress losses along the excitation ladder; ionization is increased by incorporating autoionizing levels.
- Finally, extremely low background is due to the mass separation step together with highly-efficient, low-background ion counting techniques.

A number of RIMS techniques were applied succesfully to the determination of specific radioisotopes with selectivities above 10^{10}, these have been brilliantly reviewed by Payne et al. in 1994 (14). Here we give only a brief overview and an update:

Noble Gases: The interest in noble gas ultratrace determination ranges from radiodating of groundwater, oceanic tracer studies up to surveillance of release from nuclear reactors. Major isotopes of interest are ^{3}He, ^{39}Ar and ^{81,85}Kr. The study of the two ultratrace isotopes ^{85}Kr and ^{81}Kr has been one of the pioneering concerns of RIMS since its early stages, leading to the lowest detection limits reported so far: about 500 atoms in samples of altogether 10^{23} atoms and with suppression of 15 orders of magnitude of other Kr-isotopes have been measured (15). The technique used two conventional preenrichment steps, applying successively a velocity filter and a quadrupol mass spectrometer, which have an efficiency of 10% and gave 9 orders of suppression of the interfering isotopes. Finally a three-step resonant excitation and ionization in a closed vacuum system with recuperation of atoms is used. In addition to the large number of experimental steps, a major complication of this technique is caused by the inconvenient vuv-transition of the first optical excitation from the atomic ground state at 116 nm, which can only be produced by extensive nonlinear frequency mixing of strong pulsed laser radiation. Hence, a refined approach uses one of the particular advantages of collinear fast atomic beam laser spectroscopy. Here excitation can start from metastable atomic states, which are populated in the charge exchange process applied for neutralization of the accelerated fast positive ions. From there, resonant excitation with visible light can be observed either by resonance ionization detection (16) or via multiple excitation and detection of photon bursts. The collinear RIMS approach has already been succesfully applied for the selective determination of the ultratrace noble gas isotope ^{3}He in environmental samples (17), as well as for the study of the nuclear structure of short-lived Yb-isotopes via determination of isotope shift and hyperfine structure (18). The technique of photon burst detection is still under development. It can reach extremely high isotopic selectivities, but requires efficient photon collection from the beam of the fast moving atoms (19,20).

Radiostrontium: The determination of the release of the radiotoxic isotopes ^{89,90}Sr into the environment requires isotopic selectivities above 10^{10} for suppression of the dominant stable isotope ^{88}Sr. Collinear resonance ionization has been succesful in providing a selectivity of about 10^{11} and an efficiency of 10^{-5} in the determination of ^{90}Sr. A schematic drawing of the relevant components of the collinear RIMS facility is given in Figure 2.

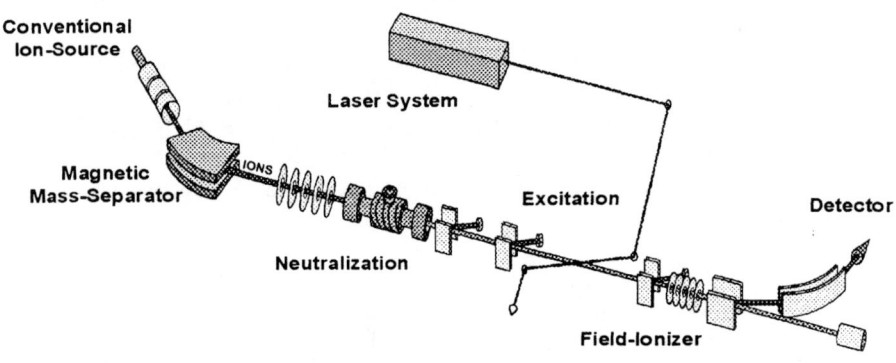

Figure 2. Schematic drawing of the relevant components in collinear RIMS

Detection limits of 2×10^6 atoms were demonstrated and a wide variety of environmental and technical samples were analyzed (21). For comparison, recent AMS results in ^{90}Sr gave selectivities of 3×10^{13} but a somewhat higher detection limit of 5×10^7 atoms (22), which is comparable to the capabilities of direct radiometric β-counting. Even though showing better efficiency of up to 10^{-3}, two-step resonant excitation RIMS with narrowband cw laser radiation on an atomic beam and subsequent quadrupole mass spectrometry could not provide sufficiently high selectivity, while recent results with this approach but using three step excitation will be presented at this meeting (23).

^{210}Pb and ^{41}Ca: The technique of multi-step narrow-band cw-laser RIMS with a quadrupol mass spectrometer has been first developed by Bushaw with primary goal of ultraselective trace determination in lead (24). A succeeding project utilizes diode lasers for determination of ^{41}Ca, where selectivities as high as 10^{15} are mandatory. Using one to three coherent narrowband excitation steps and subsequent ionization, an optical selectivity as high as 10^{12} is predicted. Maximum values of up to 10^6 have been demonstrated experimentally so far (25). In combination with the abundance sensitivity of an optimized quadrupole mass spectrometer, which has been shown to reach values of about 10^8 (26), yields extremely high overall selectivity and good overall efficiency of about 10^{-6}. A major advantage of this system is the compactness and the reliability of its components, an schematic diagram of the set-up is given in Fig. 2. Presently the technique has been demonstrated in studying bio-medical material and meteorite samples for cosmochemical studies. Also a first spectroscopic determination of a 10^{-6} enriched ^{41}Ca-sample is reported, while radiodating applications on ^{41}Ca and the extension to studies of other elements are foreseen but will require further optimizations of the optical excitation process (25).

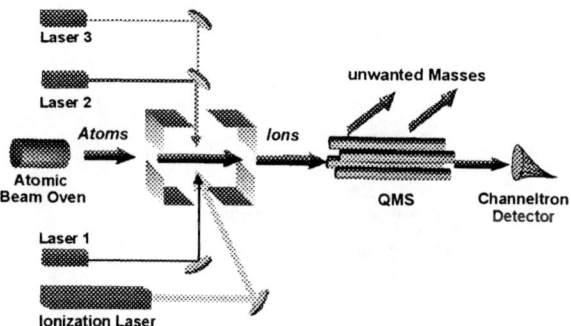

Figure 3. Schematic diagram of the RIMS setup with multi-step cw-laser excitation and ionization followed by a quadrupol mass spectrometer.

Actinides and ^{99}Tc: Due to their high radiotoxicity, ultratrace determination of actinides and ^{99}Tc in environmental samples is an important task. Having no stable isotopes, the analysis of these species does not require extremely high isotopic selectivities and thus can be carried out by efficient high-repetition rate, broad-band pulsed lasers and simple time-of-flight mass spectrometers. Routine measurements on Pu-concentrations in biomedical and environmental samples with detection limits of about 10^6 atoms, corresponding to extremely low activities of e.g. 1 µBq for ^{239}Pu, are reported (27). This detection limit is about 2 orders of magnitude better than that of competitive radiometric techniques.

CONCLUSION AND OUTLOOK

RIMS has nowadays proven to be a very valuable technique for determination of ultralow concentrations of radioisotopes with excellent detection limits, which presently lie in the range of 10^6 atoms per sample or better for practically all radioisotopes studied so far.

Isotopic selectivity as high as 10^{15} can be achieved either by multistep resonant, narrow-band cw-laser excitation or by making use of the Doppler-effect, both in combination with refined mass spectrometers. As isobaric interferences, which limit conventional MS, ICP-MS and even AMS, are fully surpressed in RIMS, widespread ultraselective applications in fundamental and applied research are expected for the near future, making RIMS a complementary more than a competitive technique to AMS. Even though ultrahigh selectivity has already been demonstrated, most successful applications so far concern ultratrace determination of radiotoxic isotopes. Here RIMS, in contrast to standard radiometric techniques, gives reliable isotopic compositions and low detection limits independent of the half-life. The additional advantage of RIMS of being a rapid technique makes it also attractive for routine surveillance and emergency use.

Still nowadays, the most important drawback of RIMS is the complicated and money- as well as man-power consuming experimental set-up, i.e. the lasers. This fact has limited the application to a small number of university and government research laboratories and has not yet permitted reasonable commercialisation of the technique. Nevertheless, it is expected that the combination of compact and reliable QMS- or TOF-mass spectrometers with small and easy-to-use laser systems, based on diode lasers or other compact solid-state lasers, is the significant and promising step toward surpassing these limitations.

REFERENCES

1. Hurst, G.S. et al., *Rev. of Modern Physics* **51**, 767 (1979)
2. Tykva, R., Sabol, J., *Low-level environmental radioactivity*, Lancaster: Technomic Publ., 1995
3. Wagner, M.J.M. et al., *Nucl. Instr. and Meth. Phys. Res.* **B 113**, 490 (1996)
4. Kutschera, W., *Nucl. Instr. and Meth. Phys. Res.* **B 50**, 252 (1990)
5. Kutschera, W., *Nucl. Instr. and Meth. Phys. Res.* **B 123**, 594 (1997)
6. Purser, K.H. et al., *Nucl. Instr. and Meth. Phys. Res.* **186**, 487 (1981)
7. Mamyrin, B.A., Tolstikhin, I.N., *Helium isotopes in nature*, New York, Elsevier Publ., 1984
8. Rokop, D.J. et al., *Anal. Chem.* **62**, 1271 (1990)
9. Altrep, M. et al., *Radiochimica Acta* **57**, 15 (1992)
10. Perrin, R.E. et al., *Int. Journal Mass Spect. Ion Proc.* **64**, 17 (1985)
11. Ross, R.R. et al., *Radioact. and Radiochem.* **4**, 24 (1993)
12. Bailey, E.H. et al., *J. Anal. Atom. Spec.* **8**, 551 (1994)
13. Contributions to the 7. Int. Conf. on AMS, *Nucl. Instr. and Meth. Phys. Res.* **B123**, 1-598 (1997)
14. Payne, M.G. et al., *Rev. Sci. Instr.* **65**, 2433 (1994)
15. Thonnard, N. et al., RIS-92, *Inst. Phys. Conf. Series* **128**, 27 (1992) and references therein
16. Aseyev, S.A. et al., *J. Phys.* **B 24**, 2755 (1991)
17. Kudryavtsev, Y.A. et al., *Appl. Phys.* **B 48**, 93 (1989)
18. Schulz, C. et al., *J. Phys.* **B 24**, 4831 (1991)
19. Lehmann, B. et al.,RIS-94, *AIP Conference Proceedings* **329**, 331 (1995)
20. La Belle, R.D. et al., *Phys. Rev.* A **54**, 4461 (1996)
21. Wendt, K. et al., *Radiochimica Acta* **79**, 183 (1997)
22. Paul, M. et al., *Nucl. Instr. and Meth. Phys. Res.* **B 123**, 394 (1997)
23. Bushaw, B.A. and Cannon, B., *Contribution to this issue*
24. Bushaw, B.A., RIS-92, *Inst. Phys. Conf. Series* **128**, 31 (1992) and references therein
25. Müller, P. et al., *Contribution to this issue*
26. Blaum, K. et al, *Int. Journal Mass Spect. Ion Proc.*, in print
27. Erdmann, N. et al., *Fresenius J. Anal. Chem.* **359**, 378 (1997)

Ultra-sensitive detection of hydrogen isotopes by the Lyman-α RIS

Y. Miyake[1], K. Shimomura[1], A. P. Mills, Jr.[2], J.P. Marangos[3], and K. Nagamine[1,4]

[1] *Meson Science Laboratory, High Energy Accelerator Research Organization (KEK-MSL), Oho, Ibaraki-305, Japan*
[2] *Bell Laboratories, Lucent Technologies, Murray Hill, NJ 07974, U.S.A.*
[3] *Blackett Laboratory, Imperial College, of Science, Technology and Medicine, Prince Consort Road, London SW72BZ, U.K.*
[4] *Muon Science Laboratory, Institute of Physical and Chemical Research (RIKEN), Wako, Saitama 351-01, Japan*

Abstract. An ultra-sensitive way of detecting hydrogen isotopes is described by utilizing a resonant ionization scheme through the $1S \rightarrow 2P \rightarrow unbound$ transition. A time-of-flight measurement by the use of the pulsed lasers coupled with a mass separation enables us to distinguish any hydrogen isotope with very low background. The detection limit of the present method is estimated to be about 10^4 atoms/cm^3

INTRODUCTION

Detection of hydrogen isotopes has been a considerable challenge in experiments dealing with low density gases and plasmas such as ion-sources [1], flames [2], gas discharges [3,4], plasma enhanced chemical vapor deposition [5] or fusion reactors [6,7]. As a result, various techniques have been invented to detect and determine the density of hydrogen and deuterium atoms, such as the laser induced fluorescence technique of Kajiwara et al. [7-10], the laser induced stimulated emission method of Bokor et al. [11,2,4,5], the laser absorption spectroscopy [1], the resonant three-photon ionization technique [12].

On the other hand, a serious investigation to extract muonium atoms (which is the lightest hydrogen isotope with 1/9 of the mass of H, consisting of a positive muon μ^+ and electron e^-; designated as Mu) has been on-going at KEK-MSL, for the purpose of generating ultra slow muons [13]. In the course of the development of the laser resonant ionization method to ionize Mu utilizing the Lyman-α laser light, it was found that this method is also very useful for detecting atoms of any hydrogen isotope. It was revealed that not only H and D, but also T (or Mu) are identified and extracted much more efficiently and with a much lower background than by any other method. In this paper, a brief description of the laser resonant ionization method using Lyman-α radiation is given.

EXPERIMENTAL ARRANGEMENT

In the present method, hydrogen isotope atoms near a hot tungsten (W) foil of 50 μm are resonantly ionized by intense laser light. Once ionized, the hydrogen isotopes are extracted by an immersion lens with an acceleration voltage of 9.0 kV, and transported to a micro-channel plate (MCP) detector located 8 m from the target. Any ions produced at the W target region can be identified either through the mass/charge (Q) ratio by adjusting the bending magnet in the ion optics, or through the time-of-flight (TOF) spectrum. The target chamber was initially evacuated down to 2×10^{-10} mbar at room temperature (R.T.). The W target was cleaned by a surface treatment in 3×10^{-7} mbar of oxygen at 1800 K for about 7 hours, and then heated up to 2200 K for the experiments.

For ionizing hydrogen isotopes efficiently, we adopted a resonant ionization scheme through the $1S \rightarrow 2P \rightarrow unbound$ transition. VUV (vacuum ultra violet) generation in the vicinity of the Lyman-α wavelength was achieved by a Sum-Difference Frequency Mixing method using two photons of 212.55 nm (ω_r) for two-photon resonant excitation of the $4P^5 5P[1/2, 0]$ state in krypton, subtracted by a photon of a tunable difference wavelength (ω_t; 820-848 nm) [14,15]. A single mode 850 nm light with a band width of 0.5-1.0 GHz which is obtained from an OPO system (Continuum Mirage800). It is amplified by a Ti-sapphire crystal (18 mm in diameter and 15 mm long) which is pumped by two synchronized frequency-doubled Nd-YAG lasers (Continuum 9020). The amplified 850 nm light as high as 300-600 mJ/p, is doubled by using the first β-Ba$_2$BO$_4$ crystal, generating 425 nm light of an intensity of 100-150 mJ/p with a good beam quality. The doubled 425 nm light is quadrupled by using the second β-Ba$_2$BO$_4$ crystal, generating ω_r of a pulse energy of 8-12 mJ/p. The difference wavelength ω_t between the $4P^5 5P[1/2, 0]$ state of krypton and the desired Lyman-α wavelength was generated by a broad-band TiS laser system (STI-LRL). By phase matching with an addition of Ar gas which has positive dispersion in the vicinity of the Lyman-α wavelength, the Lyman-α intensity can be increased by about 1.5 times higher for ω_r with an intensity of $\sim$ 10 mJ/p. For ionizing the 2P state of hydrogen atom, the third harmonic of a Nd:YAG laser of 355 nm (Spectra Physics GCR4) with an intensity of $\sim$ 30-40 mJ/pulse was introduced together with the Lyman-α light. The intensity of the Lyman-α photons reaching the W target is measured by an NO cell to be about 1.0 μJ/pulse.

DISSOCIATION OF H_2 AND D_2 OBSERVED BY RIS

In the UHV (Ultra High Vacuum) chamber if carefully treated, the main components of the residual gas is well known to be H_2 molecule. The heated W foil was used as a catalyzer to dissociate H_2 molecules into H atoms through the interaction between the molecules and the hot W surface. Therefore, the H atoms resonantly ionized correspond to neutral hydrogen atoms dissociated from H_2 molecules near the W surface, and then evaporated into the vacuum space where the Lyman-α and 355 nm laser beams are passing. It was found that H atoms dissociated from the residual H_2 molecules in the UHV chamber (3×10^{-10} mbar at 1200 K) is sufficient to give a large signal without introducing any H_2 gas. We are able to evaluate the number of hydrogen ions extracted and detected by the MCP as a ratio of the amplitude × time-width of the MCP raw signal obtained in the condition between on-resonance and off-resonance wavelength, since no more than one hydrogen ion per laser pulse is expected to reach the MCP in the case of the off-resonance wavelength. From the UHV chamber of 2×10^{-9} mbar, as much as 2000 H atoms per laser pulse were found to be extracted at 2150 K, when the Lyman-α wavelength was adjusted to be 121.57 nm (resonant value for H).

In the case of residual H_2 experiments, quantitative interpretation is difficult because of the unknown temperature dependence of the H_2 density near the W surface coupled with the reading of the pressure gauge. In order to simplify the experimental conditions, the temperature dependence of the D yield was investigated with a constant D_2 pressure of 4.0×10^{-8} [mbar], regulated by a variable leak valve (Granville Phillips). The entire measurement was done with several electrostatic quadrupole lenses turned off, resulting in the reduction of the transmission efficiency to about 1 %. Without this precaution, the D yield was easily beyond the

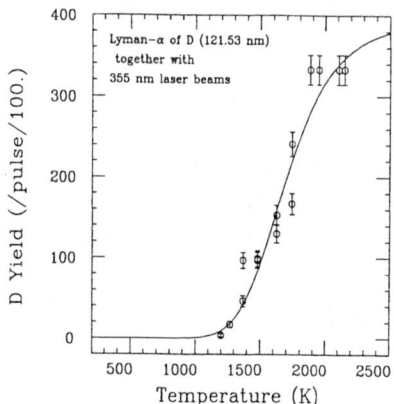

FIGURE 1. Temperature dependence of the yield of the resonantly ionized D atoms, with a constant D_2 pressure of 3×10^{-9} mbar. The W foil heated up was used as a catalyzer to dissociate D_2 molecules into D atoms through the interaction between the molecules and the hot W surface. The Lyman-α wavelength was adjusted to be 121.53 nm (resonant value for D). The fitted curves are from Eq.(1).

saturating level of the MCP. The Lyman-α wavelength was adjusted to be 121.53 nm (resonant value for D). The yield of D atoms was negligible at a temperature below 1200 K, started to increase by 1300 K, and increased monotonically with increasing temperature, leveling off at around 2000 K, as is shown in Fig. 1. According to Hickmott [16], of the H_2 molecules striking a hot W filament, the fraction dissociated is constant (at least 0.05.) for a H_2 pressure $\leq 1 \times 10^{-7}$ mbar and a filament temperature above 1475 K. If we thus assume that the probability of D_2 dissociation is constant in the range of measurements, our observed temperature dependence must be related to the emission of D atoms either from the W bulk or surface, but not directly related with the molecular dissociation rate. The solid line in Fig. 1 is a fit to the temperature dependence Eq. (1) shown below.

$$y(T) = \frac{c_1 \sqrt{T} e^{-\frac{E_a}{kT}}}{1 + c_2 \sqrt{T} e^{-\frac{E_a}{kT}}}, \qquad (1)$$

where y is the total yield of the neutral atoms, T the temperature, k Boltzman constant, E_a the activation energy of hydrogen atom in eV. The fitted activation energy E_a is 1.56 (2) eV. Hydrogen atoms are known to be bound to the W surface by about 2.97 (2) eV [17]. On the other hand, the work function of atomic Hydrogen in vacuum relative to the interior of bulk solid W is estimated to be about 1.56 eV [18]. Since our measured activation energy for D atoms agrees with the work function of hydrogen to the solid, we have experimental evidence that D atoms are not emitted into vacuum from the surface, but are rather emitted directly from the bulk at high temperatures. The dynamics of the emission of atomic hydrogenic atoms from tungsten is thus analogous to the thermionic emission of electrons from a metal.

PRESSURE DEPENDENCE OF THE D YIELD

The D yield was examined as a function of the D_2 pressure in the range between 1.6×10^{-9} and 5.0×10^{-7} mbar. Again, to prevent saturation of the MCP detector, all the measurements were taken with one of the electrostatic optics elements detuned to reduce the transmission efficiency to about 0.45 %, and for D_2 pressure lower than 10^{-6} mbar. The amount of D_2 gas introduced was regulated by a variable leak valve. The dependence of the D yield as a function of the D_2 pressure shown in Fig. 2 can be fitted well by a linear relationship. According to Hickmott [16], the rate of atomic hydrogen production is proportional to p_{H_2} below 1×10^{-6} mm, but to $p_{H_2}^{0.5}$ for $1 \times 10^{-6} \leq p_{H_2} \leq 1 \times 10^{-2}$ mm. Similarly the amount of D atoms is considered to be proportional to the pressure of D_2, in the present range of pressure. On the other hand, the D yield around the W target was experimentally found to be proportional to the D_2 pressure, as is shown in Fig. 2. We thus conclude that the present laser resonant ionization method can determine the D and D_2 density quantitatively below 10^{-6} mbar if an appropriate calibration is made.

DETECTION LIMIT

In the case of the laser induced fluorescence technique using the Lyman-α [7,8,10] or the Balmer-α [7] photons, the lowest detection limit is reported to be on the order of 10^7 -10^9 cm^{-3}, and 10^7 cm^{-3}, respectively. In the case of the two photon-induced Balmer-α emission by Bokor et al. [11], the detection limit of 3×10^9 cm^{-3} is reported to be of the same order as that by the three photons ionization by Bjorklund et al. [12]. In the case of the laser absorption spectroscopy [1], using the Lyman-β laser, line density ($\int_{path} n(l)dl$, where n is number density) of 2×10^{12} cm^{-2} is reported.

On the other hand, the detection limit of the present Lyman-α resonant ionization method can be evaluated in the following procedure. Assuming the probability of D_2 dissociation at the W surface at 2150 K to be 10% in 3×10^{-9} mbar of D_2 gas, about 1.6×10^7 D atoms exist in a volume of about 1 cm^3 which is a effective volume of extraction along the laser passage. Since we obtained an MCP signal of 5.0V with 90 ns at 2150 K, in contrast with that of 12 mV with 18 ns, we can evaluate an amount of 2100 D atoms per laser pulse are extracted at 2150 K. Therefore, the extraction and detection efficiency is calculated to be 1.3×10^{-4}. This corresponds to a detection limit of the present Lyman-α resonant ionization method is 7.7×10^3 D/cm^3, since one event detected by the MCP is sufficient to give an appropriate information about hydrogen owing to a feature of the very low backgrounds measurement. It is thus 3-5 orders of magnitude better than the other methods. In the case of detecting fluorescence, the solid angle for light collection must be limited in order to avoid stray light from the pumping laser. In contrast,

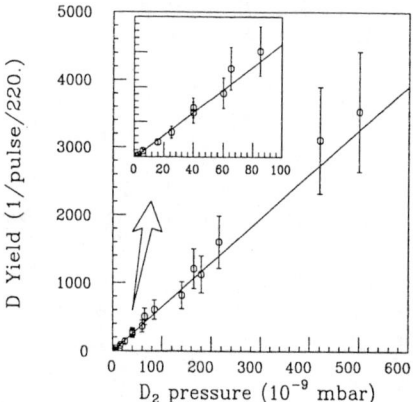

FIGURE 2. Pressure dependence of the D yield as a function of the D_2 pressure in the range between 1.6×10^{-9} and 5.0×10^{-7} mbar. It can be fitted well by a linear dependence. The W foil heated up was used as a catalyzer to dissociate D_2 molecules into D atoms through the interaction between the molecules and the hot W surface.

laser-ionized hydrogen can be extracted towards the MCP quite efficiently in the present laser resonant ionization method, although there is some loss either coming from the transport or the detection efficiency of the MCP itself. Compared with the multi-photon ionization scheme, resonant ionization through the ($1S \rightarrow 2P \rightarrow unbound$) transition naturally has a much higher ionization probability. We conclude that the resonant ionization spectroscopy method for detection of hydrogenic atoms is an ultra sensitive and efficient mass-selective technique.

REFERENCES

1. G.C. Stutzin, A.T. Young, A.S. Schlachter, J.W. Stearns, K.N. Leung, W.B. Kunkel, G.T. Worth and R.R. Stevens, Rev. Sci. Instrum. **59**, 8 1363 (1988)
2. J.E.M. Goldsmith, Appl. Opt. **29**, 4841 (1990)
3. J. Amorim, J. Loureiro, G. Baravian and M. Touzeau, J. Appl. Phys. **82**, 2795 (1997)
4. L. Tomasini, A. Rousseau, G. Baravian, G. Gousset, and P. Leprince, Appl. Phys. Lett. **69**, 1553 (1996)
5. R.C. Cheshire, W.G. Graham, T. Morrow, V. Kornas, H.F. Dobele, K. Donnelly, D.P. Dowling and T.P. O'Brien, Appl. Phys. Lett. **66**, 5 (1995)
6. G. Reinhold, J. Hackman, and J. Uhlenbusch, J. Plasma Phys. **28**, 281 (1982)
7. P. Gohil and D.D. Burgess, Plasma Phys. **25**, 1149 (1983)
8. T. Kajiwara, M. Inoue, T. Okada, K. Muraoka and M. Akazaki, Rev. Sci. Instrum. **56**, 12 (1985)
9. P. Bogen, Y. T. Lie, Appl. Phys. **16**, 139 (1978)
10. P. Bogen, Ph. Mertens, E. Pasch and H.F. Dobele, J. Opt. Soc. Am. B **9**, 2137 (1992)
11. J. Bokor, R.R. Freeman, J.C. White and R.H. Stortz, Phys. Rev. A **24**, 612 (1981)
12. G.C. Bjorklund, C.P. Ausschnitt, R.R. Freeman and R.H. Storz, Appl. Phys. Lett. **33**, 54 (1978)
13. K. Nagamine, Y. Miyake, K. Shimomura, P. Birrer, J. P. Marangos, M. Iwasaki, P. Strasser and T. Kuga, Phys. Rev. Lett. **74**, 4811 (1995).
14. J.P. Marangos, N. Shen, H. Ma, M.H.R. Hutchinson and J.P. Connerade; J. Opt. Soc. Am. B **7**, 1254 (1990).
15. Y. Miyake, J. P. Marangos, K. Shimomura, P. Birrer and K. Nagamine, Nucl. Instr. and Meth. B **95**, 265 (1995).
16. T.W. Hickmott, J. Chem. Phys. **32**, 810 (1960)
17. P. W. Tamm and L. D. Schmidt, J. Chem. Phys. **54**, 4775 (1971)
18. A.P. Mills,Jr., J. Imazato, S. Saito, A. Uedono, Y. Kawashima, and K. Nagamine, Phys. Rev. Lett. **56**, 1463 (1986).

Multi-element trace analysis of solid samples using one-photon two-step RIMS

H.H. Telle, C.J. Abraham, O.R. Jones and T. Krustev

*Department of Physics, University of Wales Swansea,
Singleton Park, Swansea SA2 8PP, United Kingdom*

Abstract. In this study we have investigated the feasibility of multi-element analysis using a simple 1+1 photo-excitation / photo-ionisation scheme. Although such schemes are usually far from ideal for optimum resonance ionisation, they are the approach of choice if one wishes to maintain a simple, easy-to-operate laser set-up which is potentially suitable for routine analysis. In addition, we only made use of the second-harmonic tuning range of a single dye. While this limits the range of elements which are accessible in the 1+1 RIS scheme it further adds to the simplicity and allows for automation of sequential multi- element analysis.

INTRODUCTION

Routine analytical examination of samples for their elemental composition exploiting the technique of SIMS is common place. The extension of the method, RIMS, which uses laser radiation to generate the ions for mass analysis has not yet been applied to multi-element analysis, despite its huge successes in detecting very small traces of specific elements (and their isotopes) down to below concentration levels of parts-per-million, and even parts-per-billion or less (see e.g. the application of SIRIS to the analysis of semiconductors (1)). This is largely due to the intrinsic property of RIS/RIMS which is element specific: the laser excitation steps only favour a single element and largely suppress the others; this results in the extraordinary selectivity and sensitivity of the method.

In this study we have investigated the feasibility whether a scheme for multi-element analysis could be devised which makes use of the technique of RIMS. Naturally, such an approach has to be implemented to sequentially interrogate the elements of interest because of the single-element selectivity of resonantly induced ionisation.

It should be noted that all-element post ionisation schemes have been described in the literature in which a powerful laser pulse, whose wavelength (usually in the UV) is not in resonance with any transition of the analyte atoms. Although an increase in signal strength over normal SIMS can be obtained, selectivity remains unchanged and mass interference problems persist.

EXPERIMENTAL PROCEDURE

As has been pointed out in the introduction, the aim of this study is the implementation of (sequential) multi-element analysis using RIMS. Throughout the work reported here simple 1+1 photo-excitation / photo-ionisation schemes were realised for the task. Although such schemes are usually far from ideal for optimum resonance ionisation, they are the approach of choice if one wishes to maintain a simple, easy-to-operate laser set-up which is potentially suitable for routine analysis. As an additional constraint, we only made use of the second-harmonic tuning range of a single dye. While this limits the range of elements which are accessible in 1+1 RIS schemes it further adds to the simplicity and potentially allows for automation of sequential multi- element analysis.

The samples were investigated in a reflectron-type time-of-flight mass spectrometer, with a potential mass resolution of 5000. Material for SIMS and RIMS analysis was sputtered from the sample using a standard Ar^+ duo-plasmatron ion gun, which was synchronised to the laser system and the ToF mass analysis. With appropriate adjustment of the timing, the extraction voltages, and the use of time-variable blanking of the mass spectrometer, it was easy to switch between SIMS and RIMS mode of the instrument, allowing for comparison of results from the two methods. A full description of the instrument is given elsewhere (2).

The laser system used throughout all these experiments was a tuneable dye laser, pumped by a Nd:YAG laser. The dye used throughout these experiments was rhodamine B (= rhodamine 610), and with second harmonic generation we could cover the wavelength range 287-301nm.

The RIMS spectra were stored in a two-dimensional array, laser wavelength against mass number. The wavelength of the dye laser was advanced in steps of 0.1nm (the line width was approximately 0.007nm); for each wavelength setting the complete ToF mass spectrum is stored, accumulated over 1,000 laser pulses.

RESULTS

We have tested this approach for the quantitative analysis for a number of metal matrix materials to demonstrate the versatility of the procedure. All samples were "certified" samples so that a rough protocol of the sample composition, specifically trace element concentrations, was available. Samples with concentration levels of a few percent down to a few parts-per-million were used in the study, to ascertain linearity of the analysis over a few orders of magnitude in concentration levels. Elements for which quantitative analysis has been realised include the majority of elements used in stainless steels. A small selection of sample materials and part of their elemental composition are summarised in Table 1.

All the elements listed in Table 1, and many others exhibit ionisation limits in the range 6.7-7.8 eV which require at the high energy end two photons at the wavelength of at least $\lambda=310$nm. Equally, all elements have intermediate energy levels accessible to the second-harmonic photons in the range mentioned above.

TABLE 1. Selected sample materials and their nominal elemental composition.

sample matrix	specified element concentration in ppm					
	Al	Cr	Fe	Mo	Ni	Ti
Ni (99.98+)	1	8	15	n/a	matrix	10
Pb (99.98+)	1	< 1	1	n/a	n/a	n/a
Ti (99.95)	500	500	300	n/a	500	matrix
Inconel 718	5,000	190,000	185,000	30,500	520,000	9,000
stainless steel SS 410	n/a	13,400	matrix	3,600	21,600	n/a

matrix = majority element in the sample, amount not specified;
n/a = information not available.

Typical mass spectra are shown in Figure 1, for the resonant excitation of Fe and Cr in the titanium sample matrix (for the concentration of trace constituents see Table 1 above). The mass spectra are shown for two consecutive steps of laser excitation, at $\lambda=297.0$nm and $\lambda=297.1$nm. The first wavelength is in close coincidence with the Fe transition at $\lambda=297.010$nm, while the second is in close coincidence with the Cr transition at $\lambda=297.110$nm. Although these transitions are not the most efficient for the implementation of RIS/RIMS, the two spectra in demonstrate clearly what care may need to be taken in selecting apparent on/off resonance wavelengths.

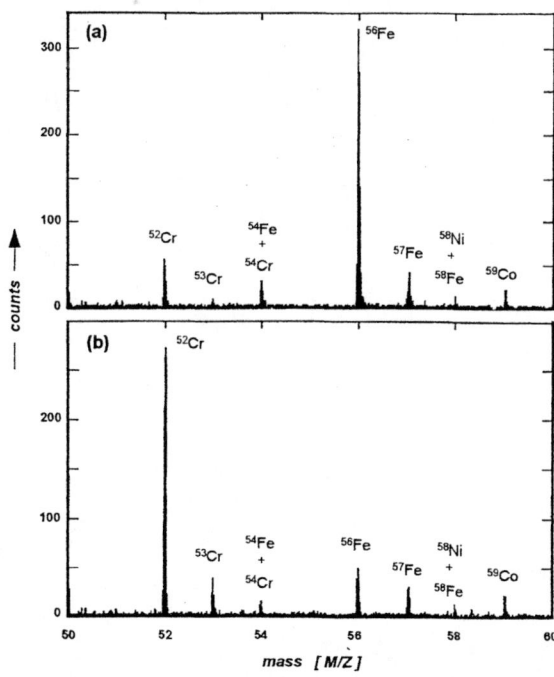

FIGURE 1. ToF RIMS spectrum of *pure* titanium sample (500ppm Cr, 300ppm Fe) in the mass range 50-60amu; (a) laser at $\lambda=297.0$nm - on Fe resonance; (b) laser at $\lambda=297.1$nm - on Cr resonance.

The RIMS response for Cr, Fe and Ni is shown in Figure 2, over the wavelength range 290-300nm; the data shown are for the most abundant isotope in each case. A number of resonances with intermediate states can be observed, as well as a broad "background" from non-resonant multi-photon excitation (its shape mimics the intensity distribution of the laser radiation). The non-resonant background contribution makes it paramount that quantitative measurements of individual elements are carried out for isotopes which do not experience mass interference from other elements.

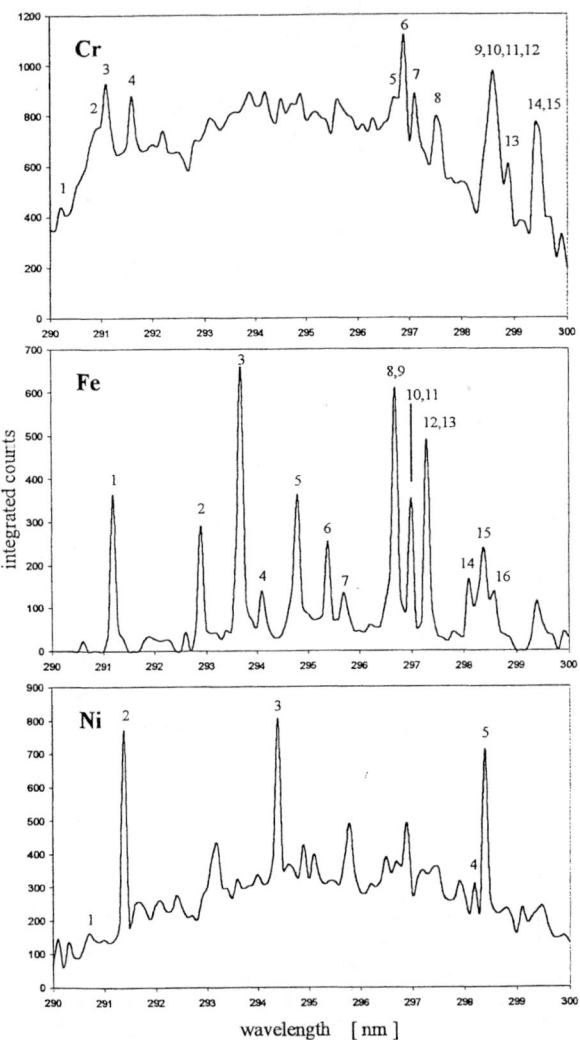

FIGURE 2. RIMS signal as a function of laser wavelength in 1+1 excitation scheme. The numbers correspond to transitions from the ground state manifolds Cr (a $^5S,^5D$), Fe (a ^5D) and Ni (a $^3D,^3F$). The non-resonant ionisation background mimics the intensity distribution of the laser radiation.

The problem of mass interference of isotopes of different elements is a common one in standard mass analysis using SIMS. This is exemplified for three common constituents in steels, namely chromium, iron and nickel who are close neighbours in the periodic table. Part of the relative isotope abundance for these elements is collected in Table 2. It is clear that for trace element analysis one would like to measure the most abundant isotope to achieve the highest sensitivity; this is feasible for Cr(52) and Fe(56) but the most abundant isotope Ni(58) experiences interference from Fe(58), and in steels Fe as the matrix element would largely swamp traces of Ni.

TABLE 2. Mass interference of selected isotopes.

element	natural isotopic abundance, in %						
	52	53	54	56	57	58	60	61
Cr	83.76	9.55	2.365					
Fe			5.9	91.72	2.1	0.28		
Ni						68.077	26.23	1.19

It is usually stated that in RIMS this problem of mass interference can be is largely overcome. However, as the spectra in Figures 1 and 2 demonstrate, in simple 1+1 RIS schemes adequate care has to be taken because of non-resonant background contributions. Furthermore, occasionally resonance wavelengths for different elements may be very close to each other.

One would argue that line co-incidences normally do not pose problems, but the example outlined in this presentation are in contradiction to this. The spacing of the two transitions addressed, Fe at $\lambda=297.010$nm and Cr at $\lambda=297.110$nm is over ten times larger than the laser linewidth $\Delta\lambda=0.007$nm. However, power broadening of the transitions may push the off-resonance transition into near-coincidence with the on-resonance transition. For example, for observed FWHM of the Fe transition changes dramatically, if the laser pulse energy density is altered (see Table 3); for further details on power broadening see (3) and (4).

TABLE 3. Power broadening of the Fe transition at 297.010nm.

laser pulse energy density [mJ/cm^2]	FWHM [nm]	counts in RIS signal
5.0±0.2	0.021±0.004	24±5
330±3	0.127±0.015	325±22 [1]

[1] count rate saturated beyond 170 mJ/cm^2.

It is clear from the data, that for high laser pulse energy density one may loose the selectivity of RIMS to a large degree. Thus a compromise has to be struck if highly selective quantitative analysis is to be carried out.

On the one hand, at low laser pulse energy density very good selectivity is maintained, since normally the line spacing of neighbouring transitions of different elements are larger than the observed transition linewitdhs (combination of Doppler and power broadening). Furthermore, non-resonant multi-photon ionisation becomes less probable so that interference from background ions is reduced.

On the other hand, it is often argued that for quantitative analysis ideally the transitions should be driven into saturation so that simple mathematical expressions can be used to determine the actual, absolute concentration values. This is at the expense of selectivity, and mass interference effects have to be considered carefully.

SUMMARY

Regardless of all the adverse effects encountered in the study we have been able to carry out sequential multi-element analysis for a wide range of trace elements (Co, Cr, Cu, Fe, Mo, Ni, Nb, Sn, Ti and V) in numerous metal samples. As has been shown for three elements in Figure 2, the RIMS response for all elements investigated in this study were recorded. Subsequently a computer controlled sequence was carried out in which the laser was tuned exactly to the most efficient 1+1 resonance excitation scheme (some of the routes had not been listed previously, see e.g. (5)), always detuneing the laser afterwards away from resonance by about five linewidths to measure the non-resonant background contribution.

Using the response from the samples with the smallest trace element concentrations we extrapolated detection sensitivities well below 1ppm for most of the studied elements. For example, for Fe we obtained detection limits in the range 6-51ppb, depending on the matrix material (for further details see (3)).

ACKNOWLEDGEMENTS

We gratefully acknowledge the financial support for T. Krustev by the Higher Education Funding Council for Wales under the *Technology Foresight* initiative.

REFERENCES

1. Downey, S.W., Kopf, R.F., Schubert, E.F., and Kuo, J.M., *Appl. Opt.* **29**, 4938-4942 (1990).
2. Abraham, C.J., "Trace element analysis of metallic and organic matrix materials exploiting RIMS", *PhD thesis*, University of Wales Swansea, 1997.
3. Telle, H.H., Abraham, C.J., and Jones, O.R., *Rapid Commun. Mass Spectrom.* **12**, 000-000 (1998), in press.
4. Perks, R.M., Jones, O.R., Grey Morgan, T., and Telle, H.H., *Rap. Commun. Mass Spectrom.* **10**, 1725-1738 (1996).
5. Saloman, E.B., *Spectrochim. Acta B,* **45**, 37-83 (1990); ibid. **46**, 319-378 (1991); ibid. **47**, 517-543 (1992); ibid. **48**, 1139-1203 (1993).

Resonance Enhanced Laser Mass Spectrometry for Process- and Environmental-Analysis: Applications and Perspectives

Ralf Zimmermann[1,2]*, Hans Jörg Heger[1,3], Ralph Dorfner[1,2],
Ulrich Boesl[3], Antonius Kettrup [1,2]

[1] *GSF Forschungszentrum Umwelt und Gesundheit GmbH, Institut für Ökologische Chemie, Ingolstädter Landstraße. 1, D-86764 Neuherberg, Germany*
[2]*Technische Universität München, Lehrstuhl für Ökologische Chemie und Umweltanalytik, 85350-Freising, Germany*
[3] *Technische Universität München, Institut für Physikalische und Theoretische Chemie, Lichtenbergstr. 4, D-85747 Garching, Germany*

ABSTRACT

Laser induced Resonance-Enhanced Multi-Photon Ionization Time-Of-Flight Mass Spectrometry (REMPI TOFMS) is a highly selective as well as sensitive analytical technique, well suited for species selective, on-line monitoring of trace-substances. In this contribution some analytical applications of a mobile REMPI-TOFMS are presented. This includes REMPI-TOMS on-line analysis of coffee roasting gas and waste incineration flue gas as well as headspace measurements of pulp processing lye or rapid analysis of polycyclic aromatic hydrocarbons from soil samples via thermal desorption.

INTRODUCTION

Recent industrial on-line process analysis often focuses on a few, easily measurable parameters. Thus relatively cheap measurement principles with restricted chemical selectivity, such as IR or UV/VIS spectroscopy or chemical sensors, are commonly used. Such approaches allow e.g. selective detection of smaller inorganic compounds such as oxygen, nitrogen monoxide, carbon monoxide, sulfur dioxide from combustion or thermal process offgases. However, the control of advanced, highly complex industrial production processes, such as chemical synthesis, semiconductor manufacturing or high performance combustion technology, requires advanced, more selective on-line process analytical instrumentation. In this contribution, a mobile process monitor based on **Resonance Enhanced MultiPhoton Ionization – Time-Of-Flight Mass Spectrometry** (REMPI-TOFMS) is described. The laser based Resonance-Enhanced Multiphoton Ionization (REMPI) technique utilizes excited molecular states for resonance-enhancement of a two-photon absorption-ionization process, thus involving laser UV-spectroscopy as analytical parameter to the ionization process. In combination with mass analysis (preferably with the affordable and highly competitive Time-of-Flight mass analyzers, TOFMS) one achieves a two dimensional analytical technique. The

---

* corresponding author, GSF-Research Center

technique was developed for applications in molecular spectroscopy and basic research in the late 1970's [1] With the availability of reliable, robust and affordable commercial laser systems, ideas for application of REMPI as highly selective and sensitive ion source for analytical mass spectrometry came up in the 1980's and early 1990's[2,3,4,5,6].

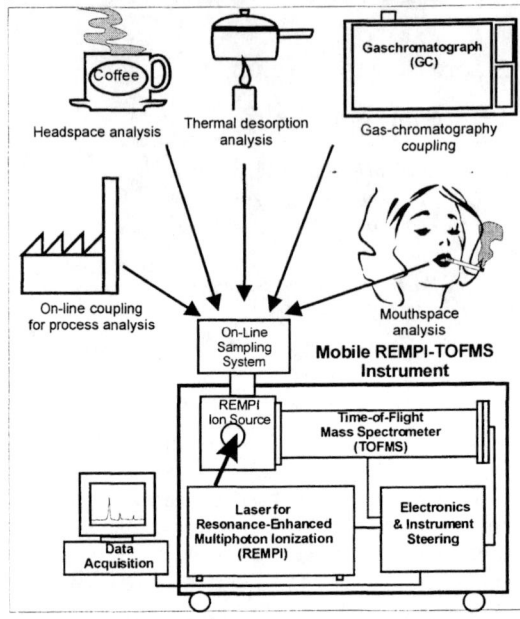

Figure 1 Schematic representation of the mobile REMPI-TOFMS instrument and the different sample inlet techniques.

In contrary to other multidimensional techniques, like gas chromatography-mass spectrometry (GC-MS), REMPI-TOFMS is a direct and ultra-fast technique and thus can be used for an on-line identification and quantification of trace components from "dirty" samples.

In the last years, REMPI-based analytical instruments have been developed. Recent applications include highly time-resolved on-line measurements of chemicals (e.g. pollutants as polycyclic aromatic hydrocarbons, PAH) in flue gases from internal combustion engines[7,8], waste incinerators[9,10] and in tobacco smoke[11] or monitoring of phenolic flavour active compounds in the coffee-roasting process-gas[12,13]. Further a gas-chromatography-REMPI-TOFMS device (three dimensional analytical instrument featuring retention time, UV-laser spectroscopy and mass spectrometry) was developed[14] Coupling to special sampling- and inlet systems allows the analysis of volatile and semi-volatile components. For example, a large sample thermal desorption interface allows a fast estimation of the contamination of soil samples with polycyclic aromatic hydrocarbons (PAH)[11].

EXPERIMENTAL

The mobile on-line REMPI-TOFMS monitoring device was designed for process analytical application at industrial locations. Therefore a compact and robust setup was required. A linear TOFMS set-up was used with an effusive sample inlet and an attached fixed-wavelength laser. The overall selectivity of the instrument still is sufficiently high for successful analysis of highly complex samples. With the laser systems used (Nd:YAG laser: 266 nm and KrF Excimer laser: 248 nm), several aromatic compounds can be detected in sub ppb-concentrations. Data acquisition and analysis is performed via a 500MHz transient recorder (*Signatec DA500A,* on board in a PC). The acquisition rate for entire mass spectra (16 Kbytes) is up to 10 Hz. Data analysis, as e.g. averaging for increased detection limits, can be performed with the storaged original data. For real-time display of the mass spectrometric information during a measurement, a digital storage oscilloscope (*Le Croy 9361*) is used. The dimension of the mobile on-line REMPI-TOFMS device (including *Lambda Physik* excimer laser MINEX™, 248 nm), is 1.15m x 0.85m x 1.45 *m*. Although the instrument is compact and robust, the flexibility of the system is high (for example a supersonic jet inlet can be mounted).

RESULTS AND DISCUSSION

The described mobile REMPI-TOFMS instrument was applied recently to a measurement campaign at a waste incineration pilot plant. Traces of polycyclic aromatic hydrocarbons were measured in the flue gas by REMPI-TOFMS as a function of process parameters. On-line monitoring of industrial flue gases require a carefully optimized sampling technique and mass spectrometric inlet system, in order to avoid memory effects or other problems related to condensation of low volatile material in the sampling lines. An all-quartz-glass sampling system, heatable to temperatures of up to 350°C, was developed. Filtration of particulate matter (fly ash, dust and aerosols) was performed via a heatable quartz-fiber filter. Alignment and calibration of the instrument as well as quantification of the measurements was performed via external standards (dynamic standards, i.e permeation and diffusion cells[15]). It was possible to observe variations of the naphthalene concentration on a some 10 pptv level directly in the post-combustion chamber. A similar sampling system was used for direct measurement of combustion byproducts in the mouth space of cigarette smokers (i.e. in the in- and exhaled air passing by). A special mouth piece was placed in the mouth of the smoker while he smoked a cigarette. Different combustion byproducts as benzene, toluene and xylenes (BTX), PAH as well as plant material pyrolysis products like phenol, styrene or indol were identified and measured highly time-resolved. Figure 2 shows two time to intensity profiles, measured in the waste incineration process (top) and in the mouth space of a smoker (bottom) respectively. The upper trace shows the naphthalene concentration in the afterburner chamber during a process change at the incinerator. The normal naphthalene concentration at this sampling point is about 50-200 ppt. Due to insufficient steering procedures, malfunctions with high emission peak concentrations can occur, as shown e.g. at the right side of the upper trace. This clearly emphasizes the use of a fast measuring technique for feed-back steering of the incinerator. However, a detailed analysis of the results obtained at the incinerator suggest, that most probably it is necessary to register a set of different

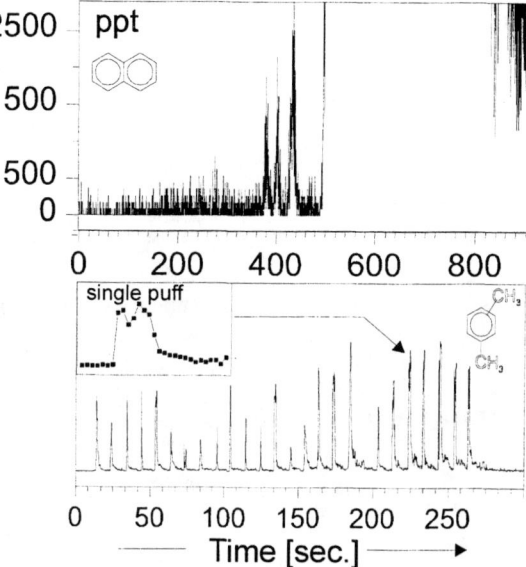

Figure 2: REMPI-TOFMS time to intensity profiles, demonstrate the sensitivity and time resolution of the approach. Top) Naphthalene concentration in the ppt concentration range in waste incineration flue gas. Bottom) Mouthspace measurement of xylene. The inset shows single puff resolution.

chemical components simultaneously. By application of fast numerical pattern recognition techniques and fuzzy logical approaches detailed process information may drawn from the observed emission patterns[10,16].

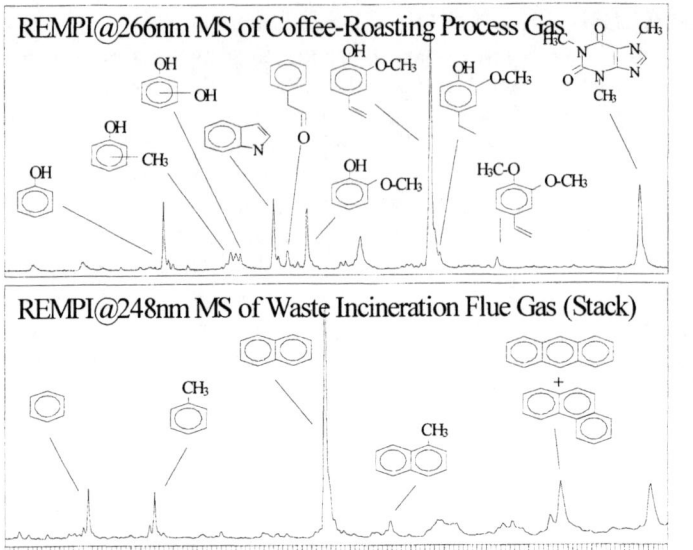

Figure 3: On-line registered REMPI mass spectra of coffee roasting gas (upper trace) and waste incineration flue gas (lower trace).

Figure 3 shows the REMPI mass spectra of highly complex process gases. The upper trace is due to coffee roasting gas. In coffee roasting gas, several hundred compounds in the mass range from 50 to 200 amu are present in ppb- to ppm- concentrations. The optical selectivity of REMPI allows direct detection of phenolic compounds as e.g. phenol, cresols, guaiacol and 4-vinylguaiacol as well as nitrogen heterocyclic components like caffeine and indol. Some of phenolic compounds belong to the key-flavour compounds of coffee and can be used as indicators for the coffee-roasting degree[17]. The lower spectrum is due to waste incineration flue gas, on-line measured close to the stack. Several aromatics can be detected in the stack gas. More details on the applications of REMPI-TOFMS for on-line measurements of waste incineration flue gases[16,9,10] and coffee-roasting gases[12,13,17] are given in the literature.

In general the analysis of liquid or solid samples require much more sample preparation effort compared to the analysis of gas phase samples. Conventional instrumental analysis of complex solid or liquid samples includes sample pre-concentration and (partly) clean-up steps prior to identification and quantification by hyphenated chromatographic and spectrometric instruments. (GC-MS or *high performance liquid chromatography*-with UV/fluorescence detection).

However, especially for applications in the field of industrial process analysis and detection of e.g. spills, a continuous monitoring technique with a fast response time would be highly desirable. For a straight forward, fast characterization of organic compounds in process fluids REMPI-TOFMS can be used with a dynamic or static headspace sampling technique (e.g. the gas phase close to the liquid surface is sampled).

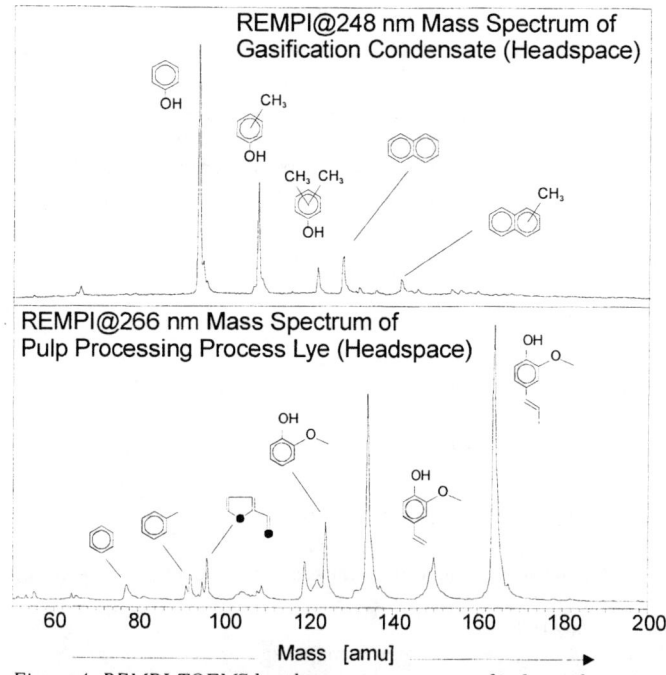

Figure 4: REMPI-TOFMS headspace mass spectra of industrial process fluids: Aqueous waste gasification condensate (upper trace) and pulp processing lye from a paper mill (lower trace).

Figure 4 shows the headspace REMPI-TOF mass spectra of aqueous waste gasification condensate (upper trace, 248 nm) and of pulp processing lye (lower trace 266 nm). The REMPI mass spectrum of the gasification condensate exhibits phenol, cresol and higher methylated phenols as well as naphthalene and its methylated derivatives. The phenolic compounds are products of wood gasification (pyrolysis), while the polycyclic aromatic hydrocarbons are combustion process related. Higher polycyclics are not volatile enough to be detected in the headspace spectrum at 20°C. The headspace REMPI mass spectrum of the pulp processing lye from a paper mill shows smaller volatile aromatics as benzene (78 amu) and toluene (92 amu) as well as typical sub-units from the lignin polymer, such as eugenol (164 amu) and furfural (96 amu). As pulp processing lye samples taken from different process stages show different headspace REMPI-TOFMS patterns, an application of headspace-REMPI-TOFMS on-line analysis may be the determination of the optimal reaction termination-time.

Future work in this field may include a combination of the REMPI-TOFMS instrumentation with a membrane inlet system[18]. The membrane inlet (MI) method allows a direct mass spectrometric analysis of volatile and semi-volatile organic compounds from water or aqueous solutions. Solid samples can be analyzed via a thermal desorption unit. Figure 5 shows a thermal-desorption REMPI-TOFMS mass

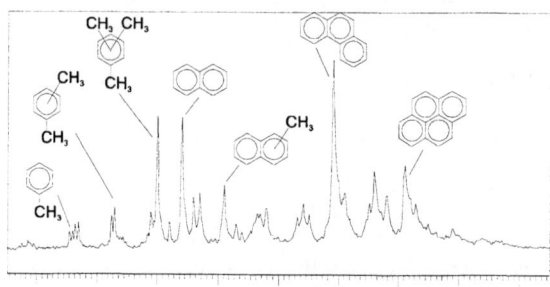

Figure 5: Thermal desorption REMPI@266nm mass spectrum of a soil sample, showing an increased PAH contamination

spectrum of a soil sample (agricultural soil, South Germany). It came out, that the soil exhibits significantly increased PAH levels in accordance with conventional measurements. The thermal desorption-REMPI-TOFMS approach allows an rapid rough classification of soil samples according to their PAH contamination.

In conclusion this work show, that there are a lot of practical application fields for (mobile) REMPI-TOFMS instruments. However, practical applications require very robust and stable instrumentation at moderate costs. The tasks for the next years in analytical laser mass spectrometry will be the construction of a black-box REMPI-TOFMS process monitor, the development of further sampling and inlet systems and hyphenated techniques as well as the investigation of new application fields

ACKNOWLEDGMENT

Financial support from the *Deutsche Bundesstiftung Umwelt*, Osnabrück, the DECHEM e.V, Frankfurt and Nestle´ Ltd. (NESTEC Ltd., Lausanne) is gratefully acknowledged.

REFERENCES

[1] U. Boesl, Neusser, E.W.Schlag, Z. Naturforsch. 33a (1978) 1546

[2] R.Tembreull, C.H.Sin, P.Li, H.M.Pang, D.M.Lubman; Anal. Chem. 57 (1985) 1186

[3] E.A.Rohlfing; 22nd Symposium (International) on Combustion, The Combustion Institute, Pittsburgh (1988)1843

[4] R.B.Opsal, J.P.Reilly, Optic News 6/1986 (1986) 18

[5] B.A.Williams, T.N.Tanada, T.A.Cool; 24th Symposium (International) on Combustion, The Combustion Institute, Pittsburgh (1992) 1587.

[6] U.Boesl, C.Weickhardt, R.Zimmermann, S.Schmidt, H.Nagel; SAE Technical Paper Series 930083 (1993) 61

[7] C.Weickhardt, U.Boesl, E.W.Schlag; Anal. Chem. 66 (1994) 1062 and U.Boesl, H.Nagel, R.Zimmermann, R.Frey; EnviroSense´97 Proceedings, Conference on Combustion Diagnostics, June 16-20, München (Germany) SPIE Series 3108 (1997) 2

[8] R.Zimmermann, D.Lenoir, A.Kettrup, H.Nagel, U.Boesl, 26th Symposium (International) on Combustion, The Combustion Institute, Pittsburgh (1996) 2859

[9] R.Zimmermann, H.J.Heger, A.Kettrup, U.Boesl, Rapid. Communic. Mass Spectrom. 11 (1997) 1095

[10] H.J.Heger, R.Zimmermann, R.Dorfner, U.Boesl, H.Griebel, M.Beckmann, A.Kettrup, submitted for publication

[11] R.Zimmermann, H.J.Heger, A.Kettrup, Annual report of the GSF Research Center for Environment and Health 1997, ISSN 0941-3847(1998) 17-28

[12] R.Zimmermann, H.J.Heger, C.Yeretzian, H.Nagel, U.Boesl, Rapid. Communic. Mass Spectrom. 10 (1996) 1980

[13] R.Zimmermann, H.J.Heger, R.Dorfner, C.Yeretzian, U.Boesl, A.Kettrup, Proceedings of the 1st International Convention on Food Engineering, FoodIng 1997, Cuneo, Italy (1997) 343

[14] R.Zimmermann, C. Lermer, K.-W.Schramm, A. Kettrup, U. Boesl, European Mass Spectrometry 1 (1995) 343 , see also: Anal. Chem. 68 (1996) 80A, R.Zimmermann, U.Boesl, H.J.Heger, E.R.Rohwer, E.K.Ortner, E.W.Schlag, A. Kettrup, J. High. Resol. Chromatogr. (HRC) 20 (1997) 461

[15] J.Namiesnik, *J.Chromatog.* 300 (1984) 79

[16] H.J.Heger, R.Zimmermann, R.Dorfner, A.Kettrup, U.Boesl, Proceedings of the 9th Int. Symposium on Resonance Ionization Spectroscopy, AIP press, this volume

[17] R.Dorfner, R.Zimmermann, U.Boesl, Proceedings of the 9th Int. Symposium on Resonance Ionization Spectroscopy, AIP press, this volume

[18] F.R.Lauritsen, T.Kotiaho, Rev. Anal. Chem. XV (1996) 237

SESSION IV
APPLICATIONS

Laser Mass Spectrometry of Chemical Warfare Agents Using Ultrashort Laser Pulses

C. Weickhardt, C. Grun, J. Grotemeyer

Lehrstuhl Für Physikalische Chemie und Analytik, BTU Cottbus,
Am Technologiepark 1, D-03099 Kolkwitz, Germany

Abstract. Fast relaxation processes in excited molecules such as IC, ISC, and fragmentation are observed in many environmentally and technically relevant substances. They cause severe problems to resonance ionization mass spectrometry because they reduce the ionization yield and lead to mass spectra which do not allow the identification of the compound. By the use of ultrashort laser pulses these problems can be overcome and the advantages of REMPI over conventional ionization techniques in mass spectrometry can be regained. This is demonstrated using soil samples contaminated with a chemical warfare agent.

INTRODUCTION

Among the various ionization techniques used in analytical mass spectrometry resonant multiphoton ionization (REMPI) has risen considerable interest during the last couple of years. This is due to the unique features of this method which allow the selective ionization and detection of the compounds of interest in even crude mixtures with high sensitivity and usually a low degree of fragmentation thus simplifying the interpretation of mass spectra. When coupled to a time-of-flight mass spectrometer REMPI enables on-line analysis with high temporal resolution as it significantly reduces or even removes the need for time consuming clean ups.

However, these features of REMPI mass spectrometry can only be obtained as long as the excited state(s) involved in the multiphoton ionization process are long lived compared to the duration of the ionizing laser pulse. If conventional nanosecond laser pulses are used this is by no means the usual case. Several relaxation processes may take place on such time scales that intermediate states are significantly depleted within this time. Depending on the nature of a certain energy redistribution mechanism its effect on the multiphoton ionization process and finally on the mass spectrum will differ. The probability for absorption of further photon(s) can be reduced because of vanishing Franck-Condon factors of vibrationally highly excited molecules which underwent Internal Conversion (IC), Internal Vibrational Redistribution (IVR), or Intersystem Crossing (ISC) processes. Furthermore, molecules containing sufficient vibrational energy may dissociate. In any case the resulting mass spectrometric measurement will lack sensitivity and be problematic as far as interpretation is concerned.

CP454, *Resonance Ionization Spectroscopy*
edited by J. C. Vickerman, I. Lyon, N. P. Lockyer, and J. E. Parks
© 1998 The American Institute of Physics   1-56396-810-X/98/$15.00

Inspecting various groups of environmentally and technically relevant substances one notices that fast relaxation processes with time constants greater than 10^9 sec^{-1} in excited states are a widespread phenomena which sets drastic limitations to RIMS using nanosecond laser pulses. A preliminary list includes nitro organic compounds (explosives), organic compounds containing heavy or metal atoms, aromatic ketones, larger biomolecules and chemical warfare agents.

A straight forward way to overcome the problems related to fast relaxation processes is the use of ultrashort laser pulses (T < 1 ps), which finish the ionization process before the energy redistribution processes can set in. In order to maintain a high level of sensitivity these short pulses have to be high in intensity. It has to be investigated whether sensitivity and easyly interpretable mass spectra can still be maintained under these conditions.

EXPERIMENTAL

The substances under investigation were cooled in a supersonic jet expansion in order to extract the major part of internal energy from the molecules and to prepare them in few initial states before ionization. The substances investigated exhibit a too low vapor pressure to be directly mixed with the carrier gas of the supersonic jet expansion. Thus, they were laser desorbed by a CO_2 laser on the low pressure side of the nozzle.

Multiphoton ionization could either be performed by the frequency doubled output of a conventional nanosecond dye laser (pulse energy ca. 200 µJ) or a dye laser system delivering 0.5 ps UV laser pulses (pulse energy ca. 40 µJ). Both laser beams could be coupled into the ion source of the reflectron time-of-flight mass spectrometer simultaneously thus ensuring equal conditions for a comparison of nanosecond and sub-picosecond RIMS.

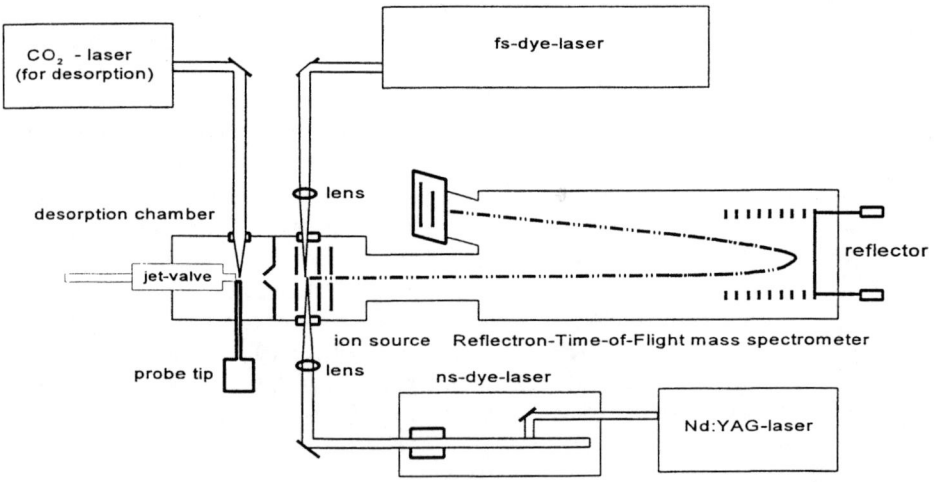

Figure 1. Reflectron TOF mass spectrometer equipped with laser desorption device and ionization lasers

RESULTS AND DISCUSSION

Comparison of ns and sub-ps RIMS

The changes in mass spectra when going from nanosecond to sub-picosecond ionization pulses depends on the nature of the relaxation process occurring in the photo excited molecule. Ns and sub-ps-REMPI mass spectra of typical examples for each of the processes fragmentation, IC, and ISC are compared in figure 2. IC and ISC leave the excited molecule in high vibrational quantum numbers which block the ionization and reduce the number of molecular ions produced. Thus, only the formation of small fragment ions is observed under nanosecond conditions in the cases of β-carotene and benzophenone. The ns-laser mass spectrum of Clark I (diphenylarsinechlorine), a warfare

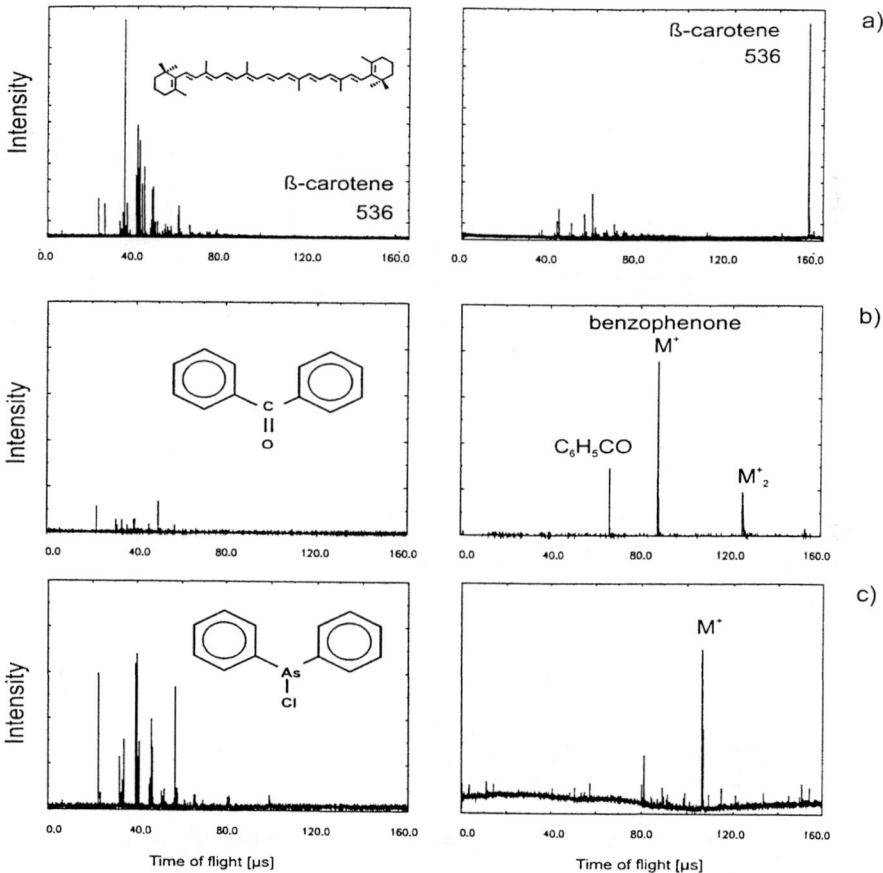

Figure 2. Comparison of ns-REMPI mass spectra (left column) and sub-ps-REMPI mass spectra of (a) β-carotene, (b) benzophenone, and (c) Clark I.

agent which undergoes very fast dissociation in its excited levels only shows the signals of fragment ions of the aromatic ring. Obviously, non of these mass spectra obtained by ns-REMPI does contain analytically useful information and permits the identification of the analyte.

When sub-picosecond laser pulses are used for REMPI the mass spectra of these example substances change drastically. The molecular ion becomes the base peak in all spectra and besides it few but structure specific fragments appear. Due to the shortness of the laser pulses relaxation processes are not able to influence the multiphoton ionization. The mass spectra obtained in this way are easy to interpret and allow the identification of compounds even in cases where several substances are ionized simultaneously.

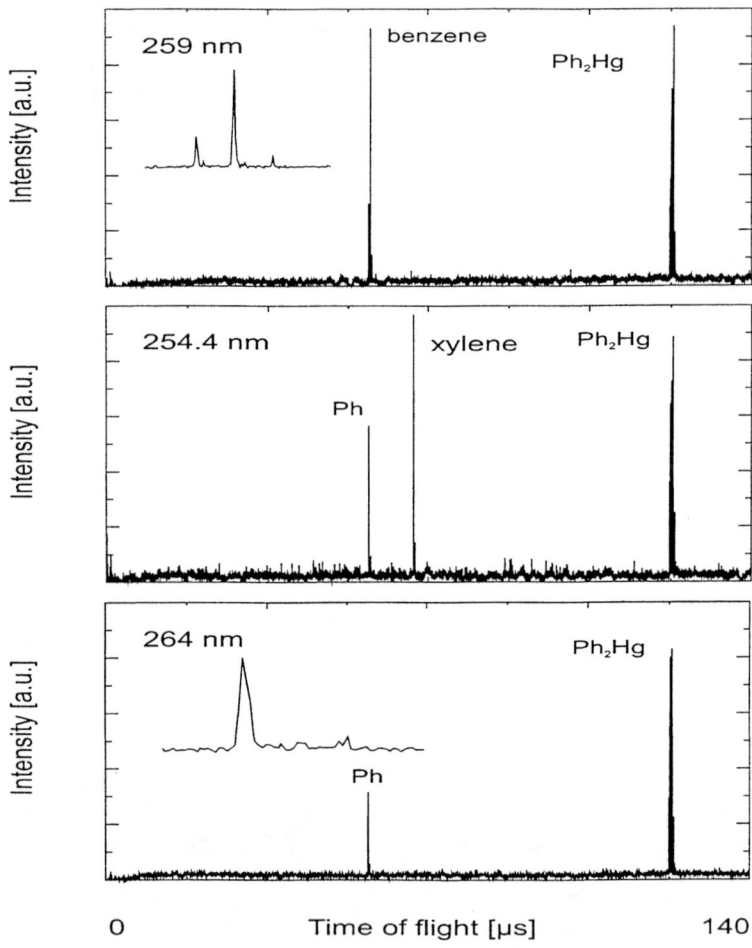

Figure 3. Sub-ps laser mass spectra of a mixture of diphenylmercury, benzene and xylene at three different wavelengths demonstrating that selectivity is maintained and non-resonant MPI is negligible.

Selective Ionization with sub-ps laser pulses

These mass spectra demonstrate that one of the major advantages of REMPI over several other ionization techniques, the small degree of fragmentation, can be maintained under the conditions of short but high intensity pulses.

However, the high light intensity (ca. 10^{10} to 10^{11} W/cm^2) and the increased spectral width of such pulses attacks REMPI's ability to selectively ionize certain compounds in mixtures. The high intensity increases the probability for non-resonant multiphoton ionization processes whereas a spectral broadening of the laser radiation raises the problem of simultaneous ionization of compounds with closely neighbouring absorption lines. While the second point has to be investigated for each specific mixture to be analyzed, the part of non-resonant ionization at the conditions used here was checked by creating a mixture of three components and observing the ion signal of two of them at a laser wavelength off their respective resonances. As an example benzene and xylene were seeded in argon to which platinum diphenylmercury was mixed by laser desorption. Figure 3 a and b show that the aromatic compounds are present and can be ionized by tuning the laser to one of their specific resonances. For the spectrum in figure 3 c the sub-ps laser was tuned to a wavelength for resonant ionization of the metal organic compound but not for the aromatic components. The result is a mass spectrum which does not exhibit any contribution of benzene or xylene demonstrating that a high degree of selectivity is still maintained under the conditions used in this experiment.

Application to Real Environmental Samples

In order to apply the technique of laser desorption (LD) combined with resonant multiphoton ionization by ultrashort laser pulses to a typical analytical problem, adamsite was detected in soil samples spiked with this chemical warfare agent. Soil samples were pressed to pallets and introduced into the mass spectrometer without any cleanup. Figure 4 shows the mass spectra obtained by LD-sub-ps-REMPI mass spectrometry for pure adamsite, "clean" soil, and "contaminated" soil. Each of the spectra was obtained within 1 sec by averaging over 20 laser shots. These measurements demonstrate the capability of the method to quickly detect pollutants in environmental samples without the need for time consuming cleanup procedures.

CONCLUSION

The use of ultrashort laser pulses in the sub picosecond range allows the extension of the advantageous features resonant multiphoton ionization offers for analytical mass spectrometry to molecules with quickly relaxing intermediate states. While up to intensities of 10^{11} W/cm^2 a high degree of selectivity is still maintained in the ionization step, the mass spectra of such molecules are characterized by a dominant molecular ion signal accompanied by structure specific fragments, which simplifies their identification. As an example for a typical analytical application the chemical warfare agent adamsite was detected in untreated soil samples.

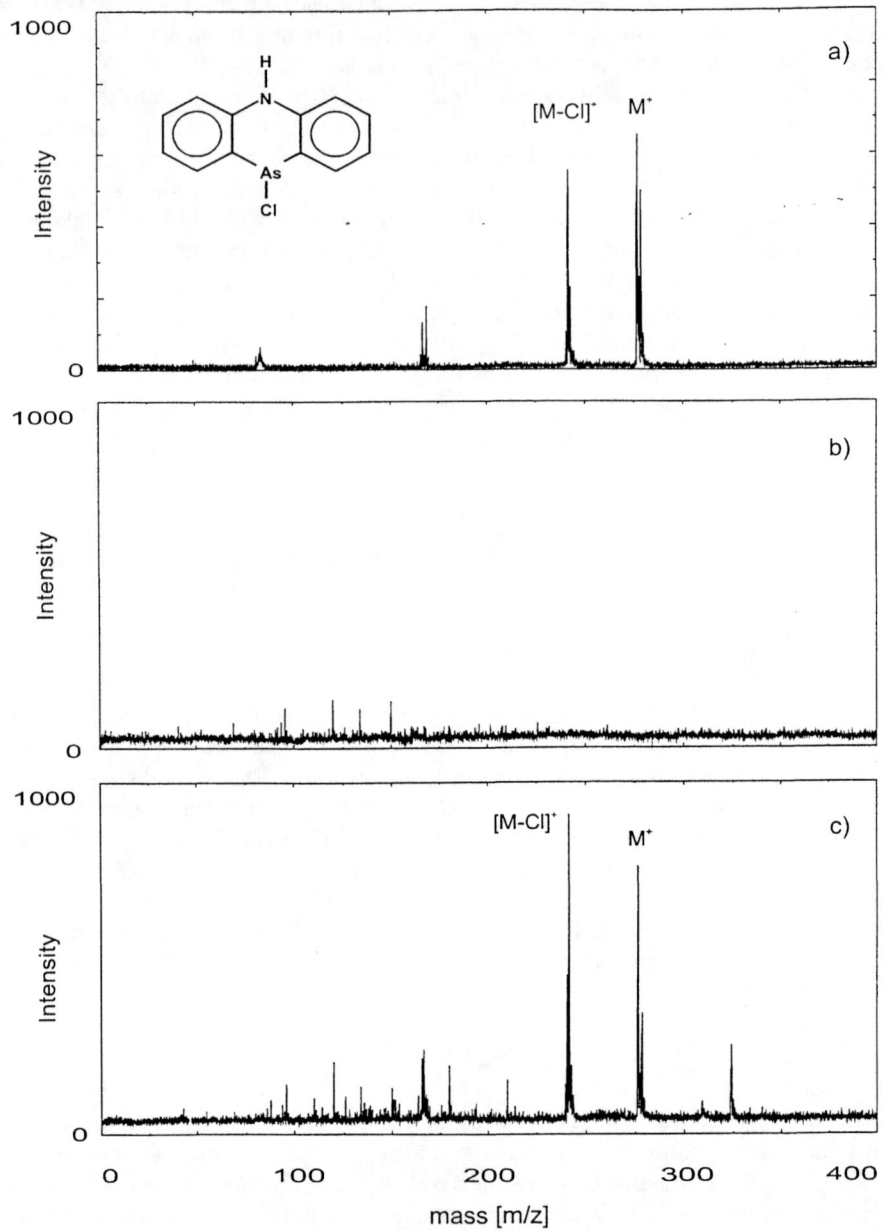

Figure 4. Detection of Adamsite in untreated soil samples. Sub-ps-laser mass spectra of (a) pure Adamsite, (b) clean soil, (c) soil contaminated with Adamsite.

Ultratrace Determination of the Long-lived Isotope ^{41}Ca by Narrowband CW-RIMS

P. Müller[†], B. A. Bushaw[*], K. Blaum[†],
W. Nörtershäuser[†], N. Trautmann[‡], K. Wendt[†]

[†]*Institut für Physik*, [‡]*Institut für Kernchemie*,
Johannes Gutenberg-Universität, D-55099 Mainz, Germany
[*]*Pacific Northwest National Laboratory, Richland, WA 99352, USA*

Abstract: Determination of odd isotopes with multi-step RIMS is usually hampered by the hyperfine structure arising from a non zero nuclear spin. The splitting of the transition strength leads to a reduced ionization efficiency and large uncertainties in the determination of isotope ratios. We present a technique to overcome these drawbacks by appropriate use of optical pumping. Applying this technique, first isotopic ratio measurements on synthetic ^{41}Ca-samples could be accomplished using double-resonance three photon ionization in a collimated atomic beam combined with a quadrupole mass spectrometer.

INTRODUCTION

Determining the abundance of the long-lived ultratrace isotope ^{41}Ca ($t_{1/2}=1\cdot10^5$ a) enables a number of interesting applications. These cover the use of ^{41}Ca as a specialized medical tracer for the study of the calcium kinetics in the human body (1), dating of geological and anthropological samples in the time window between 10^4 and 10^6 years as well as evaluation of exposure histories of extraterrestrial material (2). The difficulty in the detection of ^{41}Ca is the presence of large amounts of the stable isotopes, predominantly ^{40}Ca. Depending on the analytical application, isotopic selectivities of more than 10^{14} and overall detection efficiencies of up to 10^{-4} are required. Currently, only AMS-facilities can reach these specifications, which implies large experimental and financial expenditure (3).

We report on trace determination of ^{41}Ca utilizing the extremely high selectivity and sensitivity achievable by resonance ionization mass spectrometry, combining multi-step resonant optical excitation by narrowband lasers with quadrupole mass spectrometry (4). Isotope ratio measurements of odd isotopes must overcome the difficulties of hyperfine splitting. A technique for precise and efficient isotope ratio measurements on odd isotopes has been demonstrated and first measurements of ^{41}Ca are presented.

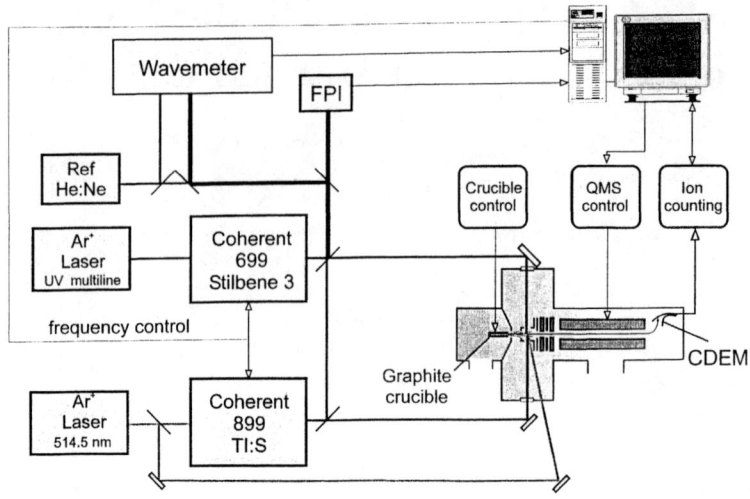

FIGURE 1. Experimental setup for doubly resonant three-photon RIMS of Calcium

EXPERIMENTAL SETUP

The experimental setup used for this work is shown in Figure 1. Calcium is atomized from either nitrate or metallic samples in a cylindrical graphite crucible, which is directly heated to temperatures of up to 1300 C. The atomic beam effusing out of the exit hole is collimated with a full-angle divergence of about 5° and intersects the laser beams perpendicular within the ionization region. The ionization region is held at a small positive potential of about 20 V, relative to the crucible, to suppress surface ions from the atom source. The resonant optical excitation uses the $4s^2\ ^1S_0 \rightarrow 4s4p\ ^1P_1 \rightarrow 4s4d\ ^1D_2$ scheme at 422.7 nm and 732.8 nm, which is driven by tunable narrowband cw lasers. For this work a cw dye laser operating with Stilbene 3 was used for the first step and a cw titanium-sapphire laser for the second step. The nonresonant photoionization is performed with an argon-ion laser operating at 514.5 nm. The resulting ions are accelerated and focused into the quadrupole mass spectrometer (QMS) by applying a weak electrical field. The QMS is operated either in the so called rf-only mode, where all masses are transmitted simultaneously, or in the mass filter mode, where only a single mass is transmitted. Ions reaching the end of the QMS are finally detected by a continuous dynode electron multiplier (CDEM). The frequencies of the two resonant lasers are controlled and stabilized by computer controlled fringe-offset-locking to a single mode He:Ne laser in a 150 MHz confocal Fabry-Pérot interferometer (FPI). The complete experiment including lasers and QMS as well as the data acquisition is controlled by a single PC.

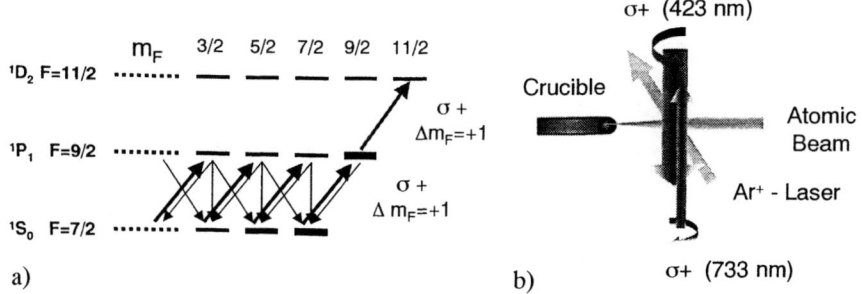

FIGURE 2. Scheme of optical pumping as applied for ^{43}Ca and ^{41}Ca in the double-resonance optical excitation

OPTICAL PUMPING

^{41}Ca as well as ^{43}Ca exhibit hyperfine structure in both excitation steps due to their nuclear spin of 7/2. This results in a complicated spectrum when tuning both lasers over the complete hyperfine structure. Therefore, precise isotope ratio measurements relative to the even isotopes are difficult, or even impossible, since the intensities of the HFS components vary greatly with alignment and polarization of the lasers due to optical pumping induced redistribution of hyperfine level and/or magnetic sublevel populations. Moreover the partitioning of transition strength into a number of hyperfine lines drastically reduces excitation efficiencies.

To obtain reproducible and precise isotope ratios of odd isotopes and to avoid a reduction of the excitation probability, one can take advantage of optical pumping in the first excitation step by using σ^+ light. Consequently, according to the selection rules, only transitions with $\Delta m_F = +1$ are allowed for the optical excitation, while the fast radiative decay back to the ground state is allowed in all Zeemann levels with $\Delta m_F = 0,\pm 1$. After several pump cycles, nearly all the ground-state population accumulates in the $m_F = 7/2$ Zeeman level. Hence, subsequent double-resonance excitation with two σ^+ photons can lead only to the F = 9/2 state in the first and the F = 11/2 state in the second excited level, as shown in Figure 2a. By using this technique, the excitation probability for the odd isotope is concentrated almost exclusively in one HFS component. Furthermore, the oscillator strength becomes equal to that of the even isotopes and hence precise isotopic ratio measurements can be accomplished, as will be further shown.

To achieve ideal experimental conditions for optical pumping, both lasers must be carefully σ^+ polarized. The first-step laser beam has to be slightly larger in diameter than that of the second-step laser and needs to be positioned in the optical excitation region with a small spatial offset relative to the center of the second-step laser as shown in Figure 2b. This setup ensures that the atoms are first exposed to the blue light and can interact long enough with the resonant laser radiation for a good polarization of the atomic beam prior to the second excitation.

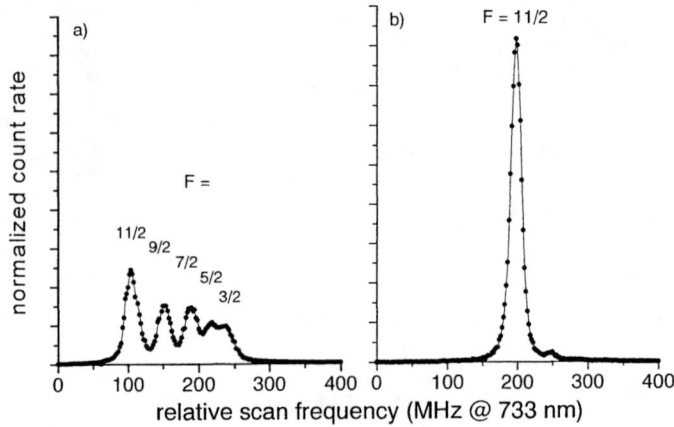

FIGURE 3. Normalized intensities of double-resonance signals of ^{43}Ca, the frequency of the first-step laser is fixed, the frequency of the second-step laser is scanned, a) two-photon transition far off-resonance from the intermediate level, b) first-step laser tuned to F=7/2 → F=9/2 transition

For a demonstration of the potential of this technique the stable isotope ^{43}Ca was used, which is an ideal test candidate due to its similarities to ^{41}Ca. Hyperfine structure patterns with and without optical pumping were recorded. The latter was realized by detuning the first-step laser from the intermediate resonance. In this case both resonant laser beams were linear polarized and carefully adjusted for maximum beam overlap. To get sufficient intensity for the off-resonant two-photon transition, we used about 40 mW at 423 nm and 500 mW at 733 nm. The first-step laser was fixed at 500 MHz above the F=7/2 → F=9/2 transition while the second-step laser was scanned over the two-photon resonance. The result of this measurement is shown in Figure 3a. All HFS components are clearly resolved.

To test the pumping efficiency, both lasers were σ^+ polarized and operated at a power of about 5 mW. The blue laser beam was expanded to a diameter of 3.5 mm while the red laser beam had a diameter of 2.1 mm. Both were aligned for optimum pumping efficiency as described above (Fig. 2b). The frequency of the first-step laser was fixed at the center of the F=7/2 → F=9/2 transition while the second laser was again scanned over the complete HFS. The result of this measurements is shown in Figure 3b. The effect of optical pumping is evident. As expected, the F=9/2 → F=11/2 transition predominates, with residual contribution from all other HFS components smaller than 4 %.

ISOTOPE RATIO MEASUREMENTS OF ^{41,43}Ca

In a second experiment, this technique was used to perform isotope ratio measurements of ^{43}Ca relative to stable even isotopes. To accomplish this task a two-dimensional frequency scan of both resonant lasers was performed, covering the resonances of ^{42,43,44}Ca. Since optical selectivities in double-resonance ionization are already sufficient to separate these isotopes, the QMS was operated in the rf-only mode. In Figure 4 the

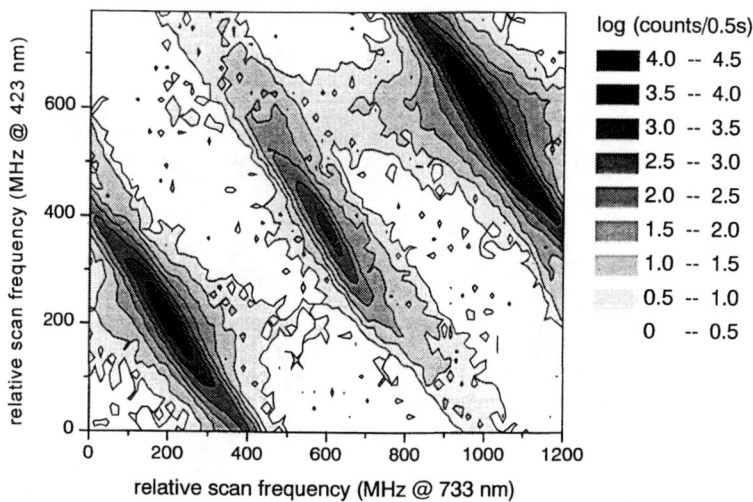

FIGURE 4. two dimensional scan of both resonant lasers over the isotopes ^{42,43,44}Ca with optical pumping, QMS operated in rf-only mode to transmit all masses.

logarithmic count rate is shown in a contour plot as a function of the detuning of both lasers. Although the middle peak is the odd isotope ^{43}Ca, all three peaks exhibit the same peak structure, which is determined by the coherent optical excitation (i.e., the prominent diagonal structure is the result of a coherent two-photon transition from the ground state to the second excited state). The ^{43}Ca peak shows only a small asymmetry on the right flank due to the residual contribution of other HFS components. From the peak heights and the well-known abundances of ^{42}Ca and ^{44}Ca, the abundance of ^{43}Ca was calculated to be 0.124(12) %, which shows excellent agreement with the literature value for the natural ^{43}Ca abundance of 0.122 %.

After successfully applying the optical pumping technique to isotope ratio measurements of ^{43}Ca, a first trace detection of an artificially enriched ^{41}Ca sample was performed using the same experimental setup. The sample was 10 μl of nitrate solution containing 62 mg Ca^{2+}/ml. The ^{41}Ca/Ca ratio was specified to be $1.0(3) \cdot 10^{-6}$, corresponding to 620 pg or 10^{13} atoms of ^{41}Ca inserted into the crucible. Similar to the measurements on ^{43}Ca, a two-dimensional scan over the resonances of ^{41,42}Ca was performed. However, operation of the QMS in the mass filter mode was necessary to suppress isotopic background originating from ^{40}Ca. Consequently, the QMS was manually switched between mass 41 and 42. The result of this measurement is shown in Figure 5. Experimental data were corrected for exponential decrease of atomization rate as well as for dead-time effects in the counting electronics. Again, the ^{41}Ca peak exhibits no HFS and peak ratios were used to determinate the isotope ratio to be $0.8(1) \cdot 10^{-6}$, which is in good agreement with the specified value. In further measurements, the overall detection efficiency of the double-resonance ionization, including atomization and mass selective detection, was determined to be $1 \cdot 10^{-6}$, which is close to the expected value.

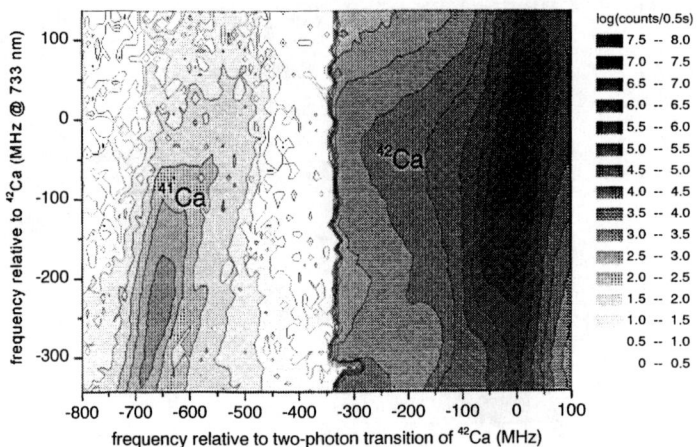

FIGURE 5. Contour plot of a two dimensional scan of both resonant lasers over the isotopes ^{41,42}Ca with optical pumping, QMS switched to transmit either mass 41 (left of dark line) or mass 42 (right of line).

CONCLUSION AND OUTLOOK

The optical pumping technique described in this work has been shown to be extremely well suited for efficient and precise isotope ratio measurements of odd calcium isotopes, including the long-lived isotope ^{41}Ca. The isotopic selectivity of the double-resonance optical excitation surpasses 10^5 while the isotopic abundance sensitivity of a refined QMS system has been shown to be 10^8, yielding a combined selectivity of greater than 10^{12}, which is fully sufficient for medical applications. To further increase selectivity and efficiency, a triple-resonance excitation scheme will be used, which has already been tested with stable even isotopes.

ACKNOWLEDGEMENTS

This work was supported by the Office of Basic Energy Sciences of the U.S. Department of Energy under Contract DE-AC06-76RLO 1830, the Deutsche Forschungsgemeinschaft under Contract Tr336/1-3, and the DAAD.

REFERENCES

1. Freeman, S.P.H.T., et al., *Nucl. Instr. Meth. Phys. Res. B* 123, 266-270 (1997)
2. Fink, D., et al., *Nucl. Instr. Meth. Phys. Res. B* 52, 572-582 (1990)
3. Dittrich-Hannen, B., et al., *Nucl. Instr. Meth. Phys. Res. B* 113, 453-456 (1996)
4. Bushaw, B.A., et al., "Multiple Resonance RIMS Measurements of Calcium Isotopes Using Diode Lasers", in *Proceedings of the Conference on Resonance Ionization Spectroscopy*, 1996, pp. 115-118

SESSION V
PHOTOELECTRONS AND PHOTODISSOCIATION

Photoelectron Spectroscopy of Reactive Intermediates

J.M. Dyke*, S.D. Gamblin*, A. Morris*, J.B. West[†], T.G. Wright*

*Department of Chemistry, University of Southampton, Highfield,
Southampton SO17 1BJ, UK
[†]CCLRC, Daresbury Laboratory, Daresbury,
Warrington WA4 4AD, UK

Abstract. The advantages of using synchrotron radiation to study reactive intermediates are described and likely future developments are outlined. Examples of studies on $O_2(a^1\Delta_g)$, OH and SO are used to show that the use of synchrotron radiation in this research area allows more information to be obtained on the molecular ions and the associated photoionization processes.

INTRODUCTION

The last ten years has seen photoelectron spectroscopy of reactive intermediates develop from a research area which uses an inert gas discharge as the photon source to a field which also uses synchrotron vuv radiation and pulsed uv lasers, utilized in resonant and non-resonant ionization modes, as the sources of ionizing radiation. This has in general led to more information being obtained on the reactive intermediate and its ionic states and on the photoionization processes involved. This paper will concentrate on the study of reactive intermediates using photoelectron spectroscopy where synchrotron radiation is used as the photon source.

The use of monochromatized synchrotron radiation in the vacuum ultraviolet region to study reactive intermediates introduces an extra degree of freedom over that available with a discrete line source. This allows more information to be obtained on the ionic states produced. In particular:

(i) measurement of photoelectron angular distributions at fixed photon energy should allow information on the photoionization dynamics to be obtained; such measurements are possible as the synchrotron source is polarized - typically linearly polarized in the horizontal direction;

(ii) a study of a photoelectron feature as a function of photon energy allows autoionization resonances to be located. Once these resonances have been identified, a photoelectron spectrum recorded with the photon energy chosen to coincide with one of these resonances can give rise to extra structure over that observed in a photoelectron

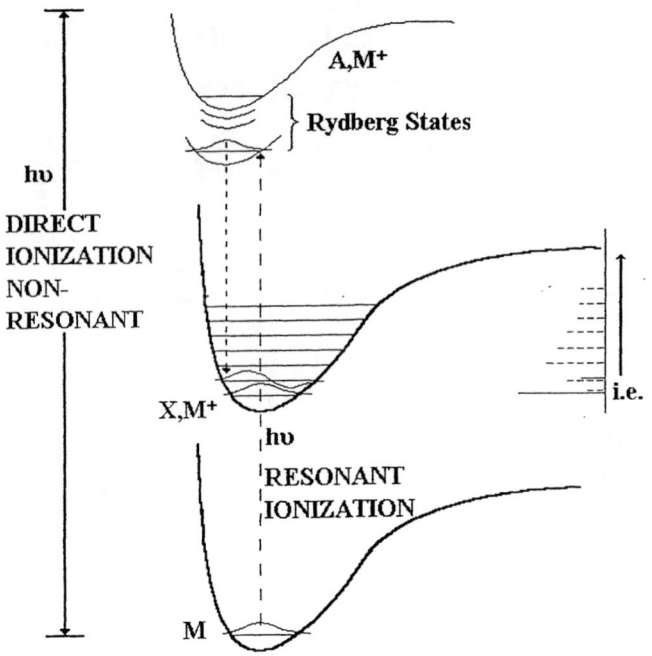

FIGURE 1. Comparison of expected vibrational band envelopes for
(a) Non-resonant ionization from a non-bonding orbital (solid lines), and
(b) Resonant ionization from a vibrational level of a Rydberg state which converges to an ionic state A,M$^+$ which has different characteristics to the ground ionic state X, M$^+$ (broken lines).

spectrum recorded off-resonance. It was this last feature which initially attracted the Southampton PES group to use synchrotron radiation because in a number of previous investigations on reactive intermediates some of the valence photoelectron bands when recorded with a HeI (21.22 eV) photon source showed an intense adiabatic component with very little intensity in other vibrational components. This was the case for the first photoelectron bands of the radicals OH, SH, N$_3$, CH$_3$ and CH$_3$O.

Such an observation for the first photoelectron band of a reactive intermediate is particularly disappointing when a lot of effort may have been made to prepare the reactive intermediate of interest and where measurement of the vibrational level separations in the ground ionic state is one of the main experimental objectives.

Synchrotron radiation can alleviate this problem. As is illustrated in Figure 1, the Franck-Condon factors for an off-resonant ionization from a molecular ground state M to an ionic state M$^+$ may be such that only the $v^+ = 0 \leftarrow v'' = 0$ component appears with any intensity. However, if the photon energy is tuned to match the energy separation from M to Rydberg states that converge to a higher ionic state, then these excited states, M*, can autoionize to the ground state of M$^+$. If the higher ionic state has a different potential curve from the first ionic state (i.e. the higher ionic state has a different shape and is displaced in the horizontal direction relative to the first ionic state), then the

Franck-Condon factors between the resonant vibronic level and the vibrational levels in the ion will be different from the Franck-Condon factors between M, $v'' = 0$ and M^+, v^+. Hence, as illustrated in Figure 1, by this autoionization route extra vibrational structure will be seen in the photoelectron band allowing vibrational level separations in the ground state of M^+ to be measured.

On-Resonance PE Spectra

- PE signal enhancement
- Vibrational profile can be extended
- FC analysis can aid CIS assignment

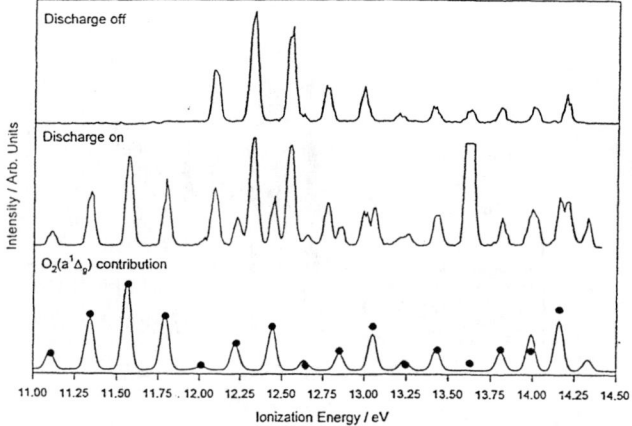

FIGURE 2. Photoelectron spectrum of molecular oxygen passed through a microwave discharge obtained at a photon energy of 14.37 eV. Black dots represent sum of direct and indirect FCF's calculated assuming autoionizing state to be $O_2(p^1\Phi_u)$ $v' = 5$.

A photoelectron spectrometer designed for studies of reactive intermediates with synchrotron radiation has been designed and built. With this apparatus, two types of experiment can be performed.
(a) The intensity of a photoelectron feature, e.g. a vibrational component in a photoelectron band, can be monitored as a function of photon energy. This is known as a constant-ionic-state (CIS) spectrum.
(b) A photoelectron spectrum (PES) can be recorded at a fixed photon energy - for example, at a photon energy corresponding to a resonance in the CIS spectrum.

Photoionization Studies of OH

$$H_2 \longrightarrow \mu wave \rightarrow 2H$$

$$H + NO_2 \rightarrow NO + OH$$

$$OH + OH \rightarrow H_2O + O$$

$$O + O + M \rightarrow O_2 + M$$

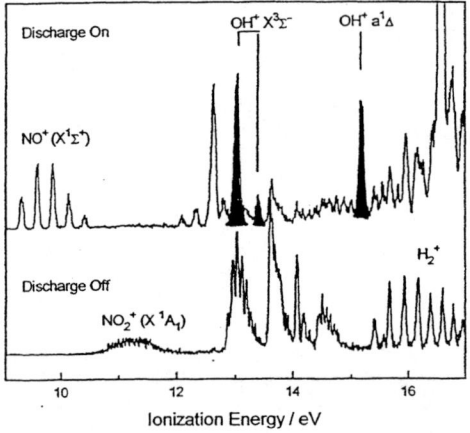

FIGURE 3. Photoelectron spectrum recorded for the H + NO$_2$ reaction. With hydrogen passed through the discharge (discharge on spectrum), the first two bands of OH can be identified and these are shaded in the diagram.

RESULTS

The yield of $O_2(a^1\Delta_g)$ obtained on passing flowing molecular oxygen ($O_2(X^3\Sigma_g^-)$) diluted in an inert gas, such as argon, through a microwave discharge is typically $\approx 15\%$. As a result, in the HeI photoelectron spectrum recorded for the discharge mixture, bands due to $O_2(X^3\Sigma_g^-)$ dominate. However, by recording CIS spectra of the first four vibrational components of the first photoelectron band of $O_2(a^1\Delta_g)$, resonant photon energies can be identified at which the $O_2(a^1\Delta_g)$ contribution to the photoelectron spectrum is enhanced. An example occurs at $h\nu = 14.37$ eV where an intense signal in the CIS spectrum is observed, corresponding to excitation of $O_2(a^1\Delta_g)$ $v'' = 0$ to the Rydberg state $O_2(^2\Phi_u, 3s\sigma_g)^1\Phi_u$ in the vibrational level $v' = 5$. As can be seen from

Figure 2, recording photoelectron spectra at this photon energy led to a clear enhancement in the signal associated with $O_2(a^1\Delta_g)$. Ionization to seventeen vibrational components of the ionic state, $O_2^+ \; X^2\Pi_g$, can be observed compared to only four components in an off-resonant spectrum. As can also be seen in Figure 2, the calculated Franck-Condon factors from $O_2(^1\Phi_u)$ $v' = 5$ to the vibrational levels of the ground ionic state, $O_2^+ \; X^2\Pi_g \; v^+$, match the experimental vibrational relative intensities very well. If the vibrational numbering in the excited state is changed for example to $v' = 4$ or $v' = 6$ the agreement of the experimental relative intensities and the computed Franck-Condon factors is very poor, providing support for the vibrational assignment in the resonant state.

$$\text{CIS of OH}^+ \; (X^3\Sigma^-) \; v^+ = 0 \leftarrow \text{OH} \; (X^2\Pi) \; v'' = 0$$

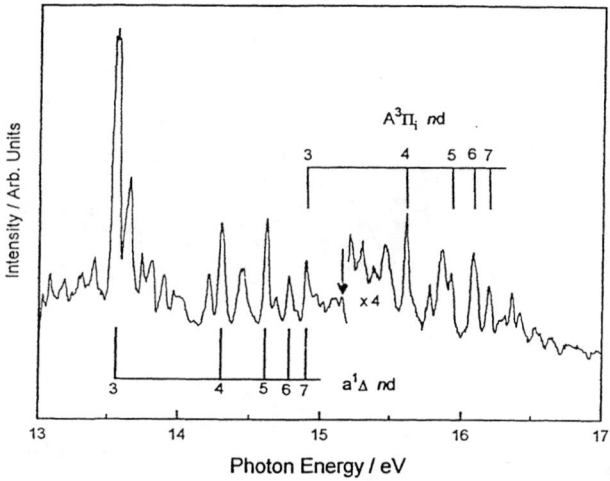

FIGURE 4. CIS of the first band of OH recorded in the 13.0-17.0 eV photon energy region.

As well as making short-lived molecules by microwave discharge of a flowing gas mixture, reactive intermediates can be prepared for study by rapid atom-molecule reactions. Figure 3 shows a photoelectron spectrum recorded for the $H + NO_2$ reaction at ≈ 1 ms reaction time. Bands due to NO can be clearly seen as well as two sharp bands associated with the $OH^+(X^3\Sigma^-, a^1\Delta) \leftarrow OH(X^2\Pi)$ ionizations. Both of these OH bands arise from ionization from the outermost 1π non-bonding molecular orbital and hence exhibit typical non-bonding envelopes. Only the first band shows an observable extra vibrational component (ionization to $v^+ = 1$) which is much weaker than the $v^+ = 0$ component. The CIS spectrum recorded for the $OH^+ \; X^3\Sigma^-, v^+ = 0 \leftarrow OH \; X^2\Pi, v'' = 0$ ionization is shown in Figure 4. This is a highly structured spectrum which is still undergoing analysis. Nevertheless, as shown, two Rydberg series can be identified, one which converges to OH^+ $a^1\Delta$ and the other which converges to OH^+ $A^3\Pi$, an ionic state obtained on ionization from the highest occupied σ orbital, which is bonding in character. Photoelectron spectra have been recorded at all the main resonances in Figure 4 and extra vibrational structure has been observed in the first band of OH in most of these spectra, compared with that recorded in an off-resonant spectrum. An example of this is shown in Figure 5 where six vibrational components are observed in the first band of OH, recorded at 15.15 eV photon energy, and eight vibrational components are observed for OD recorded at 15.08 ev photon energy.

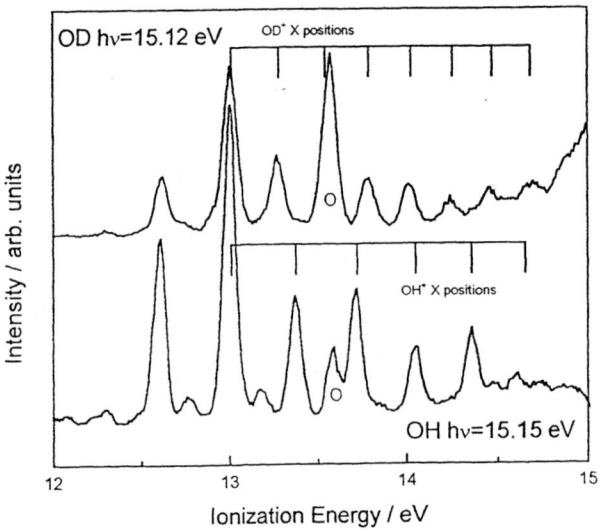

FIGURE 5. On-resonance photoelectron spectra obtained for OH and OD. 15.15 eV corresponds to the vertical arrow on Figure 4.

In the example outlined in Figure 1, the Rydberg states converging to the second ionic state are assumed to have different characteristics from the ground ionic state. If however these Rydberg states have similar characteristics to a lower ionic state then useful information can also be obtained. An example of this occurs in CIS spectra recorded for the fourth ionic state of SO, the $b^4\Sigma^-$ state, recorded at photon energies between the fourth and the fifth ionization energies. Because the fourth and fifth states of SO^+ (the $b^4\Sigma^-$ and the $B^2\Sigma^-$ states) arise from the same $(7\sigma)^{-1}$ SO ionization, the potential curves of the SO^+ $b^4\Sigma^-$ and the $B^2\Sigma^-$ states are very similar. As a result

Franck-Condon factors from the resonant Rydberg states, which are parts of series which converge to SO^+ $B^2\Sigma^-$, to the SO^+ $b^4\Sigma^-$ state are essentially diagonal. Hence in the CIS spectra obtained $\Delta v = 0$ transitions between the resonant excited state and the lower ionic state are strongly favoured. Figure 6 shows CIS spectra recorded for $SO^+(b^4\Sigma^-)$ and as expected the $v^+ = 0$ CIS spectrum shows transitions only from $v' = 0$ and the $v^+ = 1$ CIS spectrum shows transitions only from $v' = 1$ in the resonant state. Two strong series may be identified in the $v^+ = 0$ CIS spectrum, one of which consists of negative window resonances while the other is composed of positive window resonances. Both series converge to (16.43 ± 0.01) eV, the fifth adiabatic ionization energy of SO. Similar series are observed in the $v^+ = 1$ CIS spectrum, with the shift in the resonant positions for a given principal quantum number being the $v' = 0\text{-}1$ separation in the resonant Rydberg state. Recording photoelectron spectra at a photon energy corresponding to one of these resonances allows a vibrational component in the fourth photoelectron band to be selectively enhanced.

The results of these initial investigations demonstrate that the ability to change the photon energy using monochromatized synchrotron radiation is very valuable in that it allows control over the extent and relative intensities of the vibrational components observed in a photoelectron band.

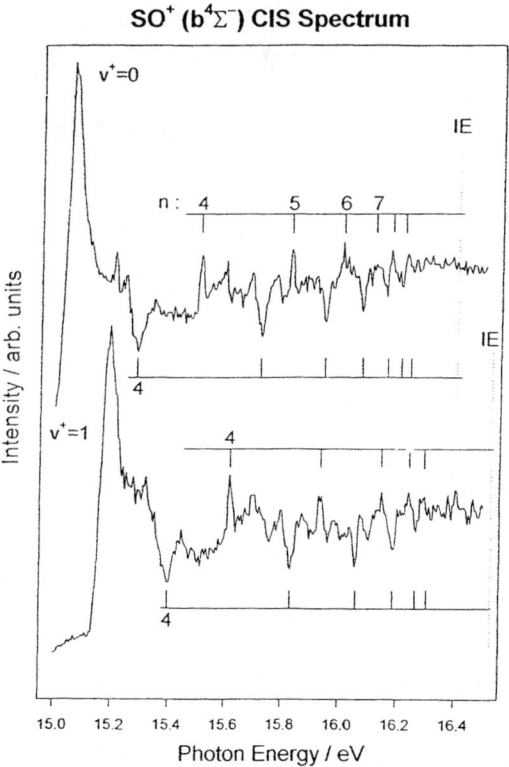

FIGURE 6. CIS spectra recorded for $SO^+(b^4\Sigma^-)$ in the photon energy region 15.0-16.4 eV.

FUTURE DEVELOPMENTS

As a result of these initial studies, a number of future developments of this work can be clearly identified. These can be summarised as follows.

(i) It is well known that the angular distribution parameter, β, can change dramatically at positions of resonances, and this is one of the reasons why angular distribution measurements have emerged as a powerful way of studying the dynamics of photoionization in stable molecules. It is planned to extend these measurements to reactive intermediates.

(ii) Overlapping bands can be a serious problem in the study of reactive intermediates with PES. This can be conveniently avoided by use of the photoelectron-photoion coincidence (PEPICO) method.

(iii) The present Southampton synchrotron-PES apparatus is limited to studying radicals and other reactive intermediates with lifetimes ≥ 1 ms. It is proposed to improve the apparatus to allow shorter-lived species to be studied by up-grading the pumping efficiency of the system. Other instrumental improvements should allow short-lived molecules to be produced by methods other than microwave discharge or atom-molecule reactions, such as pyrolysis or photodissociation.

ACKNOWLEDGEMENTS

The authors would like to thank EPSRC, NERC and the Leverhulme Trust for support of this work.

Femtosecond Photodissociation Dynamics of $Cr(CO)_6$

T. Kono[1,2], V. Vorsa[1], K. F. Willey[3], N. Winograd[1]

[1]Department of Chemistry, The Pennsylvania State University
184 Materials Research Institute Building, University Park, PA 16802, USA
[2] Permanent address ; Analytical Research Laboratory, Asahi Chemical Industry Co., Ltd.
2-1 Samejima, Fuji-city, Shizuoka 416, Japan
[3]Atom Sciences, Inc.
114 Ridgeway Center, Oak Ridge, Tennessee 37830-8810 USA

Abstract. Pump-probe techniques are used in conjunction with time-of-flight mass spectrometry to investigate the ultrafast photodissociation dynamics of $Cr(CO)_6$ following excitation at 200 nm and 267 nm. Time-resolved experiments reveal the first loss of CO ligand(s) occurs within 140 fs.

INTRODUCTION

Metal carbonyls exhibit unique photochemical behavior that is fundamentally different from small organic and inorganic molecules (1). Photoexcitation of metal carbonyls results in dissociations of multiple bonds, while small inorganic and organic molecules typically exhibit rupture of one bond. This is attributed to the lower bonding energies in metal carbonyls. Beginning with the studies of $Fe(CO)_5$, $Cr(CO)_6$, and $Mo(CO)_6$ (2), photoionization of metal carbonyls with nanosecond (ns) laser pulses in the ultraviolet have predominantly resulted in the production of bare metal ions. Several mechanisms for photodissociation have been proposed (1), although the operative mechanism(s) is still under debate.

The rapidly expanding field of femtosecond chemistry is emerging as an ideal forum for the study of the early time behavior of chemical processes (3, 4). Several groups have recently reported time-resolved studies on the ultrafast photodissociation of metal carbonyls in the gas phase. In this study, the researchers probed the real-time dissociation of $Fe(CO)_5$ (5) and $Mn_2(CO)_{10}$ (6). Previously, our laboratory has investigated the photoionization mechanisms for $Cr(CO)_6$ using high intensity laser pulses in the near-IR in which it was proposed that ionization occurs via a barrier suppression mechanism (7). In this paper we describe the ultrafast photodissociation dynamics of $Cr(CO)_6$ following excitation at 200 nm and 267 nm.

EXPERIMENTAL METHOD

The experiments are performed by coupling a reflectron-based time-of-flight (TOF) mass spectrometer with a femtosecond Ti:sapphire laser system. A variable leak valve introduces $Cr(CO)_6$ into the extraction region of the mass spectrometer where it intersects the laser beams and is pulse-extracted into the mass spectrometer. The 1 kHz Ti:sapphire femtosecond laser system (3,7) (Clark-MXR, Inc.) produces 100 fs pulses at 800 nm containing 3.5 mJ of energy per pulse. This output is directed into a harmonic generator to produce 400, 267, and 200 nm wavelengths. The pulse widths for these wavelengths are ~ 200, 250, and 400 fs, respectively. In the pump probe experiment, the power of pump and probe beam are attenuated so that each pulse yields no significant ion signal. These laser intensities are 1.8×10^9, 2.9×10^{12}, 2.8×10^{11} and 7.9×10^{11} W/cm^2 for 267, 800, 200, 400 nm, respectively.

RESULTS AND DISCUSSION

Single-color, single-pulse photoionization of $Cr(CO)_6$ with femtosecond laser pulses at 400 nm, 267 nm and 200 nm are shown in Fig. 1. At threshold laser intensities, 400 nm and 267 nm excitation yield only the molecular ion, while 200 nm excitation produces predominantly $Cr(CO)_3^+$ and $Cr(CO)_2^+$. This is because 2 photons of 267 nm and 3 photons of 400 nm excites $Cr(CO)_6$ to an energy just above that of $Cr(CO)_6^+$, whereas 2 photons of 200 nm excites it to above threshold energy for $Cr(CO)_2^+$ (refer to Fig. 1 (d)). At higher laser intensities, additional photons are absorbed leading to further fragmentation. Since the calculated Keldysh parameter is much greater than 1 for the laser intensities used here, the ions are most likely generated through a multiphoton ionization (MPI) mechanism. These results imply that $Cr(CO)_6$ dissociates on the fs time scale.

Shown in Fig. 2 are the ion yields for all species obtained with 267 nm (pump) and 800 nm (probe) pulses (parallel polarization). The 267 nm pump pulse excites the $^1A_{1g} \rightarrow c^1T_{1u}$ transition. Since 267 nm (4.65 eV) excitation can only dissociate a maximum of 4 M-L bonds, the strong $Cr(CO)^+$ and Cr^+ signals are indicative of fragmentation from the probe pulse. The normalized data (dots in Fig. 3) clearly illustrate the unique fs scale time-dependent behavior of fragment ions following excitation at 267 nm.

In order to explain these results, we attempted a simulation incorporating fragmentation from the probe pulse. We examined several models including a concerted process and one that consists of 2 relaxation scales where the 2nd relaxation occurs from each state in the 1st sequential process. The best fit is obtained with a model incorporating 7 states where each state except the last one relaxes sequentially. The agreement between simulation and experiment is illustrated in Fig. 3. The lifetimes of these initial and intermediate states are 20, 20, 25, 25, 30,

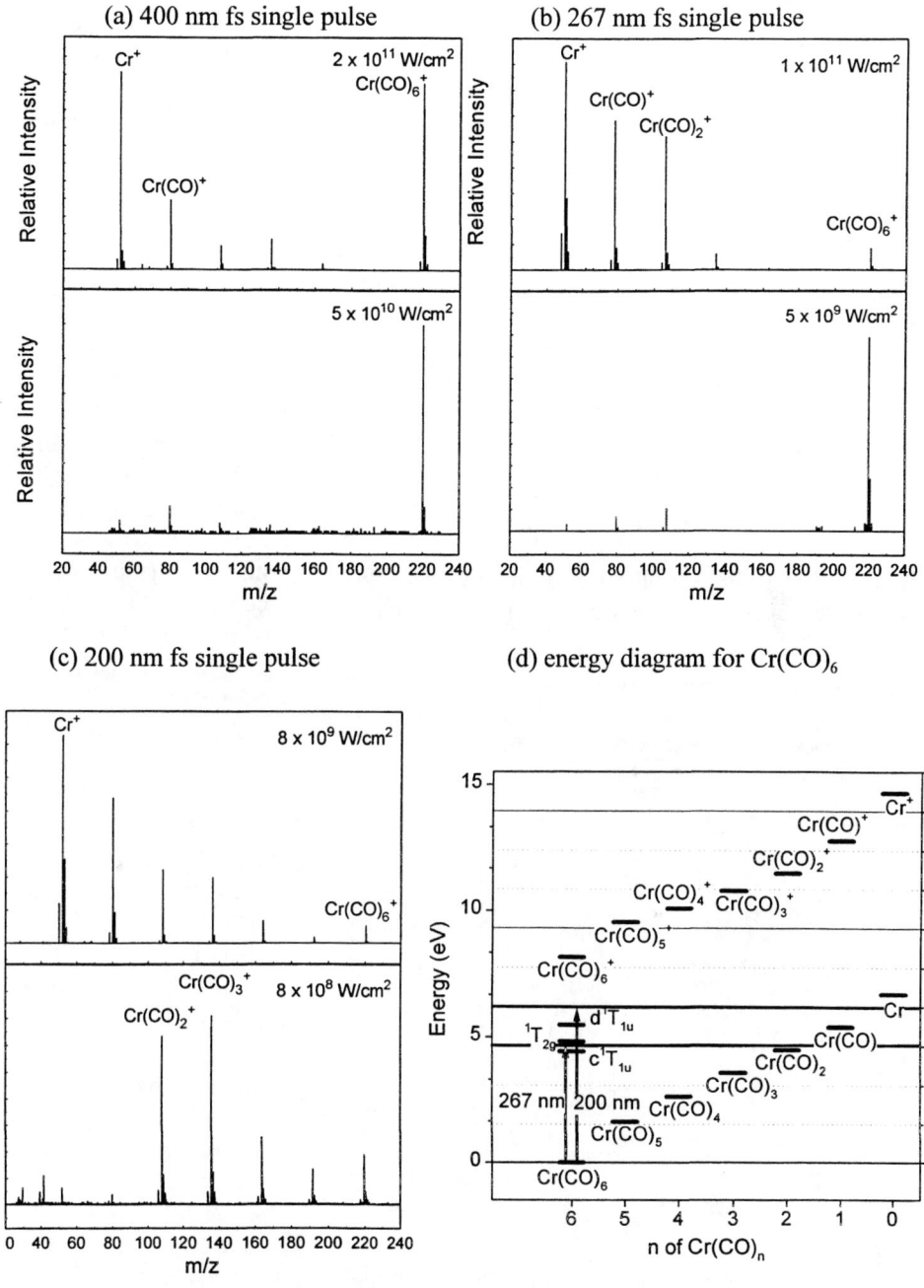

Figure 1. Femtosecond Photoionization Spectra of $Cr(CO)_6$ at (a) 400 nm, (b) 267 nm and (c) 200 nm and (d) energy diagram for $Cr(CO)_6$ and its fragments.

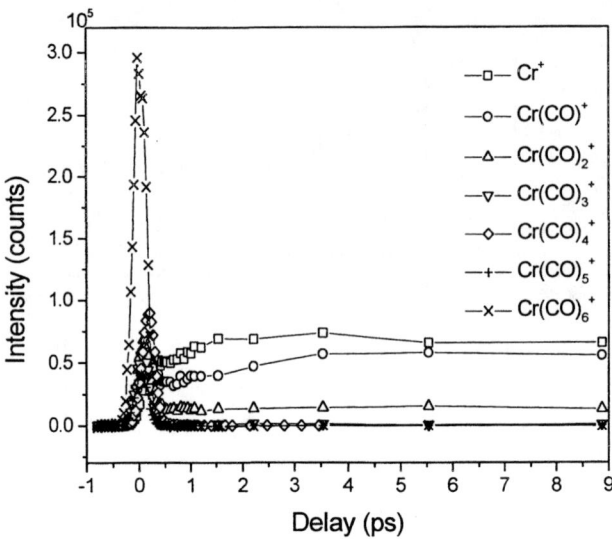

Figure 2. Ion yields for all species obtained with 267 nm (pump) and 800 nm (probe) pulses (parallel polarization).

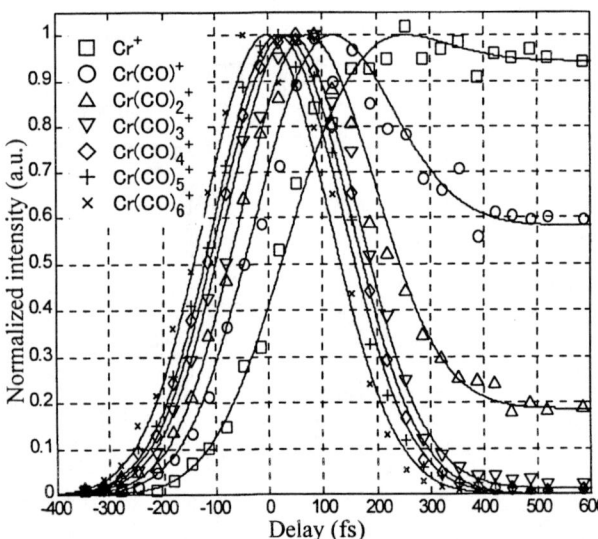

Figure 3. Normalized ion yields for 267 nm - 800 nm. Dots show experimental data and solid lines show simulation result.

20 fs respectively and the overall time that is needed for completing these fs sequential processes is 140 fs. The lifetime of the last state is in the time range of 10 ps or longer. When an individual lifetime was changed by times as small as 10 fs, significant error was introduced into the fit. Recently, Trushin, et al. reported similar results (8) for this system. Our results are in reasonable agreement with their study which incorporated a 5 level sequential process.

A constant signal from $Cr(CO)_n^+$ (n = 0~3) is observed between 1 and 10 ps. This signal is proposed to originate from the last long lived state of the sequential process. Since this constant signal is observed in $Cr(CO)_n^+$ with n = 0~3, this long-lived state must belong to $Cr(CO)_n^+$, where n = 3~6. Moreover, the initial loss of CO ligand(s) from $Cr(CO)_6$ should occur during the 140 fs associated with the 7-state sequential process. This conclusion is further supported by our previous observation

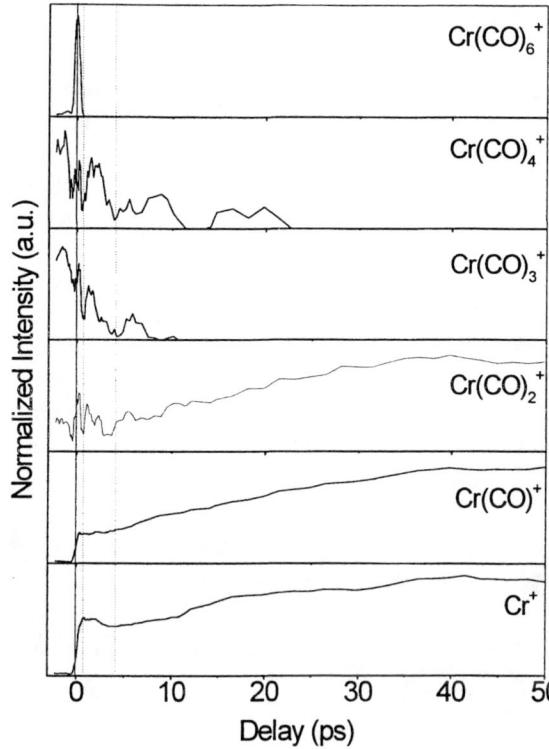

Figure 4. Normalized ion yields for all species obtained with 200 nm (pump) and 400 nm (probe) pulses (parallel polarization).

where it was found that the $Cr(CO)_6^+$ ion is not observed when $Cr(CO)_6$ is irradiated with a 2 ps pulse. The initial CO loss yields $Cr(CO)_n$ where n = 3~5 as the product species.

Normalized ion yields for all species obtained with 200 nm (pump) and 400 nm (probe) pulses (parallel polarization) are shown in Fig. 4. The data have been smoothed with 5-point averaging. The $Cr(CO)_5^+$ signal is not displayed due to its extremely low signal level.

In the 200 nm - 400 nm pump - probe experiments, the peak at zero time is observed strongly for $Cr(CO)_6^+$ but weakly for other fragment ions. This is presumably because the ground states of $Cr(CO)_n^+$ (n = 2~5) that are generated by the 400 nm probe have high vibrational energy and undergo rapid decomposition. To verify this hypothesis, we are planning a 200 nm - 800 nm pump - probe experiment which would enable us to access lower vibrational ionization states and avoid fragmentation by the probe pulse.

The peaks near zero time shift to longer times as the molecule sequentially loses CO ligands, as was also found for the 267 nm - 800 nm pump - probe experiment. This implies there is a similar type of mechanism that is operative. However, since the absolute intensity of these peaks for $Cr(CO)_n^+$ (n = 2~4) is extremely low, further experiments are needed to examine the near zero time trend more precisely.

A 2nd component is observed in the region of ~ 5 ps with a life-time of approximately 2 ps. This time scale corresponds to rotational dephasing of the fragmented metal carbonyl complex. A 3rd component was also seen in the 5 ~ 50 ps region for $Cr(CO)_n$ (n <= 2) showing a rise time of around 20 ps. This indicates there is a dissociation process with yet a longer time scale.

CONCLUSIONS

We have completed experiments on the ultrafast ionization and dissociation dynamics of $Cr(CO)_6$ using femtosecond laser pulses. Multiphoton ionization of $Cr(CO)_6$ yields large molecular ion signals. Pump-probe experiments with 267 nm excitation reveal the presence of a fast sequential process in which rapid dissociation of Cr – CO bond(s) occur. The last product of this fs sequential process is considered to be one of $Cr(CO)_n$ (n = 3 ~ 5). In the ultrafast experiments using 200 nm for the pump pulse, however, the intensity of the near zero time scale is extremely low or not observed for $Cr(CO)_n^+$ (n = 2 ~ 5). This is presumably because the ground states of $Cr(CO)_n^+$ (n = 2 ~ 5) generated by the 400 nm probe has high vibrational energy and undergoes rapid decomposition. The longer time scale component is also seen in $Cr(CO)_n$ (n <= 2) for up to 50 ps. This indicates the existence of a dissociation process with longer lifetime following the fs sequential mechanism. Further experiments using other probe wavelengths are planned.

ACKNOWLEDGEMENTS

We are grateful for helpful discussions with Dr. Sergei Trushin. We also acknowledge the financial support of the National Institute of Health and the National Science Foundation.

REFERENCES

1. Jackson, R. L., *Acc. Chem. Res.* **25**, 581 (1992).
2. Duncan, M. A., Dietz, T. G., Smalley, R. E., *Chem. Phys.* **44**, 415 (1979).
3. Zewail, A. H., *Femtochemisitry: Ultrafast Dynamics of the Chemical Bond*, Singapore: World Scientific, 1994.
4. *Femtochemistry 97, J. Phys. Chem. A*, **102** (1998)
5. Banares, L., Baumert, T., Bergt, M., Kiefer, B., Gerber, G., *J. Chem. Phys.* **108**, 5799 (1998).
6. Kim, S. K., Pedersen, S., Zewail, A. H., *Chem. Phys. Lett.* **233**, 501 (1995)
7. Willey, K. F., Brummel, C. L., Winograd, N., *Chem. Phys. Lett.* **267**, 359 (1997).
8. Trushin, S. A., Fuss, W., Schmid, W.E., Kompa, K. L., *J. Phys. Chem. A*, **102**, 4129 (1998)

Time-resolved photoelectron emission from Rydberg atoms

L. D. Noordam

Noordam@amolf.nl
FOM-Institute for Atomic and Molecular Physics,
Kruislaan 407, 1098 SJ Amsterdam, The Netherlands

Abstract. The dynamics of a Rydberg Stark wavepacket above the saddlepoint of the combined Coulomb-electric field potential is investigated by conventional pump-probe techniques and an atomic streak camera. This new device records when the electron is ejected from the autoionizing system. With some modifications this streak camera can also be used to measure the shape of infrared laser pulses with a picosecond time resolution.

1. Introduction

In a Rydberg atom [1,2] the loosely bound electron moves in a large Kepler orbit around the atomic nucleus. The system is very sensitive to external perturbations; for instance, a moderate electric field drastically influences the behaviour of the Rydberg electron. A static field of a few kV/cm is sufficient to change the bound Rydberg atom into a system in which the electron can escape. Within a few picoseconds (10^{-12} s) the atom falls apart. It is an experimental challenge to detect how this decay actually happens. Does the electron come out at once, or are there signatures of the quantum nature of the system which are made manifest by subsequent bursts of probability that the atom emits an electron?

The behaviour of the decaying electron has been studied for more than a decade using spectroscopy, either in the frequency domain [3,4] by measuring the absorption spectrum in the continuum, or in the time domain [5-8] by following the motion of an electronic wavepacket created by a short optical pulse. Each of these techniques probe the atom when the excited electron is near the core. If the electron does not escape immediately but returns at least once to the core, this is seen in the frequency domain as a resonance in the absorption spectrum. The same situation can be probed in the time domain by a delayed second pulse that stimulates the electron back to a bound state.

However, spectroscopy does not tell the full story. What if the electron does not leave the atom immediately but also does not returns to the core. A spectroscopist will not observe a recurrence and conclude that the electron has escaped, which is an incorrect conclusion since the electron is still in the atom but just out of the scope of the spectroscopist. An atomic streak camera [9] has been constructed that provides additional information: the camera makes pictures of when the electron leaves the atom [10].

To our surprise, the electron usually makes first a few orbits around the nucleus before escaping. After each orbit there is some probability of escaping, or in more quantum mechanical terms, some fraction of the electron wavefunction is emitted. The camera directly sees how much electron wavefunction leaks out after each orbit. In this contribution we discuss in detail the dynamics of electron emission of a highly excited Rydberg state. Moreover, we present some applications of an atom ejecting an electron upon photoexcitation to an autoionizing Rydberg state.

CP454, *Resonance Ionization Spectroscopy*
edited by J. C. Vickerman, I. Lyon, N. P. Lockyer, and J. E. Parks
© 1998 The American Institute of Physics   1-56396-810-X/98/$15.00

2. Dynamics of a Rydberg electron in an electric field

The ionization dynamics of Stark states above the saddle point have been studied [7,11] using short optical pulses to excite a coherent superposition of the 'quasi continuum' states. The decay of the wave packet [12] is monitored [13] by measuring the amount of wave function returning to the core region as a function of time. In these experiments a ps optical pulse excites a wave packet above the saddle point energy. After a delay a second pulse is applied to probe the amount of wave function that has returned to the core. In this way the evolution of the wave packet *near the core* has been observed.

Two types of motion of the autoionizing wavepacket above the saddlepoint of the potential are seen. **Firstly**, the radial motion [14] of the wave packet starting near the core and moving towards the outer turning point and back to the core. This motion is due to beating of eigenstates with different principal quantum number n. Since the energy difference of Rydberg states is given by $dE/dn = 1/n^3$ the beating time is given by $\tau = 2\pi/\Delta E = 2\pi n^3$. This radial oscillation time of the wave packet corresponds to the Kepler motion of an electron around the nucleus. For $n = 20$ this orbit time is 1.2 ps.

Secondly, the angular momentum of the electron is not conserved in an electric field F. The wavepacket exhibits oscillations in the angular momentum between $l = 0$ and $l = n$ [5], starting from the initial angular momentum state $l \sim 0$ (see fig.1).

These angular momentum oscillations are due to the beating of different Stark states k within a n-manifold. Since the energy difference of Stark states is given by $dE/dk = 3Fn$, the beating time is given by $\tau = 2\pi/\Delta E = 2\pi/3Fn$. For $n = 20$ and a field of 2 kV/cm, the angular momentum oscillation time corresponds to 6 ps. Recurrences in the optical pump-probe studies are seen whenever the angular momentum and the radial distribution are the the initial value at the time the packet was launched by the pump pulse: $r \sim 0$ and $l \sim 0$.

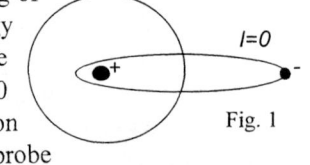

Fig. 1

A limitation of using optical techniques to study the ionization dynamics is the fact that the dynamics of the Rydberg electron can only be probed near the core. Using the atomic streak camera, we are able to identify when the electron escapes from the ionic potential (see fig. 2).

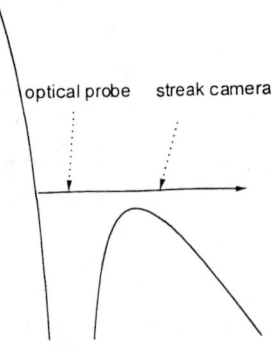

Fig. 2: The sterak camera probes the electron while escaping over the saddlepoint, while optical techniques probe recurrences near the core.

3. Atomic Streak camera

We present a new device, the atomic streak camera [9], which is able to measure the ionization dynamics (i.e., the escape of the electron from the atomic potential) with ps resolution. The design of the atomic streak camera (see fig. 3) is based on a conventional streak camera. In a conventional streak camera, [15] the optical pulse is transformed into an electron pulse by means of a photocathode. In the atomic streak camera, instead of a photocathode, a low pressure atomic gas is used to create the electrons. These atoms, which are in a static electric field, are excited by a short laser pulse.

Let us consider the case where the wave packet dynamics are such that the electrons leave the atoms in two bursts. The double electron pulse is accelerated by the electric field and passes the slit in the anode. After the slit the electrons enter the deflection region. The voltage applied to one of the two deflection plates is ramped. Therefore, the electron pulse arriving first between the deflection plates will be deflected less than the second one, resulting in a different impact position on the detector. Thus from the position on the detector, the time resolved ionization spectrum of the atoms is retrieved.

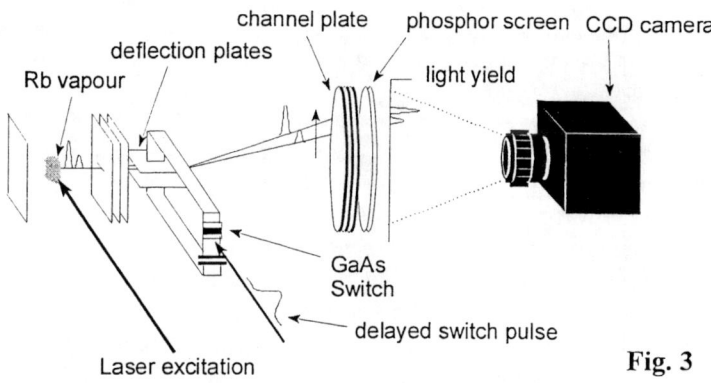

Fig. 3

4. Experimetal observation of an autoionizing Rydberg wavepacket

We have monitored the evolution of a Rydberg wavepacket in two ways [10]: near the core with a pump-probe technique and near the saddlepoint with the atomic streak camera. In the experiment rubidium atoms were excited in a static electric field of 2.0 kV/cm by a short optical pulse. The excitation energy was chosen to be above the saddle point energy, thus creating an autoionizing wave packet. In Fig. 4 a time resolved electron emission spectrum of rubidium excited with a 4-ps pulse around 0.87 E_c is shown, where E_c is the saddle point energy.

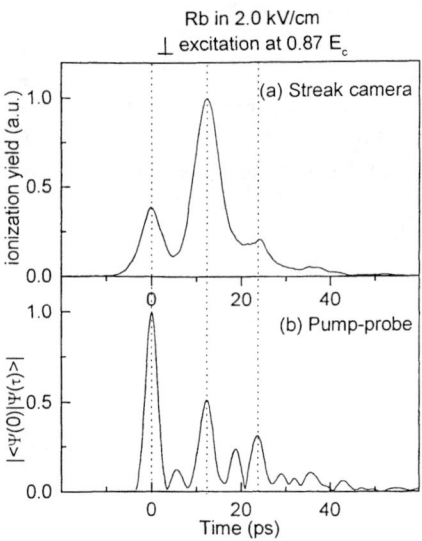

Fig. 4: In the upper spectrum the electron is probed at the saddle with the atomic streak camera, whereas in the lower spectrum the electron is probed near the core point by a conventional pump-probe technique.

Inspection of figure 4 shows that instead of observing an exponential decaying type of wave packet the main ionization was surprisingly delayed 12 ps. The polarization of the laser used to excite the wave packet was chosen perpendicular to the electric field. The wave packet will therefore be located perpendicular to the electric field. Since the electron can only escape in the direction of the electric field it is still bound in the directions perpendicular to this axis. The wave packet needs to reorient itself in the direction of the saddle point to escape. This reorientation can occur by scattering from the core electrons. The scattering in turn, depends on the average value of the angular momentum of the wave packet. For this particular case, the excited wave packet makes an oscillation in angular momentum in 6 ps. Initially excited to a low angular momentum state, the wave packet begins to increase its angular momentum. When the wave packet is in a high angular momentum state it will be located far away from the core electrons ($r_{min}= l(l+1)$) making it impossible to scatter. From Fig. 4(a) we observe that the wave packet ionizes dominantly at the second angular recurrence. In Fig. 4(b) the corresponding recurrence spectrum is plotted. This spectrum is a measure for the amount of wave function that comes back to the $l=1$ state as a function of time. We see that the amplitude is rather low after the first oscillation in angular momentum at 6 ps. The second angular recurrence gives a high amplitude indicating that a larger fraction of the wave packet returns to low angular momentum states. This causes the scatter event with the core electrons leading to the large ionization at 12 ps observed in fig. 4(a). These spectra have been reproduced by a MQDT calculation [16]. In the calculation one observes the scatter event by the transfer of population from a closed channel (bound states) into an open channel (ionizing states).

5. Time-gated spectroscopy

We describe a novel spectroscopic tool which provides the timeresolved electron emission as a function of the excitation wavelength: Time-Gated photoionization Spectroscopy (TGS [17]). Using the atomic streak camera, we record the time of electron emission after laser excitation. By setting time gates for the outgoing electron flux and recording the yield in a time gate versus excitation wavelength we obtain a time gated spectrum.

In Fig. 5 the relative ionization yields are measured for time gates centered around the prompt peak, the first and the second angular recurrence time of the wave packets. The underlying mechanism responsible for the observed oscillations can be explained in terms of the commensurability between angular and radial periods of the electron dynamics [8]. Within the laser bandwidth, k-states belonging to different n-manifolds are excited. The total wave packet is built up by two angular wave packets. Depending on the energy of the nk-states with respect to the energy of the $(n+1)k$-states the wave packet will either constructively (same energies) or destructively (interleaved energies) interfere near the nucleus after the first angular oscillation period.

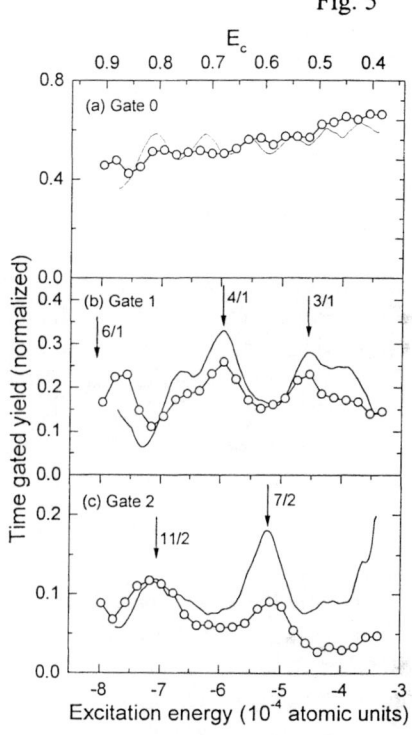

Fig. 5

The classical equivalent of this can be explained in terms of angular and radial oscillations of the wave packet. When $E_{nk} = E_{(n+1)k}$, both oscillation periods are in phase at the first angular recurrence and the wave packet returns to the core. At this time the Rydberg wave packet may scatter off the core since it is the core electrons that break the parabolic symmetry for the valence electron. This scattering event can change the direction of the wave packet and cause it to travel down-field giving rise to the observed peaks in the ionization spectrum. When the energies are interleaved the periods are out of phase at the first angular oscillation time, so when the wave packet is in a low angular momentum states it is not located near the core.

The intrinsic large bandwidth of the ps laser pulse does not permit study of individual resonances, but discloses for a small energy range at which recurrence of the wavepacket to the core the electron is scattered into an outgoing channel.

6. Application of a Rydberg photocathode: a FIR streak camera

Temporal characterization of infrared radiation is severely limited by the lack of fast IR detectors. Whereas the temporal characterization of IR pulses is cumbersome, visible light pulses are nowadays routinely characterized using a streak camera. The spectral range at which conventional streak cameras can be operated is limited by the spectral response of the photocathode. The work function of the photocathode material determines the long wavelength limit of a conventional streak camera to 1.6 μm.

Recently, a new type of streak camera has been constructed in our laboratory. Here the photocathode used in conventional streak cameras is replaced by a sample of gas-phase atoms in a Rydberg state. The low binding energy of the electrons in Rydberg atoms and the large photoionization cross section [18] make the Rydberg atom photocathode a very sensitive detector for infrared radiation [19-21] and thus very suitable for use in an IR streak camera. The operation of this streak camera was demonstrated in the infrared at wavelengths ranging from 2.6 μm [22] up to 85 μm [23], well beyond the spectral range of any conventional streak camera.

An essential requirement for proper operation of a streak camera is that the photoelectron emission is prompt. Experiments on the far-infrared ionization behavior [20] of Rydberg atoms in an electric field have shown that states with a negative Stark effect (for which the energy of the states decreases with increasing electric field strength) ionize on a sub-picosecond time scale, independent of the wavelength and polarization of the ionization laser. For states with a positive Stark effect (the energy of the states increases with increasing electric field strength) this is not true. Here the time scale for ionization depends on the wavelength as well as on the polarization of the light and can be as long as 100 ps.

The time resolution of the camera system was tested [23] at a wavelength of 35 μm by using the lowest Stark component of the n=16 manifold in Cs as intermediate level. After absorption of the 35 μm photon the initial kinetic energy of the electrons is only 10 meV. Figure 6 shows the measured temporal profile for a single laser shot. The line is a fit of a Gaussian profile to the data points. The width (FWHM) of the Gaussian profile was found to be 1.4 ps. Note that the oscillation time of the laser field at 35 μm is already 0.12 ps!

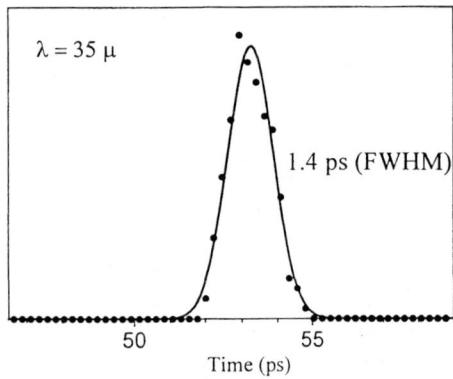

Fig. 6

7. Conclusions

From the comparison of the recurrence spectra, obtained by pump-probe spectroscopy, with the streak spectra the conclusion can be drawn that the lifetime as measured by an optical technique is not the same as the time it takes the electron to leave the atom. The Rydberg electron can be far away from the core, invisible for optical techniques, but still be captured in the attractive force of the parent ion. When the electron does not pass the core before ionizing, the electron appears to be ionized for an optical technique, but in fact has not yet escaped from the atom. Probing the dyamics of aRydberg electron in an electric field with half-cycle pulses [24,25] provides also additional information on the Rydberg electron dynamics.

We have described a novel spectroscopic tool called Time-Gated photoionization Spectroscopy (TGS). Information in both the frequency and time domain is obtained simultaneously with this technique. TGS provides additional information compared to conventional absorption spectroscopy. We have experimentally demonstrated the technique on the photoionization of Cs wave packets in a static electric field. It is found that the time-gated spectra reveal information on the commensurability between radial and angular oscillation periods of the excited wave packets. An enhancement in the ionization yield is found whenever the wave packet is in a low angular momentum state *and* has radially returned to the core, since core scattering is the relevant ionization mechanism. The experimental data is confirmed by MQDT calculations.

We have also demonstrated the operation of an atomic streak camera in the mid to far-infrared (2.6-85 μm), well beyond the spectral range of any conventional streak camera. The temporal resolution of the camera system was found to be in the order of 1 ps over its useful wavelength range.

ACKNOWLEDGMENTS

The work described in this article is performed in collaboration with M. Drabbels, D. I. Duncan, J. H. Hoogenraad, G. M. Lankhuijzen, C. W. Rella, F. Robicheaux, and R. B. Vrijen. It is a pleasure to acknowledge C. W. Rella for critical reading the manuscript. LDN is supported by the "Stichting voor Fundamenteel Onderzoek der Materie (FOM)" which is financially supported by the "Nederlandse Organisatie voor Wetenschappelijk Onderzoek (NWO)".

REFERENCES

[1] T. F. Gallagher, Rydberg atoms, Cambridge University Press, Cambridge 1994
[2] J.-P. Connerade, Highly excited atoms, Cambridge University Press, Cambridge 1998
[3] R. R. Freeman, N. P. Economou, G. C. Bjorklund, and K. T. Lu, Phys. Rev. Lett. **41**, 1463 (1978)
[4] G. M. Lankhuijzen and L. D. Noordam, Phys. Rev. A **52**, 2016 (1995)
[5] L.D. Noordam, A. ten Wolde, A. Lagendijk, and H.B. van Linden van den Heuvell, Phys. Rev. A **40**, 6999 (1989).
[6] B. Broers, J. F. Christian, J. H. Hoogenraad, W. J. van der Zande, H. B. van Linden van den Heuvell, and L. D. Noordam, Phys. Rev. Lett. **71**, 344 (1993)
[7] J. F. Christian, B. Broers, J. H. Hoogenraad, W. J. van der Zande, and L. D. Noordam, Opt. Commun. **103**, 79 (1993)
[8] M. L. Naudeau, C. I. Sukenik, and P. H. Bucksbaum, Phys. Rev A **56**, 636 (1997)
[9] G. M. Lankhuijzen and L. D. Noordam, Optics Commun. **129**, 361 (1996)
[10] G. M. Lankhuijzen and L. D. Noordam, Phys. Rev. Lett. **76**, 1784 (1996)
[11] L. D. Noordam, D. I. Duncan, and T. F. Gallagher, Phys. Rev. A **45**, 4734 (1992)
[12] R. R Jones, and L. D. Noordam, Advances in Atomic and Molecular Physics **38**, 1 (1997), Academic Press, San Diego
[13] L. D. Noordam and R. R. Jones, Journal of Modern Optics **44**, 2515 (1997)
[14] A. ten Wolde, L. D. Noordam, A. Lagendijk, and H. B. van Linden van den Heuvell, Phys. Rev. Lett. **61**, 2099 (1988).
[15] W. Knox, and G. Mourou, Opt. Commun **37**, 203 (1981)
[16] F. Robicheaux and J. Shaw, Phys. Rev. Lett. **77**, 4154 (1996)
[17] G. M. Lankhuijzen, F. Robicheaux, and L. D. Noordam, Phys. Rev. Lett. **79**, 2427 (1997)
[18] J. H. Hoogenraad, and L. D. Noordam, Phys. Rev. A **57**, 4533 (1998)
[19] J. H. Hoogenraad, R. B. Vrijen, P. W. van Amersfoort, A. F. G. van der Meer, and L. D. Noordam, Phys. Rev. Lett. **75**, 4579 (1995)
[20] G. M. Lankhuijzen, M. Drabbels, F. Robicheaux, and L. D. Noordam, Phys. Rev. A **57**, 440 (1998)
[21] J. H. Hoogenraad, R. B. Vrijen, and L. D. Noordam, Phys. Rev. A **57**, 4546 (1998)
[22] M. Drabbels, and L. D. Noordam, Optics Lett. **22**, 1436 (1997)
[23] M. Drabbels, G. M. Lankhuijzen, and L. D. Noordam, submitted to IEEE-QE
[24] R. R. Jones, N. E. Tielking, D. You, C. Raman, and P. H. Buckbaum, Phys. Rev. A **51**, 2687 (1995)
[25] R. B. Vrijen, G. M. Lankhuijzen, and L. D. Noordam, Phys. Rev. Lett. **79**, 617 (1997)

Formation of $O_2(b^1\Sigma_g^+)$ in the Ultraviolet Photolysis of O_3

P.K. O'Keeffe, T. Ridley, S. Wang, K.P. Lawley
and R.J. Donovan

*Department of Chemistry, The University of Edinburgh
West Mains Road, Edinburgh, EH9 3JJ, Scotland, U.K.*

Abstract: The formation of $O_2(b\ ^1\Sigma_g^+)$ in the ultraviolet photodissociation of O_3 (335-352 nm) has been observed for the first time using (2 + 1) resonance enhanced multiphoton ionisation. Measurement of the photofragment kinetic energy confirms that the primary photodissociation step involves the absorption of a single photon.

Introduction

The ultraviolet photodissociation of ozone has been the subject of numerous experimental and theoretical investigations (1-7). In particular, the intense bell shaped Hartley absorption band between 220 and and 310 nm (λ_{max} = 254 nm: σ = 1.159 x 10^{-19} cm^2 $molecule^{-1}$) has been well studied due to the central role that photodissociation of ozone plays in stratospheric chemistry. It is generally accepted that the Hartley band arises from an electronically allowed transition from the ground state ($\tilde{X}\ ^1A_1$) to the repulsive wall of the 1^1B_2 state, resulting in prompt (~150 fs) dissociation of the O_3. Photodissociation in this energy region is believed to take place *via* two main spin-allowed channels with the following quantum yields and thermodynamic thresholds:

$O_3 + h\nu \rightarrow O(^1D) + O_2(a\ ^1\Delta_g),$ Φ = 0.85 - 0.9; $\lambda_{threshold}$ = 310 nm   [1]

$O_3 + h\nu \rightarrow O(^3P) + O_2(X\ ^3\Sigma_g^-),$ Φ = 0.1 - 0.15; $\lambda_{threshold}$ = 1180 nm   [2]

In the long wavelength tail of the Hartley continuum (λ > 310 nm) diffuse vibrational structure becomes apparent in a region known as the Huggins band (2). In spite of the reduced absorption coefficient at these wavelengths ($\sigma_{254} \approx 10^4 \times \sigma_{340}$) the photolysis of O_3 at $\lambda > 310$ nm has a profound effect on the photochemistry of ozone in the troposphere. Until recently, it was assumed that the sole photodissociation channel operative in this region was the triplet spin-allowed process [2] as this is the only energetically accessible spin-allowed channel. Recently however, a number of groups (3-6) have reported that the following spin-forbidden channels also participate in the Huggins band photodissociation of ozone:

$O_3 + h\nu \rightarrow O(^3P) + O_2(a\,^1\Delta_g)$, $\lambda_{threshold} = 612$ nm   [3]

$O_3 + h\nu \rightarrow O(^1D) + O_2(X\,^3\Sigma_g^-)$, $\lambda_{threshold} = 411$ nm   [4]

There is also a fifth photodissociation channel that is energetially accessible *via* the Huggins band, i.e.

$O_3 + h\nu \rightarrow O(^3P) + O_2(b\,^1\Sigma_g^+)$, $\lambda_{threshold} = 463$ nm   [5]

In this paper we communicate the first experimental evidence that channel [5] is operative in the region 335-352 nm.

Experimental

The experimental arrangement consisted of two pulsed and independently tunable dye lasers (a Lambda Physik FL3002 and a Lambda Physik FL2002) pumped by a XeCl excimer laser (a Lambda Physik EMG 201 MSC). One laser was used to photolyse the ozone and the other to probe the products. The counter propagating pump and probe beams intersected a molecular beam of O_3 in He. The resulting ions were then collected by a linear time-of-flight mass spectrometer (TOF-MS).

The kinetic energy release experiments required the TOF-MS to be operated in a pulsed mode where the fragments were allowed to separate spatially under the initial recoil velocities imparted by the photodissociation event. This separation was achieved by allowing the fragments to recoil under field-free conditions during a time delay, τ (typically 300 ns), between their formation by the laser pulse and acceleration by a pulsed field (risetime < 50 ns). The time-of-flight of the resulting unfocussed ions, is related to the position of the ions in the extraction region when the field is applied; this can then be converted to the initial recoil velocities of the neutral fragments by solution of the time-of-flight equation. Core sampling collection could be achieved with our experimental arrangement. The 40 mm diameter detector is usually sufficient to collect the entire ion packet, however, the central core of the ion packet can be selectively detected by reducing the active surface of the detector to a diameter of 8.5 mm, thus reducing the collection angle subtended by the detector from 4° to 0.87°. The TOF profiles could be recorded with the dissociation laser polarisation either parallel or perpendicular to the detector axis. Rotation of the laser polarisation was achieved by passing the beam through a Soleil Babinet prism.

Ozone gas mixtures were prepared by passing the output of a commercial ozonizer through a silica gel trap at 196 K. The trap was then placed on a vacuum line and residual O_2 pumped away to a pressure of 1 Torr. The ozone was then desorbed into a Pyrex bulb to give a pressure of 10 Torr, followed by mixing with He gas to a pressure of 1 atm. The gas mixture was then introduced into the photolysis chamber *via* an all glass feed-line and a General Valve pulsed nozzle with a 250 μm diameter aperture.

Results and Discussion

Photodissociation of O_3 was studied in the region 335-352 nm (i.e. the Huggins band). The pump laser was scanned through this region and the production of $O(^3P_J)$, $O_2(a^1\Delta_g)$ and $O_2(b^1\Sigma_g^+)$ was monitored using (2 + 1) REMPI techniques. Both $O(^3P_J)$ and $O_2(a^1\Delta_g)$ have been observed previously following photolysis of O_3 at shorter wavelengths (3,10). Our results confirm previous findings and extend them to longer wavelengths. However, to our surprise, new transitions which are assigned to $O_2(b^1\Sigma_g^+)$ were observed and these observations are the main concern of the work presented here.

The formation of $O_2(b^1\Sigma_g^+)$ was observed *via* the Rydberg transition $d^1\Pi_g \leftarrow \leftarrow b^1\Sigma_g^+$. The upper state has been characterised *via* transitions from both the $(a^1\Delta_g)$ and $(X\ ^3\Sigma_g^-)$ states (3,11). The positions of vibrational transitions for the d←←b system are thus readily calculated. Experimentally, the two most convenient vibrational transitions in the d←←b system are the (2, 0) and (1, 0) bands. A spectrum of the (2, 0) band is shown in figure 1. By fixing the probe laser on this transition and scanning the pump laser, the yield of $O_2(b^1\Sigma_g^+)$ throughout the Huggins band region, was obtained. This excitation spectrum corresponds closely with the structured absorption spectrum of O_3, showing that $O_2(b^1\Sigma_g^+)$ is a primary product throughout this region. The quantum yield for $O_2(b^1\Sigma_g^+)$ is difficult to quantify but is clearly significant.

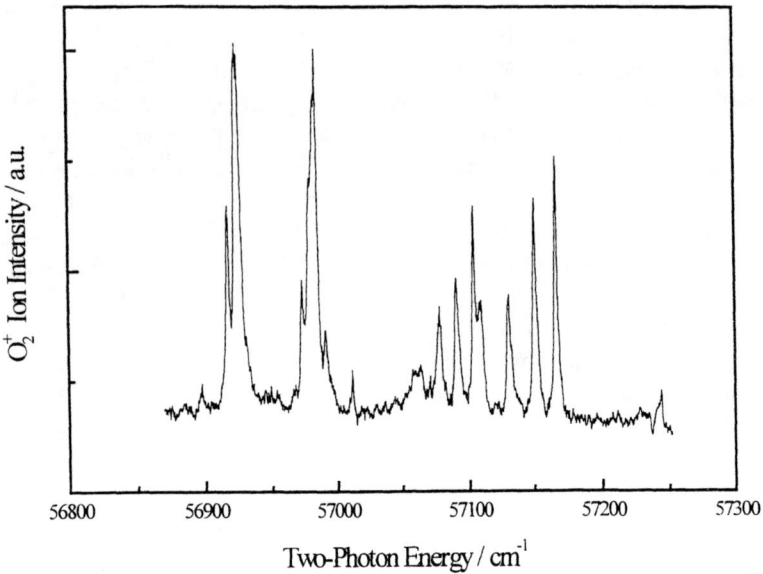

FIGURE 1. The (2,0) band of the $O_2(d^1\Pi_g \leftarrow\leftarrow b^1\Sigma_g^+)$ transition recorded by probing the molecular photofragment produced in the 340 nm (Huggins band) photolysis of ozone.

The presence of vibrational structure in the Huggins bands raises the question as to whether two photons are absorbed by O_3, in a resonantly assisted process, before fragmentation takes place. Indeed, $O_2(b^1\Sigma_g^+)$ has been observed as a primary product in the photolysis of O_3 at both 157.6 and 193 nm(8,9). The dependence of the $O_2(b^1\Sigma_g^+)$ signal on pump power was therefore investigated and was found to be linear (the power index n = 0.95 ± 0.09). Further evidence that the primary photolysis step only involves absorption of a single photon comes from a study of the kinetic energy release in the fragments. Using the technique briefly outlined above we have measured the kinetic energy of the $O_2(b^1\Sigma_g^+)$ fragment and this is illustrated in figure 2.

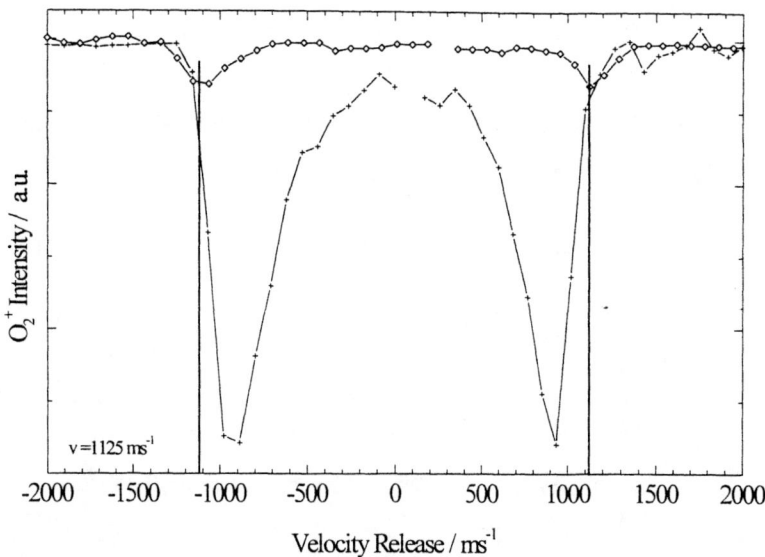

FIGURE 2. The velocity profile of the $O_2(b^1\Sigma_g^+)$ photofragment produced by photolysis of ozone with vertically polarised light (with respect to the detection axis) at $\lambda = 351.35$ nm. The $O_2(b^1\Sigma_g^+)$ photofragments were detected by (2 + 1) ionisation *via* the $d^1\Pi_g$, ($v' = 2$) intermediate level. The crosses and diamonds indicate data for the total ion packet and the central core of the ion packet, respectively.

The double peaked nature of the profile is a result of preferential recoil of ions towards and away from the detector which then experience different flight times due to the different times spent in the applied extraction field. A detailed analysis of the kinematics involved in the photodissociation step allows the internal energetic state of the fragments to be determined. These results again support a single photon process being the main primary step in the formation of $O_2(b^1\Sigma_g^+)$.

REFERENCES

1. Molina, L.T. and Molina, M.J., *J. Geophys. Res* **91**, 14501 (1986).

2. Katayama, D.H., *J. Chem. Phys.* **71**, 815 (1994).

3. Ball, S.M., Hancock G., and Winterbottom, F., *Faraday Discuss.* **100**, 215 (1995).

4. Talukdar, K., Longfellow, C.A., Gilles, M.K. and Ravishankara, A.R., *Geophys. Res. Lett.* **25**, 143 (1998).

5. Takahashi, K., Taniguchi, M., Matsumi, Y., Kawasaki, M. and Ashfold, M.N.R., *J. Chem. Phys.* **108**, 7161 (1998).

6. Ball, S.M., Hancock, G., Martin, S.E. and Pinot de Moira, J.C., *Chem. Phys. Lett.* **264**, 531 (1997).

7. Banichevich, A., Peyerimhoff, S.D. and Grein, F., *Chem. Phys.* **178**, 155 (1993).

8. Taherian, M.R. and Slanger, T.G., *J. Chem. Phys.* **83**, 6246 (1985).

9. Stranges, D., Yang, X., Chesko, J.D. and Suits, A.G., *J. Chem. Phys.* **102**, 6067 (1995).

10. Takahashi, K., Matsumi, Y. and Kawasaki, M., *J. Chem. Phys.* **100**, 4084 (1996).

11. Sur, A., Friedman, R.S. and Miller, P.L., *J. Chem. Phys.* **94**, 1705 (1991).

Multiphoton Ionization/Dissociation of Cyclopentanone at the lower Rydberg States

John G. Philis, Constantine Kosmidis and Paraskevas Tzallas

Department of Physics, University of Ioannina, GR45110 Ioannina, Greece

Abstract. The 2-photon excitation of the 3p and 3d Rydberg states in jet-cooled cyclopentanone has been investigated by resonance enhanced multiphoton ionization (REMPI) in a time of flight mass spectrometer. The three $3p_{x,y,z}$ components are clearly resolved while the case for the $3d_i$ excitations is obscure due to the S_1 one-photon resonance. The ns laser induced mass spectra are characteristic of hard ionization while the fs laser induced mass spectrum is very similar to the Electron Impact one.

1. INTRODUCTION

The photochemical properties of carbonyl-containing compounds have been a topic of extensive investigations from the earlier systematic studies of molecular dissociation up to now. Cyclic ketones were initially investigated by Norrish and co-workers (1), while a direct femtosecond study of the transition state structures has been presented recently by the group of Zewail (2). Particularly, the study of the dissociation processes of cyclopentanone (CP) by means of different experimental techniques have been reported in several articles.

CP decomposes or rearranges upon excitation to the $^1(n,\pi^*)$ state ($\lambda \sim 320$ nm) as follow:

$$C_5H_8O + h\nu \rightarrow 2C_2H_4 + CO \quad (1)$$

$$\rightarrow C_4H_8 + CO \quad (2)$$

$$\rightarrow CH_2=CH-CH_2-CH_2-CHO \quad (3)$$

The above reactions have been found to depend on the temperature and the wavelength of the radiation used and on the pressure for gas-phase experiments (3). Bacerra and Frey (3) have suggested that, for reactions (1) and (2), a simple mechanism may be sufficient to explain the observed photolysis involving a vibrationally excited singlet electronic state.

The multiphoton ionization (MPI) technique has been also used for the study of CP. Baba et al (4) have presented the mass spectra of CP using laser light at 193 and 248 nm. Although the 193 nm MPI processes were greatly affected by population leakage from the intermediate state, the mass spectra have been interpreted (4) to originate preferably by the following direct two-photon processes [three-photon for eq. (11)]:

$$C_5H_8O \rightarrow C_5H_8O^+ \text{ (m/z=84)} + e \quad IP = 9.26 \text{ eV} \quad (4)$$

$$\rightarrow C_4H_8^+ \text{ (56)} + CO + e \quad \Delta H^0 \geq 9.86 \text{ eV} \quad (5)$$

$$\rightarrow C_4H_7^+ \text{ (55)} + HCO + e \quad \Delta H^0 \geq 10.1 \text{ eV} \quad (6)$$

$$\rightarrow C_2H_3CO^+ \text{ (55)} + C_2H_5 + e \quad \Delta H^0 = (10.5\pm0.5) \text{ eV} \quad (7)$$

$$\rightarrow C_2H_3CHO^+ \text{ (56)} + C_2H_4 + e \quad \Delta H^0 = 11.6 \text{ eV} \quad (8)$$

$$\rightarrow C_3H_5^+ \text{ (41)} + CH_3CO + e \quad \Delta H^0 = 11.6 \text{ eV} \quad (9)$$

$$\rightarrow C_2H_4^+ \text{ (28)} + C_2H_3CHO + e \quad \Delta H^0 = 12.1 \text{ eV} \quad (10)$$

$$\rightarrow C_2H_3^+ \text{ (27)} + C_2H_5CO + e \quad \Delta H^0 = (13.1\pm0.5) \text{ eV} \quad (11)$$

Of particular importance is the work of Pendensen et al (2) where, by pump-probe fsec excitation at 320/640, 310/620 and 280/560 nm, the formation of the tetramethylene diradical has been studied. The time scale for CP decarbonylation (100-200 fs) was found to depend on the state excited, while the lifetime of the diradical was in the range of 500 fs.

In this contribution we study the REMPI spectra of the 3p and 3d Rydberg states as well as the multiphoton ionization-dissociation of CP by means of TOF mass spectrometry.

2. EXPERIMENTAL

CP seeded in Ar (2-4 atm) was expanded into a vacuum chamber through a pulsed valve. The pulse duration of the valve (0.5 mm orifice diameter) was 200 μs. The vacuum chamber was evacuated by a rotary-backed diffusion pump and a liquid nitrogen trap was used to ensure an oil free chamber. The experiments were performed by crossing the molecular beam jet with the laser beam (20 - 150 μJ, 10 ns pulses at 7 Hz, focused by an f = +17 cm lens) at about 4 cm downstream from the orifice. The dye laser (Lambda Physik, LPD3000) was pumped by a XeCl excimer laser (Questec 2320).

The time-of-flight (TOF) mass spectrometer was of the conventional linear Wiley-McLaren design. Ions formed by the REMPI process were introduced into a field-free region (L = 1.10 m), after a two-step acceleration through the fields of E_1 = 500 V/cm, E_2 = 2000 V/cm, and were detected by a pair of microchannel plates. The mass resolution was typically 250 at 100 Da. TOF mass spectra were accumulated and averaged over 256 laser shots by connecting the ion detector output through a fast preamplifier to a digitizing storage oscilloscope (Hewllet-Packard 54210A). The ion signal corresponding to a specific mass was processed by a gated integrator as a function of laser wavelength.

3. RESULTS AND DISCUSSION

3.1. Cyclopentanone Spectroscopy

The study of the VUV absorption spectrum of cyclopentanone (5) confirms that the n → 3p and n → 3d Rydberg excitations are in the 178.5 nm (56020 cm^{-1}) and 165.6 nm (60390 cm^{-1}) regions respectively. Detection of these transitions as two-photon resonances requires laser light around 360 nm and 330 nm respectively.

Figure 1 is the 2+1 REMPI n → 3p spectrum of jet cooled cyclopentanone obtained by monitoring specific mass peaks in the time-of-flight mass spectrometer. The parent ion (m/z = 84) is extremely weak in the nanosecond pulsed laser excitation. All the masses yield identical MPI spectra over the entire wavelength range. This suggests that these spectra reflect the REMPI spectrum of a common precursor. We have attributed the observed structure (Fig. 1 and Table 1) to two-photon resonant transitions of neutral CP taking into consideration (a) the VUV absorption spectrum of cyclopentanone (5), (b) the two-photon static cell REMPI spectrum (6), (c) the one-photon $S_0 \to S_1$ (nπ*) jet-cooled fluorescence excitation spectra of Laane et al (7), Baba and Hanazaki (8) and the room temperature absorption spectrum of Howard-Lock and King (9), (d) the energetic ordering A_2 (3p$_x$), A_1 (3p$_y$), B_2 (3p$_z$) of all three members of the "3p complex" of acetone, a molecule with the same chromophore (10).

The n → 3d 2+1 REMPI spectrum of cyclopentanone becomes complicated, because (a) only one component out of the five expected appears in the VUV absorption spectrum (λ = 165.63 nm) (5) and (b) the laser light for this (2+1) process induces also the 1+2 REMPI $S_0 \to S_1$ transition. The complex spectrum is displayed in Fig. 2 and it is mainly due to $S_0 \to S_1$. Owing to the one-photon S_1 resonance, we would expect the 3d signal to be very strong as in the well known case of the 3s REMPI spectrum of acetaldehyde. However, the MPI signal is weak, implying the existence of dissociating channels in the CP molecule.

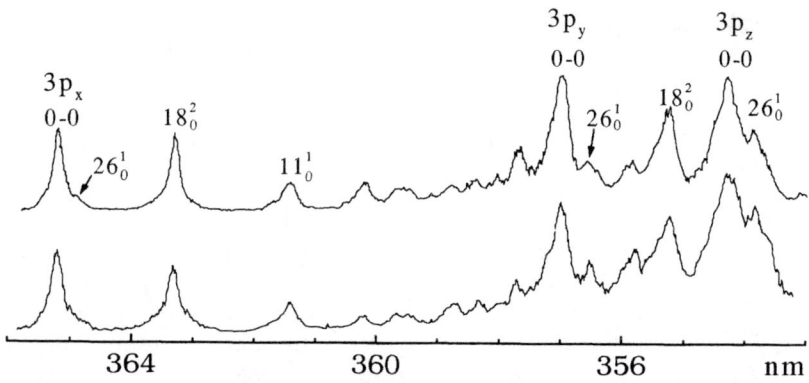

FIGURE 1. The n → 3p 2+1 REMPI spectrum of jet-cooled cyclopentanone with linearly polarized laser light (carrier gas argon, 2 atm). Upper spectrum: m/z = 26 mass channel. Lower spectrum: m/z = 39 mass channel. The 150 μJ dye laser pulses become weaker at λ < 355 nm.

TABLE 1. Summary of some spectroscopic characteristics for CP. Values of vibrations are in cm^{-1}

Transition	Origin (cm^{-1})	v_{11} ring angle bending (a$_1$)	v_{18} ring twisting (a$_2$)	v_{25} C=O out-of-plane wagging (b$_1$)	v_{26} ring bending (b$_1$)
S$_0$ → S$_1$ (A$_2$)^a	30276	532	2v_{18} = 238	2v_{25} = 309^d	91
n → 3s (B$_2$)^b	50068	671	2v_{18} ≈ 210	366	95
n → 3p$_x$ (A$_2$)	54750	574	2v_{18} = 284	-	48
n → 3p$_y$ (A$_1$)	56015	633	2v_{18} = 274	-	76
n → 3p$_z$ (B$_2$)	56446	-	2v_{18} = 280	-	66
n → 3d	60362	-	-	-	-
Ground State^c		705	2v_{18} = 238	446	95

a: From Ref. 7, b: From Ref. 11, c: From Ref. 9, d: The carbonyl wagging has a double minimum potential function in S$_1$ and the v_{25} = 0, v_{25} = 1 levels are degenerate.

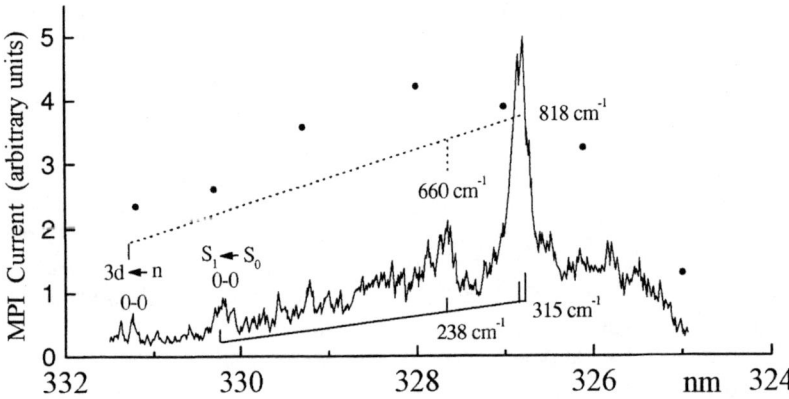

FIGURE 2. Laser light in the 330 nm region induces the n → 3d (2+1 REMPI) and S$_0$ → S$_1$ (1+2 REMPI) transitions. The dots represent the dye profile (left 35 µJ, maximum 65 µJ, right 20 µJ).

3.2. Fragmentation of cyclopentanone

Pedersen et al (2), using fs lasers, have proposed that masses 28 and 41 are due to parent ion fragmentation. In the present ns work (360 nm region, f = 17 cm lens) we have observed an extremely weak parent ion signal, a very weak signal for m/z = 28, and no signal for m/z = 41 (Fig. 3). Using an f = +56 cm lens, the 41 mass shows up very weakly. The m/z = 41 peak is also very weak in the mass spectra recorded at the 327 nm region. The m/z = 27 peak is the strongest one in all the wavelengths. Reactions (9) - (11) are taking place and it is clear that the ionization-dissociation (ID) mechanism is involved. The ns laser induced mass spectra show characteristics of hard ionization. Furthermore, the neutral fragments generated from the intermediate Rydberg states of CP are not expected to have a significant, if any, contribution to the

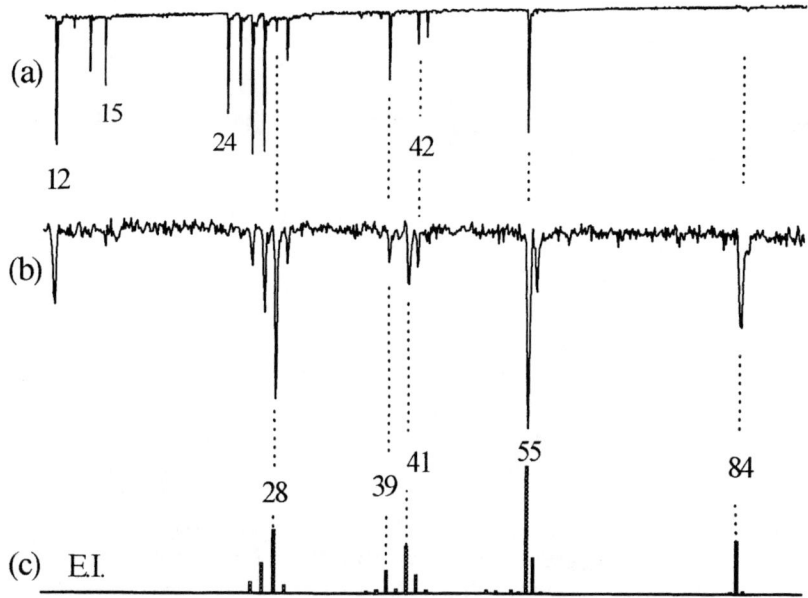

Figure 3. Fragmentation mass pattern of cyclopentanone induced by (a) nanosecond pulsed laser $I \approx 10^7$ W/cm^2, λ = 363.3 nm, (b) femtosecond pulsed laser $I \approx 10^{15}$ W/cm^2, λ = 375 nm, (c) 70 eV Electron Impact.

recorded mass spectra, since their multiphoton ionization is not resonantly enhanced at these wavelengths. The time of flight wavelength dependence of the mass spectra have shown that acetaldehyde is not a product of CP (6).

The mass spectrum of CP induced by a 50 fsec laser, at 375 nm (12) has been included in Fig. 3. It is known (13-16) that, using fsec laser pulses, the dissociation from electronic states of the neutral molecule can be "bypassed". Furthermore, the parent and the m/z = 56, 41 and 28 ion peaks appear clearly in the fsec induced mass spectrum. It is worth to note the striking similarity between fsec induced and E.I. mass spectra. This seems to be true for a large number of molecules and it will be discussed in details in a forthcoming paper.

ACKNOWLEDGMENTS

We are grateful to Dr. P. Tsekeris for his scientific contribution. The fsec mass spectra have been recorded at Rutherford Appleton Laboratory (RAL), in collaboration with the Glasgow's LIS group. C.K. and P.T. would like to express their thanks for the excellent facilities, assistance and hospitality there.

REFERENCES

1. Bamford, C. H., and Norrish, R. G. W., *J. Chem. Soc.* 1421-1428 (1938).
2. Pedersen, S., Herek, J. L., Zewail, A. H., *Science* **266**, 1359-1364 (1994).
3. Bacerra, R., and Monty Frey, H., *J. Chem. Soc. Faraday Trans.2* **84**, 1941-1949 (1988).
4. Baba, M., Shinohara, H., and Nishi, N., *Chem. Phys.* **83**, 221-233 (1984).
5. O' Toole, L., Brint, P., Kosmidis, C., Boulakis, G. and Tsekeris, P., *J. Chem. Soc. Faraday Trans.* **87**, 3343-3351 (1991).
6. Kosmidis, C., Boulakis, G., Bolovinos, A., Tsekeris, P. and Brint, P., *J. Mol. Structure* **266**, 133-140 (1992).
7. Zhang, J., Chiang, W-Y. and Laane, J., *J. Chem. Phys.* **98**, 6129-6137 (1993).
8. Baba, M. and Hanazaki, I., *J. Chem. Phys.* **81**, 5426-5433 (1984).
9. Howard-Lock, H. E. and G. W. King, G. W., *J. Mol. Spectrosc.* **36**, 53-76 (1970).
10. Xing, X., McDiarmid, R., Philis, J. G. and Goodman, L., *J. Chem. Phys.* **99**, 7565-7573 (1993).
11. Cornish, T. J. and Baer, T., *J. Am. Chem. Soc.* **109**, 6915-6920 (1987).
12. The fsec mass spectra have been recorded at Rutherford Appleton Laboratory (RAL), in collaboration with the Glasgow's LIS group.
13. Ledingham, K. W. D., Kosmidis, C., Georgiou, S., Couris, S., Singhal, R. P., *Chem. Phys. Lett.* **247**, 555-563 (1995).
14. Ledingham, K. W. D., Kilic, H. S., Kosmidis, C., Deas, R. M., Marshall, A., McCanny, T., Singhal, R. P., Langley, A. J. and Shaikh, W., *Rapid Commun. Mass Spectrom.* 9, 1522-1527 (1995).
15. Kosmidis, C., Ledingham, K. W. D., Kilic, H. S., McCanny, T., Singhal, R. P., Langley, A.J. and Shaikh, W., *J. Phys. Chem. A* **101**, 2264-2270 (1997).
16. Smith, D. J., Ledingham, K. W. D., Kilic, H. S., McCanny, T., Peng, W. X., Singhal, R. P., Langley, A. J., Taday, P. F., Kosmidis, C., *J. Phys. Chem. A* **102**, 2519-2526 (1998).

SESSION VI
MOLECULAR RIS

Charge Transfer and Charge Localization in Extended Radical Cations: Investigation of Model Molecules for Peptides

Rainer Weinkauf, Florian Lehrer

Institut für Physikalische und Theoretische Chemie der Technischen Universität München, Lichtenbergstr. 4, 85747 Garching, Germany

Abstract. Molecules consisting of a flexible tail and an aromatic chromophore are used as model systems to understand the situation of a single chromophore in a small peptide. Their S_0 - S_1 resonant multiphoton ionization (REMPI) spectra show, that in neutral molecules the tail-chromophore interaction is weak and electronic excitation is localized at the chromophore. For molecules, where the ionization energy of the tail is considerable higher than that of the chromophore, by high resolution REMPI photoelectron spectroscopy we find the charge to be localized on the aromatic chromophore. This scheme also in suitable peptides allows local ionization at the aromatic chromophore. An estimate for various charge positions in peptide chains, however, shows, that most of the amino acids electron hole positions in the nitronen and oxygen "lone pair" orbitals of the peptide bond are nearly degenerate. REMPI photoelectron spectra of phenylethylamine, which as a model system contains such two degenerate charge positions, show small energetic shift of the ionization energy but strong geometry changes upon electron removal. This result is interpreted as direct ionization into a mixed charge delocalized state. Consequences for the charge transfer mechanism in peptides are discussed.

INTRODUCTION

Charge transfer is a very important process in nature, such as in photosynthesis and respiration. Usually in charge transfer model systems the electron donor and acceptor are connected by a non-conductive bridge. After photoactivation charge transfer is supposed to occur by a tunneling process through the bridge barrier (1). In neutral molecular systems in solvents the energetic and the dynamic of this process is complicated because of the Coulomb attraction between the opposite charges and its partial shielding by solvent reorientation. Hence, the donor-acceptor distance, solvent properties such as dipole moment and polarizability and also the solvent temperature are important parameters.

In contrast to this, in peptide radical cations in the gas phase charge transfer is a "charge shift". The simplified energetics in gas phase allows an identification of the electronic states involved in the charge transfer process. As confirmed by this work in suitable tailor-made peptides charge can be localized at the aromatic chromophor by REMPI. As found previously (2-4) by resonant multiphoton excitation of the radical peptide cations by visible light, charge transfer can be activated in some peptides but not in others. These effects are energetically well described in a landscape of local ionization potentials: Dependent on the energetics of the hole positions in the individual amino acids either a downleading staircase-

CP454, *Resonance Ionization Spectroscopy*
edited by J. C. Vickerman, I. Lyon, N. P. Lockyer, and J. E. Parks
© 1998 The American Institute of Physics   1-56396-810-X/98/$15.00

like arrangement of electronic states is formed or a barrier blocks charge migration. We find that in radical cations charge transfer is a stepwise through bond multistep hole transfer mechanism (2-4).

In this zero order description for peptide cations many electronic sates, corresponding to different charge sites, are found whithin an energy range of 1 eV. Hence, in a next step, large interstate coupling is expected between charge sites neighboring in space and energy. In this work we investigate the degree of local photoexcitation and ionization in extended model molecules. The degree of local excitation is shown by REMPI spectroscopy. The degree of charge localization is investigated by high resolution REMPI photoelectron spectroscopy.

EXPERIMENTAL

REMPI spectroscopy and high resolution REMPI photoelectron spectroscopy are performed in the same apparatus. In short, the apparatus consists of three parts: the supersonic expansion chamber, the time-of-flight photoelectron spectrometer and the linear time-of-flight mass spectrometer. The sample molecules are brought into gas phase by heating the pulsed nozzle up to 50-130 °C according to sample vapour pressure. After supersonic expansion the neutral beam is skimmed and transferred into a µ metal shielded time-of-flight photoelectron spectrometer. Much care was taken on electrical and magnetical shielding. Laser ionization is performed by two excimer-pumped frequency doubled dye lasers of 6 ns pulse width. The excess energy for the ionization step was chosen to be below 300 meV in order to optimize the electron energy resolution to be 4-8 cm^{-1} (0.5-1 meV) at an absolute accuracy of 8 cm^{-1} (1 meV). In this apparatus especially no photoelectron energy calibration is necessary.

The REMPI spectra have been taken in the same apparatus. This allows mass control and finding ideal adjustments for the supersonic expansion also for the photoelectron experiment. For detection of the ions formed in the REMPI process, we wait until the ions drift out of the photoelectron spectrometer into the extraction plates of a linear time-of-flight mass spectrometer. The REMPI spectra are recorded by mass selective ion detection in dependence on laser wavelength.

RESULTS AND DISCUSSION

In the following we investigate model molecules consisting of a chromophore and a flexible tail: (i) REMPI spectra can be used to show that the electronic photoexcitation is local at the chromophore. (ii) Structural changes upon electron removal monitored by Franck-Condon effects can be used for detection of charge localisation. For this REMPI photoelectron spectroscopy is applied. As model molecules we use butylbenzene, phenylethanol, phenylpropionic acid, phenylethylamine and phenylpropylamine.

REMPI Spectra of Flexible Molecules

In Fig. 1 the S_0 - S_1 REMPI spectra of butylbenzene (a), phenylethanol (b), phenylpropionic acid (c), phenylethylamine (d) and phenylpropylamine (e) are shown. Because of the very efficient coupling of low frequency modes to the supersonic expansion (5) contributions of transitions from vibrational excited molecules (hot bands) to the spectra can be excluded. Some small peaks can be assigned to vibrational progressions. The remaining peaks are contributed to different conformers. In agreement with previous investigations two conformers are found for butylbenzene (6), two conformers for phenylpropionic acid (7) and four conformers for phenylethylamine (8). For phenylethanol and phenylpropylamine two conformers are identified. The S_0 - S_1 origins of the conformers of all molecules lye within 150 cm^{-1} around a center wavelength of 37 600 cm^{-1}. This fact shows that the chromophore property is not very much changed neither by the different functional groups at the tail nor by their geometric position to the aromatic ring.

The conformer structure analysis can be perfomed by REMPI photoelectron spectroscopy (see below). In general two structures are possible: Linear *trans* structures and folded *gauche* structures. For butylbenzene (Fig. 1 a) the folded *gauche* structure is stabilized by a interaction of one hydrogen of the third carbon atom of the alkyl chain with the π-System of the aromatic chromophore. This -C-H--π interaction is expected to be increased in the S_1 state because of the increase of the size of the electron cloud in S_1. This agrees well with the observed red shifted S_0 - S_1 origin. For the folded conformer in phenylethanol the hydrogen of the hydroxy group is supposed to be in a similar position to the chromophor as the H in the *gauche* conformer of butylbenzene. This is confirmed also by the red shift observed in the S_0 - S_1 spectrum (Fig. 1 b). The second conformer is an extended *trans* conformer.

In phenylpropionic acid in comparison to phenylethanol the chain is one -CH$_2$ - link longer. In the folded structure the chain is obviously now too long and the hydroxy group does not find a suitable position to bind to the chromophore. As a result in stead of the hydrogen the oxygen of the -C=O group which is supposed to be slightly negative comes close to the chromophore. By increasing the electron cloud of the π-system in S_1 this interaction is expected to become less attractive in agreement with the observed blue shift of the *gauche* conformer with respect to the *trans* conformer (Fig. 1 c). For butylbenzene, phenylethanol and phenylpropionic acid this conformer assignment can be supported by the REMPI photoelectron spectra (see below).

For phenylethylamine in addition to the *trans* and the folded conformer the rotation of the -NH$_2$ group allows further conformer possibilities. For our consideration of charge localization we are mostly interested in throgh bond effects and therefore only are interested in the *trans* conformers. However for phenylethylamine and phenylpropylamine a complete assignment of the conformers by our spectra is not possible. Nevertheless the red shifted origins in the S_0 - S_1 spectra are assigned to the folded conformers.

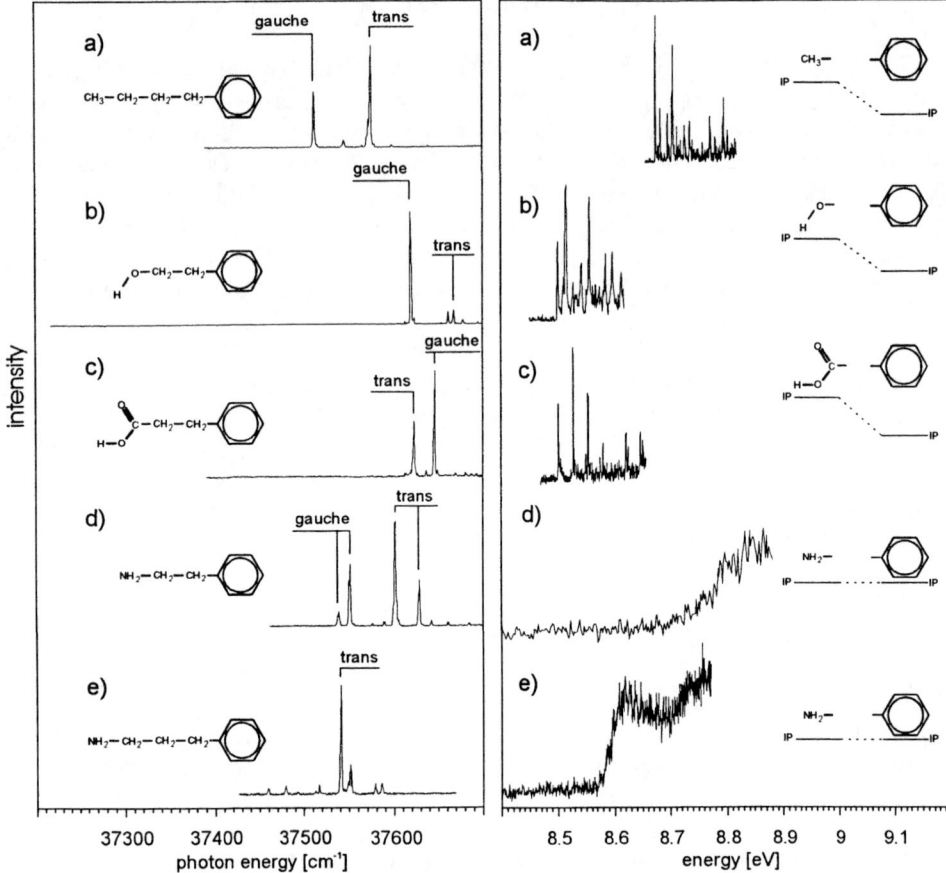

FIGURE 1: $S_0 - S_1$ REMPI spectra of butylbenzene (a), phenylethanol (b), phenylpropionic acid (c), phenylethylamine (d) and phenylpropylamine (e). Note the occurence of different conformers for each molecule. All spectra are situated within a small energy range of 150 cm^{-1} centered around 37600 cm^{-1} which shows that the property of the chromophore is neither much influenced by the functional end group nor by its position to the chromophore.

Figure 2: REMPI photoelectron spectra of the *trans* conformeres of butylbenzene (a), phenylethanol (b), phenylpropionic acid (c), phenylethylamine (d) and phenylpropylamine (e) are shown. The corresponding energy scheme is given as a sketch at the right side. Beside shifts in the ionization energy, Fig. 2 a-c show sharp structured spectra (charge localized) and Fig. d,e broad features (charge delocalized). For explanation see text.

The $S_0 - S_1$ origins of the conformers of all molecules lie within 150 cm^{-1} around a center wavelength of 37600 cm^{-1}. This shows that the chromophore property is not very much changed by the different functional groups at the tail and even by their position to the aromatic ring. Hence, the electronic $\pi - \pi^*$ excitation is localized at the aromatic chromophore. The electronic states corresponding to a tail excitation are either due to a $\sigma - \sigma^*$ or a n - σ^* excitation and are situated at very high excitation energies. This and our experimental results suggests a picture in which for large molecules electronic states at different charge sites can be treated as "local" states. Such a local state concept is widely used in liquid spectroscopy of large molecules. If we transfer this concept, a local photoexcition behavior can be expected also in peptides containing a single aromatic chromophore.

REMPI Photoelectron Spectroscopy: Charge Localization

Charge localization is investigated by high resolution REMPI photoelectron spectroscopy. The basic idea is that strong charge delocalisation should cause large geometry effects in the ionization process (Fig. 3). This should show up in large vibrational progressions or even an unstructured photoelectron spectrum. We are especially interested in „through bond" charge delocalization effects and therefore we here concentrate on linear conformers. The folded confromers have been also investigated but are not shown here. Using suitable REMPI S_0 - S_1 origin transitions exclusively molecules of *trans* or folded conformers structures can be excited. As model molecules we use molecules for which the energies for charge positions in the chromophore and the tail are (i) well separated by more than 1 eV (see Table 1) or (ii) nearly degenerate. For the latter the use of a polar end group can not be avoided. Therefore for the molecules of type (i) we investigate also non-polar (butylbenzene) or polar end groups (phenylethanol, phenylpropionic acid) in order to distinguish monopole-dipole interaction and charge delocalization effets.

Table 1: Ionization potential of the functional groups of the model molecules

functional group	ethyl-benzene	n-butane	ethanol	propionic acid	ethyl-amine	n-propyl-amine
ionization potential [eV]	8.7727[a]	10.5[b]	10.2[b]	10.4[b]	8.8[b]	8.7[b]

[a] taken from (5), [b] taken from (7)

As for butylbenzene, the REMPI photoelectron spectrum of the *trans* conformer shows besides some typical chromophor modes a characteristic chromphore-tail mode with 249 cm^{-1}. These choromphor-tail mode is attributed to the symmetric chromophore-tail bending mode. The photoelectron spectrum of the *gauche* structure of butylbenzene is also sharp (not shown here). It shows a low frequency mode of 38 cm^{-1}, which is interpreted as the fundamental frequency of the chromophore-tail twist mode. This is consistent with the lack of symmetry in the *gauche* conformer. The intensities, which are correlated with Franck-Condon factors, indicate small geometry changes by electron removal in the ionization process. Because the tail is nonpolar our observation can be directly correlated with a very small degree of charge delocalization into the tail. The extensive analysis of structures, vibrational frequencies, symmetries and *ab initio* calculations of the alkylbenzenes is presented in a separate work (5).

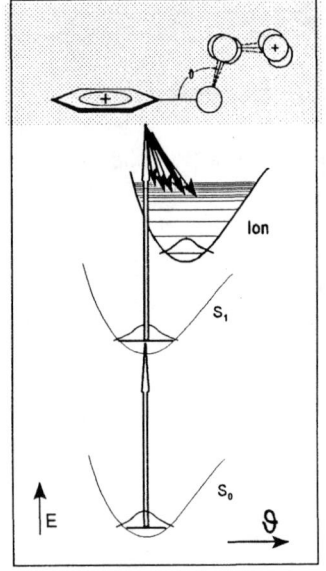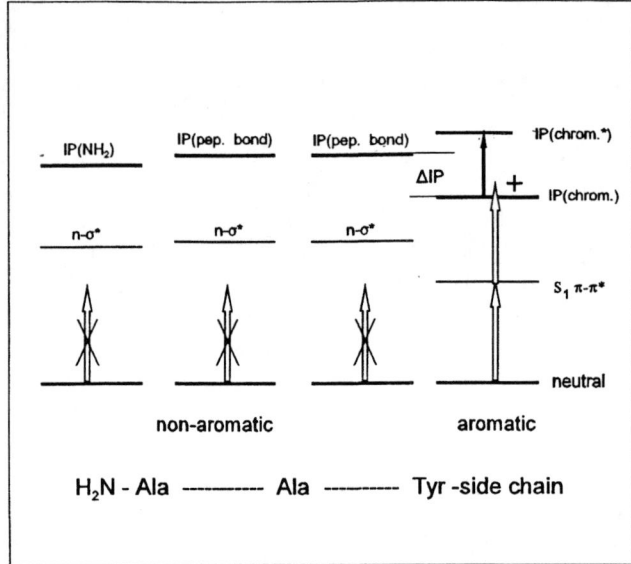

Figure 3: Detection of structural changes upon electron removal by REMPI photoelectron spectroscopy

Figure 4: In tailor-made peptides local excitation and local ionization is possible if the ionization potential of the aromatic chromophore is much lower than the ionization potentials of all peptide chain charge sites (IP(pep. bond)).

As for phenylethanol, the REMPI photoelectron spectrum of the *trans* conformer is well structured and shows low frequency tail-chromphore modes. The ionization energy shift and the vibrational intensities agree well with the expected monopole-dipole attraction. The second conformer found in the REMPI-spectrum 1 c shows a broad unstructured REMPI-photoelectron spectrum. Such behavior is expected for a folded structure, where the polar group and the chromophor come very close to each other and therefore the dipole-chromophore interaction changes strongly upon electron removal.

For phenylpropionic acid also the REMPI photoelectron spectrum of the *trans* conformer is sharp and shows low frequency tail-chromphore modes. The situation is very similar to phenylethanol: The ionization energy shift and the Franck-Condon intensities agree also well with the expected monopole-dipole attraction. Also the second conformer shows a broad spectrum as expected for a folded structure. In this case the negatively charged oxygen is close to the chromophore and therefore the chromophore-tail attraction increases strongly upon ionization i.e. electron removal from the chromophor π-system. These examples shows that by means of REMPI photoelectron spectroscopy the assignment of different conformer structures is possible.

To investigate the "through bond" charge delocalization effect between neighboring, isoenergetic charge sites we investigate phenylethylamine and phenylpropylamine. Here the two charge sites, the amine group and the phenyl group are spaced by a $-CH_2-CH_2-$ bridge like in peptides. They have very similar ionization energies of 8.77 eV (ethylbenzene (5)) and 8.8 eV (ethylamine (7)). The REMPI photoelectron spectra of all conformers of the composed molecule show both broad structures. This result can be interpreted as a strong geometry change upon electron removal. Because for the *trans* conformer this can not be a simple charge-dipole interaction (analogous to the lack of such effects for *trans* phenyethanol and *trans* phenylpropionic acid) this change in geometry is attributed to a charge delocalization in the cation over both isoenergetic sites. This is qualitatively in agreement with *ab initio* calculations predicting high charge density on both isoenergetic charge sites. The strong shift in ionization energy to higher values and the smooth onset are attributed to a lack of Franck-Condon overlap from neutral S_1-state to the vibrational ground state of the cation.

In phenylpropylamine some spectral structrures are again observed and the onset is somewhat sharper. This on one hand can be understood in terms of a larger distance between the charge sites, which reduces the coulomb repulsion between the positive partial charges and on the other hand possibly by less charge delocalization. Further investigations and *ab initio* calculations are required to clarify this aspect. Nevertheless the results of Fig. 2 d,e show that already during the ionization process the charge is instantaneously delocalized. This means that for molecules with degenerate and not too far spaced charge sites local ionization fails and the local electronic state picture is invalid. Beeing aware of this, in our peptide experiments (2-4) we, therefore, avoided any isoenergetic charge positions for the ionization process to asure a local ionization. Fig. 4 shows schematicly the local photoexcitation and the local ionization in the tripeptide alanyl-alanyl-tyrosine (Ala-Ala-Tyr). As for the isoenergetic charge sites in the peptide chain (lone pair orbitals of oxygen and nitrogen, see Fig. 4), however, complete charge delocalization over two or more sites is possible.

CONCLUSION

By high resolution REMPI photoelectron spectra of model systems we could investigate charge localisation in extended model molecules. On one hand for isoenergetic but in space well separated charge sites we found strong charge localization for charge sites which differ strongly in energy. On the other hand for isoenergetic charge sites strong charge delocalization is observed.

The results of this paper support and confirm our former local ionization concept for tailor-made, spezial selected peptides. However, they also contribute new important information concerning charge delocalization in case of the presence of energetically degenerated charge sites. Our spectroscopic evidences indicate for charge delocalization a strong dependence on bridge length. Hence for the degenerate case and short bridges the local electronic state concept is no more appropriate, the electronic states become a superposition of local states as discussed previously (4). Nevertheless for discussion of the energetics of the charge transfer the local ionization potential picture is still a very useful description.

Obviously in radical cations charge can be easily shifted even through short saturated carbon bridges. Therefore in radical cations hole transfer can be assumed to be a very fast and efficient process which even might be faster than vibrational time scales. As for mass spectrometry charge mobility is also highly correlated with radical cation reactivity.

ACKNOWLEDGMENTS

We gratefully acknowledge the support by E.W. Schlag and the 'Deutsche Forschungsgemeinschaft'.

REFERENCES

1. N.M. Paddon Row, *Acc. Chem. Res.* **27**, 18-29 (1994)
2. R. Weinkauf, P. Schanen, D. Yang, S. Soukara, and E.W. Schlag, *J. Phys. Chem.* **99**, 11255-11265 (1995)
3. R. Weinkauf, P. Schanen, A. Metsala, E.W. Schlag, M. Bürgle, and H. Kessler, *J. Phys. Chem.* **100**, 18567-18585 (1996).
4. R. Weinkauf, E.W. Schlag, T.J. Martinez, and R.D. Levine, *J. Phys. Chem. A* **101**, 7702-7710 (1997).
5. F. Lehrer, L. Lehr, R. Weinkauf, E.W. Schlag, P. Hobza, to be submitted to *J. Am. Chem. Soc.*
6. P.M. Mayer, and T. Baer, *Int. Mass. Spectrom. Ion Proc.* **156**, 133-139 (1996)
7. K. Kimura, S. Katsumata, Y. Achiba, T. Yamazaki, and S. Iwata, *Handbook of HeI Photoelectron Spectra of Fundamental organic Molecules*, Tokyo: Japan Scientific Press, 1981, p. 49, p. 106, p. 165, p. 115, p. 116

An attempt to study LiH and Li$_2$ molecules by high resolution pulsed laser spectroscopy

Nadia Bouloufa, Louis Cabaret, Patrice Cacciani, <u>Pierre Camus</u>, Boris Pitcheev and Raymond Vetter

†*Laboratoire Aimé Cotton, CNRS II, Bât 505, Campus d'Orsay, 91405 Orsay Cedex, FRANCE*

Abstract. As we start a program to study alkali hydrides and dimers, we have developed a two-step photoionisation experiment on Li$_2$ molecules based on the use of an atomic beam and two pulsed dye lasers. The first resonant step which excites the A$^1\Sigma_u^+$-X$^1\Sigma_g^+$ Li$_2$ dimer systems is a home-made cw-seeded DCM dye laser with a laser linewidth of 55 MHz (FWHM) and near the Fourier transform limit. The second step is a larger width fixed frequency UV laser which allows the photoionisation of the selectively excited molecules. The three ^{6}Li$_2$, ^{6}Li^{7}Li and ^{7}Li$_2$ spectra are recorded simultaneously by the use of a doubly-accelerating time-of-flight ion analyser. Comparison between recorded and calculated absorption spectra using Dunham parameters found in the litterature is satisfactory. To develop similar pulsed high-resolution investigations in LiH, we have characterized our molecular beam by using the laser induced fluorescence (LIF) technique with a cw blue dye laser. Two Franck-Condon LiH Doppler-free resonances have been observed.

INTRODUCTION

Single mode CW dye and solid state lasers have been previously used in high resolution molecular spectroscopy but their commercially available tunable ranges (which are generally limited to a few wavenumbers) are not easy to operate for extended wavelength range observations. More over due to their continous working conditions only low excited systems can been investigated efficiently and analysed using the modulated detection of the resonant induced fluorescence.

---

†This laboratory is associated with the Université Paris Sud

At the opposite, the use of REMPI technics with commercially available pulsed lasers and time-of-flight (TOF) analysis of the produced ions and electrons have considerably extended the knowledge of the properties of their excited states but, in general, at the detriment of the spectral resolution. The recent two-colour coherent laser manipulation of molecules named Stimulated Raman Adiabatic Passage (STIRAP) (1) which can excite selectively a vibrational state in a molecular beam has pointed out the necessity to develop single mode pulsed amplified laser system (2) if we want to spread its use for investigating properties of molecules outside of their equilibrium.

LiH INVESTIGATION

Looking at the litterature for making the choice of a small molecule to be studied, we have found that LiH which is one of the simplest heteronuclear molecules, has been the bench of extended theoretical and experimental studies (3) but surprisingly is still poorly known experimentally. Only recently, pulsed optical-optical double resonance fluorescence depletion (4) has allowed the identification of the C $^1\Sigma^+$ excited electronic state which is the second one to be known in spite of extended ab-initio predictions of the potential curves (5). Higher predicted Rydberg states and ionisation limits are ever unknown. Let's mention there that the knowledge of this molecule should be revived because of its strong astrophysical interest in the formation of the early Universe (6). Evidently if we want to investigate those unknown excited states by the way of a two or three-step photoionisation, we need at least three lasers: two single mode pulsed amplified ones (one at ≈350 nm for the first step and a second at 500-650 nm for the second step) plus a less resolving power UV one for the last photoionisation step. At that time, we are in possession of only one of these two high resolution pulsed laser which is an old home-made cw seeded red dye laser (8) and the second which will be a single mode pulsed amplified Ti: sapphire is under construction at the laboratory.

So during this time, we have performed two specific experiments. First in order to develop a very well collimated LiH molecular beam we have observed laser induced fluorescence with a cw single mode blue laser to analyse the species in the molecular beam. Second to demonstrate the usefulness of high-resolution pulsed laser excitation associated with TOF mass analysis of light ions for the LiH study we have developed a two-colour photoionisation experiment on the Li_2 dimers by including our cw seeded dye pulsed laser as the first step to get high spectral resolution .

CW LASER INDUCED FLUORESCENCE OF LiH

We have designed a set-up where a flow of hydrogen gas is passed through a powder LiH filled oven at 1000°K. To analyse the molecular beam species, induced fluorescence by a right angle crossing beam cw blue dye laser is collected. In spite of the fact that the available ≈435 nm of this cw single mode blue laser is very far from the center of the A $^1\Sigma^+$ - X $^1\Sigma^+$ LiH system at 355 nm, two Franck-Condon unfavourable LiH resonances from the 1-8 band have been observed Doppler-free and identified as many others lines belonging to the B $^1\Pi_u^+$ - X $^1\Sigma_g^+$ ^{7}Li^{7}Li and ^{6}Li^{7}Li species present in the beam. A part of the spectrum is given in Figure 1.

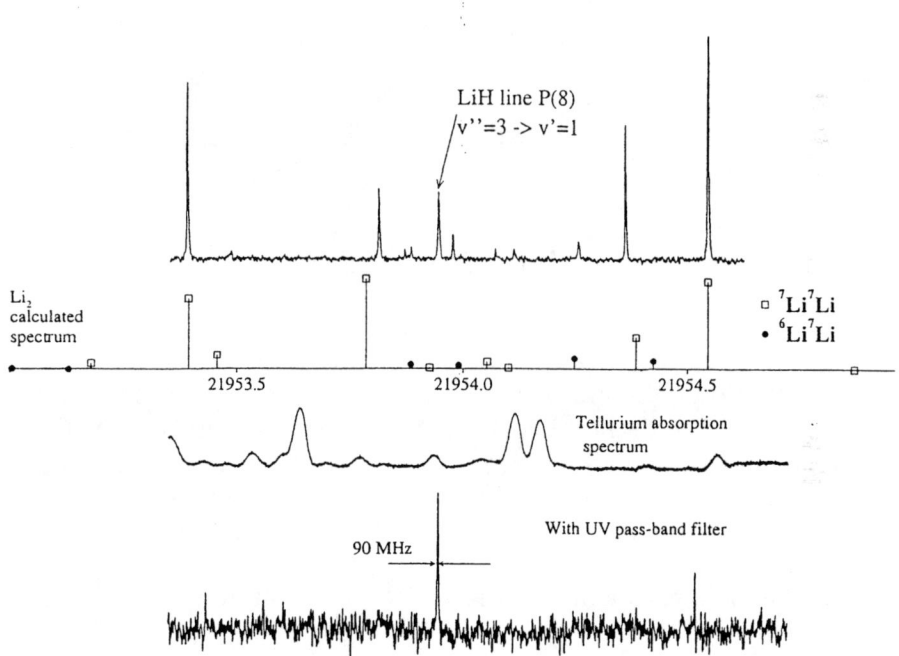

FIGURE 1. High resolution LIF spectrum of the molecular beam with a single mode cw blue dye laser.

The observed linewidth is 90 MHz (FWHM). At this high resolution deviations between the observed and calculated Li$_2$ dimer spectra using mass corrected Dunham parameters are visible on Figure 1.

RESONANT TWO-PHOTON IONISATION OF Li$_2$

First, we have improved a two-accelerating voltage TOF analyser to separate easily very light ions. The first zone is 1 cm wide with a 20V dc applied voltage and the second is 3 cm with a 200 V dc voltage. The field free zone is 100 cm long. The ion detector is a tandem of microchannel plates. By focusing only the 291 nm UV photoionisation step laser on the Li effusive atomic beam which is passing in the middle of the first zone, we have recorded the time-of-flight ion spectrum given in Figure 2.

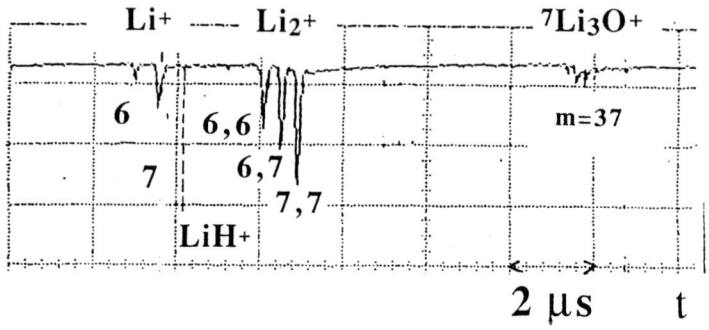

FIGURE 2. Time arrival of the different ions in the two-accelerating zone TOF analyser.

The three dimer ion peaks correspond to an adequatly composed metallic mixture of ^{6}Li and ^{7}Li in the electrically heated oven bored with a 0.2 mm hole. The dashed line indicates the TOF calculated position of the ^{7}LiH$^+$ ion. The resulting large peak separation between successive ions will allow further the use of a sweeping repulsive voltage applied in front of the microchannel plates to prevent their saturation by arrival of faster undesirable ions.

In this R2PI experiment the spectral high resolution is achieved by scanning the first step laser which is the home made cw-seeded DCM dye laser with a 10 ns Nd-YAG pulse duration. Its linewidth around 55 MHz (FWHM) is near the Fourier transform limit. Around 650 nm, it excites the A $^1\Sigma_u^+$ - X $^1\Sigma_g^+$ Li$_2$ dimer systems and the second step laser, at a fixed frequency in the UV around 350 nm, is a conventional doubled frequency 3 GHz DCM dye laser pumped by a Nd-YAG which allows the photoionisation of the selectively excited molecules. The isotopic dimer Li$_2^+$ ions are then simultaneously recorded.

Figure 3 shows a part of the 30 cm^{-1} observed energy range which includes the v"=0 to v'=5 bandhead of the A-X transition for $^7Li_2$.

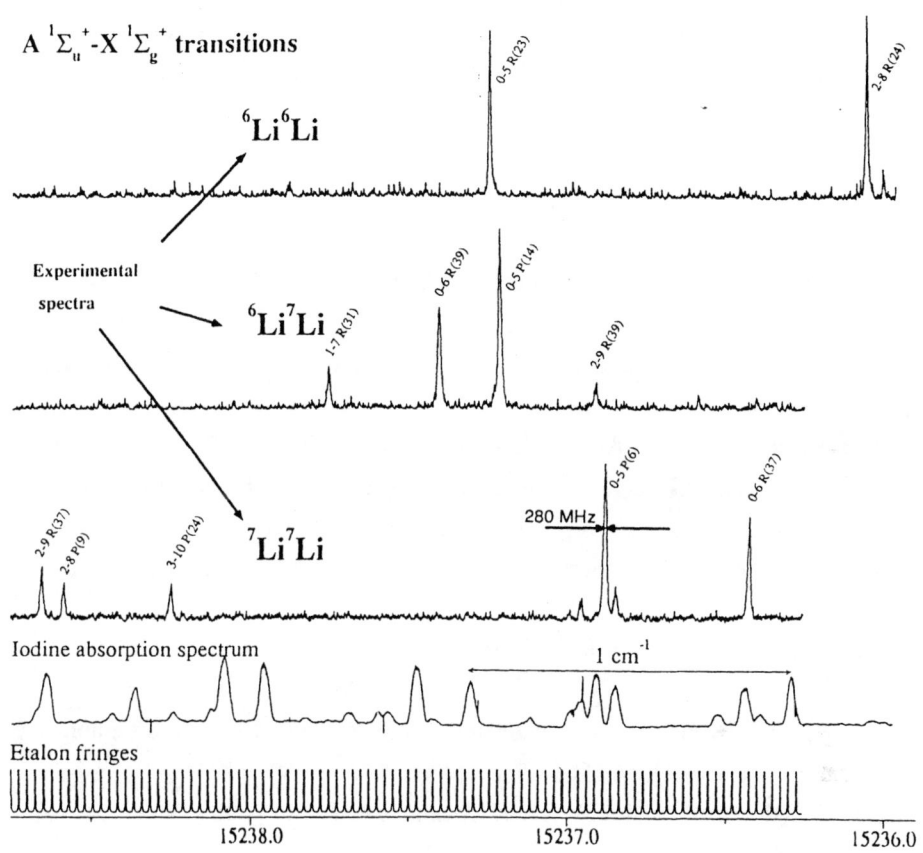

FIGURE 3. R2PI high resolution three dimer spectra of Li_2.

Here the observed linewidth is limited to 280 MHz (FWHM) by residual Doppler effect due to low collimation of the molecules in our Li atomic effusive beam. The $^7Li_2$ measured lines shows significant shifts with calculated spectrum using Dunham parameters (9). Even larger discrepancies for $^6Li^7Li$ resonances can be well taken in account by the use of a different mass correction for a heteronuclear diatomic molecule (10).

CONCLUSION

We hope next that the use in a R3PI experiment of the second single mode pulsed amplified Ti:sapphire laser under construction with the two lasers presented above will allow us to observe more data in small molecules with a very high resolution and particularly to investigate higher excited states of LiH even by using STIRAP to populate selectively vibrationally excited levels of the $X\ ^1\Sigma^+$ electronic ground state.

REFERENCES

1. Gaubatz, U., Rudecki, P., Schiemann, S., and Bergmann, K., *J. Chem. Phys.* **92**, 5363-5376 (1990).
2. Schiemann, S., Kuhn, A., Steuerwald, S., and Bergmann, K., *Phys. Rev. Lett.* **71**, 3637-3640 (1993).
3. Stwalley, W.C., and Zemke, W.R., *J. Phys. Chem. Ref. Data* **22**, 87-112 (1993).
4. Lin, W.C., Chen, J.J., and Luh, W.T., *J. Phys. Chem. A* **101**, 6709-6711 (1997).
5. Boutalib, A., and Gadéa, F.X., *J. Chem. Phys.* **97**, 1144-1156 (1992).
6. Puy, D., Alecian, G., Le Bourlot, J., Léorat, J., and Pineau des Forêts, G., *Astron. Astrophys.* **267**, 337-346 (1993).
7. Neuhauser, R., Sussmann, R., and Neusser, H.J., *Phys. Rev. Lett.* **74**, 3141-3144 (1995).
8. Cabaret, L., Thèse de Docteur Ingénieur, Université de Paris Sud, 22 décembre 1986.
9. Kusch, P., and Hessel, M.M., *J. Chem. Phys.* **67**, 586-589 (1977).
10. Vidal, C.R., and Stwalley, W.C., *J. Chem. Phys.* **77**, 883-898 (1982).

Dibenzo-p-dioxin and its chlorinated congeners: Molecular Geometry and Electronic Structure

Ralf Zimmermann

GSF Forschungszentrum Umwelt und Gesundheit GmbH, Institut für Ökologische Chemie, Ingolstädter Landstraße. 1, D-86764 Neuherberg, Germany and Technische Universität München, Lehrstuhl für Ökologische Chemie und Umweltanalytik, D- 85350-Freising, Germany

The dibenzo-p-dioxin (DD) molecule and some of its chlorinated derivatives (PCDD) are investigated by means of resonance-enhanced multiphoton ionization - time-of-flight mass spectrometry (REMPI-TOFMS). The 2+2 four-photon REMPI spectrum of the $S_1(^1B_1)/S_2(^1A_1)$ systems of DD is presented and compared with the corresponding 1+1 REMPI spectrum. Additionally the 1+1 REMPI spectra of 2-monochloro- and 2,3-dichloro-p-dibenzo-p-dioxin are presented. The spectra are interpreted regarding electronical and geometrical aspects. New concepts for a fast, REMPI-based PCDD analysis are discussed

INTRODUCTION

The tricyclic dibenzo-p-dioxin (DD) molecule consists of two aromatic benzoic moieties which are coupled via two neighboring ether bridges. The central ring, which bears the two oxygen atoms, is non-aromatic and separates the two benzoic π-systems. Therefor a slightly folded, non-planar structure of the molecule is possible. However, the oxygen's lone pairs allow some electronic resonance interaction between the π-systems of the benzene moieties. Thus the DD molecule represents a model system for a weak electronic coupling of two benzoic π-systems via hetero atom bridges.

A powerful technique for examination of electronically weakly coupled aromatic-systems is supersonic jet laser spectroscopy and analysis of the vibronic structure of the low-lying UV-transitions. Upon the optical excitation, the coupling strength between the π-systems changes. This causes ge-

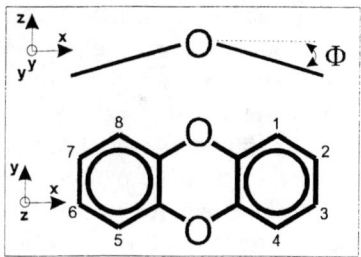

Figure 1: Structure of Dibenzo-p-dioxin and definition of folding angle Φ

ometry changes as e.g. the change of the twisting angle between the phenyl groups of the biphenyl molecule[1] or the alternation of the bondlengths in the central four-ring of the partial antiaromatic biphenylene molecule[2]. The investigation of the vibronic structure of electronic transitions of jet cooled molecules can reveal informations about the induced geometry changes as well as about the electronic structure of the molecule (e.g. via a *Franck-Condon* analysis or optical selection rules). Together with additional data, e.g. from calculations or further spectroscopic experiments often meaningful conclusions can been drawn. In this work, the jet-spectroscopy of the S_1 and S_2 band sys-

CP454, *Resonance Ionization Spectroscopy*
edited by J. C. Vickerman, I. Lyon, N. P. Lockyer, and J. E. Parks
© 1998 The American Institute of Physics 1-56396-810-X/98/$15.00

tems of DD and some of its chlorinated congeners is studied, using resonance-enhanced multiphoton ionization time-of-flight mass spectrometry (REMPI-TOFMS). The present investigation focuses on two topics, the molecular geometry of DD in its ground and first excited singlet state in the gas phase and the influence of chlorination on the electronic structure and the vibronic pattern in the REMPI spectra. The experimental setup used for recording of REMPI spectra is described elsewhere [1]. Briefly, an excimer laser pumped tunable dye laser (frequency doubled for 1+1 REMPI) and a homebuilt compact reflectron time-of-flight mass analyzer with a supersonic jet inlet system was used

RESULTS AND DISCUSSION

The folding angle Φ (~ folding angle about the axis that connects the two oxygen atoms, see figure 1) is determined by an equilibrium between electronic and sterical forces. The geometry of DD in the solid state (crystal) and in solution is known from experimental studies to be planar and slightly folded, respectively[3]. *Ab initio* calculations (6-31G**) predict a slightly folded structure (12°) of DD in the gas phase[4] (ground state), while semiempirical calculations (AM1-CI) suggest a planar DD structure in the first excited singlet state. Semiempirical calculations (this work) on INDO/S-CI level also show, that the first two excited singlet states are close together, separated by only 600 cm^{-1}. The oscillator strength of the S_1 transition ($^1B_{3u}$, for a planar structure or $^1B_1$ for a folded structure, INDO/S-CI: 34490cm^{-1}) is predicted to be weak (f= 0.05) for both, planar and folded geometry. The second singlet transition (S_2, $^1A_{1g}$ or $^1A_1$, INDO/S-CI: 35090 cm^{-1}, f= 0!) is electronically forbidden for planar (D_{2h}) geometry. If the geometry is folded (C_{2v}), the S_2 transition becomes electronically allowed character. The INDO/S calculations predict a very weak oscillator strength (f= 0.001 for 18° folding angle). As for both states, S_1 and S_2, the two photon absorption is electronically allowed, a two-photon resonant, four photon ionization REMPI spectrum of DD was recorded (2+2-REMPI). For recording of the 2+2-REMPI spectrum, the output of the dye-laser (Rhodamin 6G, 7 mJ) was focused (f = 100 mm) onto the skimmed supersonic molecular beam in the ion source of the TOFMS. High power densities are required for the non-resonant two-photon absorption step. This induces a massive fragmentation of the formed DD ions due extensive photon absorption. The mass spectrum shows C^+ (m=12 amu) as base peak and only some minor peaks of larger fragments. Therefore the 2+2 REMPI spectrum was recorded via the C^+ trace. The one-photon resonant, two-photon ionization spectrum (1+1-REMPI) was recorded via the molecular mass peak using the frequency doubled, unfocussed output of the dye laser[5]. Figure 2 shows the 1+1 (lower trace) and 2+2 REMPI spectra of DD (upper trace). The closely neighbored S_1 and S_2 electronic systems leads to partially disturbed vibronic [6]. The activity of the low frequency butterfly mode (see below) is remarkable. For example some vibronically forbidden modes appear in the spectrum. For an analysis of the 1+1 REMPI spectrum see ref. [6]. The upper trace shows the 2+2 REMPI spectrum. The first intense peak (~33800 cm^{-1}) is due to the origin region of the (two-photon allowed) S_1 system while the second intense peak (~34300 cm^{-1}) represents the origin of the two-photon allowed (but one photon forbidden) S_2 system.

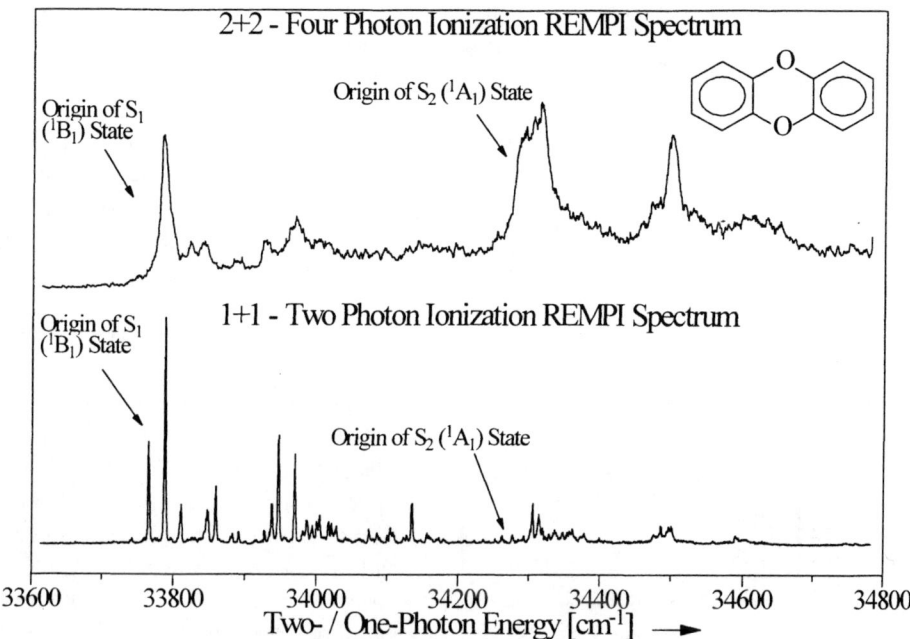

Figure 2: *Two photon resonant, four photon ionization REMPI spectrum (2+2 REMPI, top) and one photon resonant, two photon ionization REMPI spectrum (1+1 bottom) of the S_1 and S_2 region of dibenzo-p-dioxin.*

However, due to the very high laser intensities necessary for recording of the 2+2 REMPI spectrum, large power broadening effects occur, making a detailed analysis of the vibronic structure impossible. In the 1+1 spectrum, one can recognize some small peaks in the S_2 origin region, suggesting a very weak oscillator strength for the S_2 one-photon transition. Within the framework of the *Born-Oppenheimer* approximation this is a hint for a slightly folded geometry of DD in the S_0 ground state. The same conclusion can be drawn from the short progression of the butterfly mode visible in the 1+1 REMPI spectrum (i.e. the first three peaks with a spacing of 23 cm^{-1}). The progression-activity of this mode indicates a change of the folding angle upon the transition to the S_1 state. Furthermore, no anharmonicity can be detected. If the structure of the DD molecule in the S_1 would be non-planar, a very flat double-minimum potential for the butterfly motion would exist. In this case the short butterfly mode progression should exhibit a distinct anharmonicity (e.g. as in the case of the pentacene-molecule[7]). This argument for a planar structure in the S_1 state is supported by the fact, that basic MO consideration as well as semiempirical calculations clearly depict, that the π-bonding order along the C-O bonds increases upon $S_1 \leftarrow S_0$ excitation (π-bonding order according to a AM1-CI calculation, using 36 singly excited determinants: $S_0 = 0.079$, $S_1 = 0.174$). In conclusion there is a strong possibility, that DD is planar in its first excited singlet state. From the two arguments, on the one hand the observed geometry change upon the excitation (butterfly mode progression) and on the other hand the direction of the induced motion (planar DD in S_1), can be further concluded that DD in the gas phase exhibits a folded structure (C_{2v}). In Figure 3 jet-1+1-REMPI spectra of the $S_1 \leftarrow S_0$ transitions of analogous three-ring compounds are compared. Anthracene is planar in the ground and excited state, while 9,10-dihydroanthracene (DAH) exhibits a folded structure in ground) an a less folded structure in the excited state ($\Phi(S_0) = 35°$, $\Phi(S_1) = 16°$, $\Delta\Phi(S_1-S_0) = 19°$)[8]. Although a direct analysis of the *Franck Condon* pattern of the DD spectrum has not been performed jet, the comparison with the spectrum of the analogous DAH shows, that the $S_1 \leftarrow S_0$ geometry shift along the folding angle Φ for the DD molecule is considerably smaller than the 19° observed in the case of DHA.

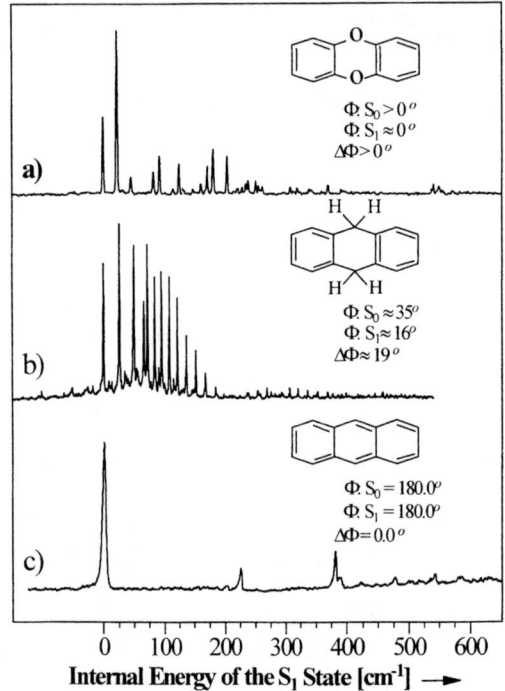

Figure 3: REMPI spectra of dibenzo-p-dioxin (DD, 9,10-dihydroanthracene (DAH) and anthracene (ANT): Comparison of the vibronic structure of the S_1 transition.

Chlorine substitution has a strong influence on the optical transitions of aromatics components. This is due to the resonance interaction between the d-/p-orbitals of the chlorine atoms and the aromatic π-system. Semiempirical calculations (INDO/S-CI hamiltonian using 33 singly excited determinants and $P_{(\pi-\pi)}=0.585$) predict relatively high oscillator strengths for all (poly)chlorinated dibenzo-p-dioxins (PCDD), bearing different numbers chlorine substituents at their two benzoic moieties, respectively. For PCDD with an equal number of chlorine atom at the two rings, the (symmetry-) disturbance of the π-system is not high enough for inducing considerably enhanced transition moments. The REMPI spectrum of the 2,3-dichlorodibenzo-p-dioxin (2,3-DCDD) is depicted in figure 4. The asymmetrical disturbance of the π electronic system induces a strong $S_2 \leftarrow S_0$ transition moment, reflected by the relatively intense S_2-transition in the REMPI spectrum. The 2,7- and 2,8-dichlorodibenzo-p-dioxin congeners (which exhibit one chlorine atom at each ring) do not show intense S_2-transitions. Similar results are also observed for the monochlorinated PCDD congeners. For example the S_1 and S_2 band systems are observable with an intensity relation of about 3:1 (INDO/S: 5:1) in the REMPI spectrum of 2-monochorodibenzo-p-dioxin. Surprisingly it was not possible to achieve a sharply resolved 1+1 REMPI spectrum of the S_1 and S_2 transitions of the 1-monochlorodibenzop-dioxin (1-MCDD). Only two extremely weak and broadened bands, according to the S_1 and S_2 transitions, are observed in the REMPI spectrum. Apparently, the chlorine substitution in ortho-position to the ether-bridge quenches the REMPI process. This "ortho-effect" may be due to an intermolecular interaction between the chlorine substituent that causes either a rapid deactivation of the S_1 excitation (inter system crossing, *ISC* or internal conversion, *IC*) and/or a large geometry shift upon excitation. If this effect is general for PCDD, the consequence may be, that PCDD that are chlorine-substituted in the 1,4,6 or 9 position can not be efficiently ionized via the S_1 and S_2 states. If the deactivation is caused by rapid inter system crossing a two color ionization scheme as suggested in reference [9] may be applied.

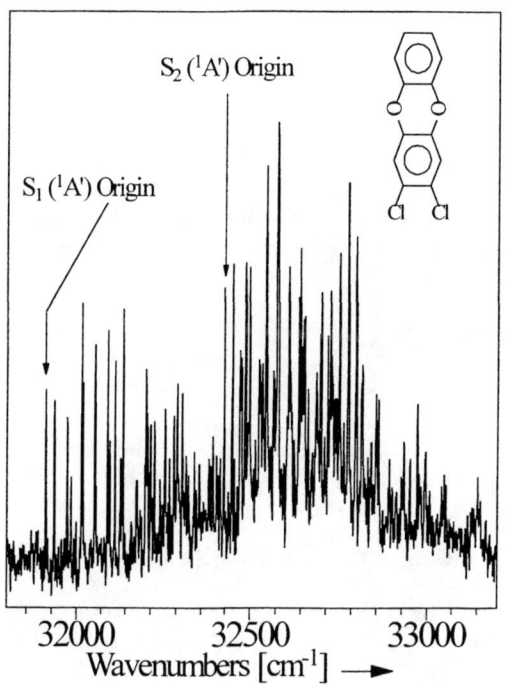

Figure 4: *1+1-REMPI spectrum of 2,3-dichlorodibenzo-p-dioxin, showing the S_1 and S_2 band systems*

The analysis of chlorinated dibenzo-p-dioxins and -furans (PCDD/F) and similar chlorinated xenobiotica is a very important field in analytical chemistry. The PCDD/F are formed as trace products in combustion or thermal processes (e.g. metallurgical processes) and are emitted to the environment. Some polychlorinated dibenzo-p-dioxins (PCDD) are extremely toxic and due to their persistence and lipophilic character they are easily bio-accumulated in the nutrition chain. Even very low emission and imission concentrations of these compounds exhibit ecotoxicological effects and thus are considered to be hazardous. Due to the low concentrations that need to be monitored and the necessity of isomer selective quantification, PCDD analysis is very challenging. For the conventional PCDD analysis, extensive and time consuming sample pre-concentration and clean-up steps are required prior the instrumental analysis step via high resolution gas chromatography-mass spectrometry. Therefore the first REMPI results on DD and the dichlorinated DD's [5] also were considered according due to the question if laser mass spectrometry can be applied for PCDD analysis. However, recent experimental and theoretical studies on the spectroscopy of these compounds pointed out, that an straight foreword analysis of PCCD is not possible with the REMPI technique [4,6,9]. This conclusion is also supported by the above mentioned results concerning the effect of chlorine substitution in the 1,4,6 and 9 position of the DD molecule. Alternative concepts, however, have been developed. The first approach is based on on-line REMPI-detection of suitable surrogates in the flue gas of incineration plants[10], utilizing the so called indicator parameter relation between chlorinated benzenes and PCDD in the combustion flue gases[11]. A second, off-line, approach is based on a gas chromatography (GC)-REMPI-TOFMS coupling. By using an in-line post-GC-column derivatization technique, that quantitatively dechlorinates all PCDD to DD, in principle all PCCD congeners can by detected[12]. Here the GC information is used for isomer identification, while the REMPI-TOFMS spectrometer is used as „dioxin"-selective detector, suppressing potential interferences. In conclusion promising concepts laser mass spectrometric methods for analysis of chlorinated aromatics as PCDD have been developed, but still are in an early experimental stage.

ACKNOWLEDGMENTS

The author thanks Dr. U.Boesl, Prof. E.W.Schlag and Prof. A.Kettrup for the ongoing support and interest. Financial support of the fundamental and applied studies by the DECHEMA e.V., Frankfurt (D), the Deutsche Bundesstiftung Umwelt, Osnabrück (D) and the Nestle´ Research Center, Lausanne (CH) is gratefully acknowledged. The author wish also to thank the coworkers in the analytical laser mass spectrometry laboratory at the GSF research center.

REFERENCES

[1] R.Zimmermann, C.Weickhardt, U.Boesl, E.W.Schlag: *J. Molec. Structure* 327 (1994) 81

[2] **a)** R.Zimmermann, *J. Molec. Structure* 377 (1996) 35. **b)** R. Zimmermann, 8th Resonance Ionization Spectroscopy Symposium 1996, *AIP-Conference Proceedings* 388, AIP-Press, New York (1997) 399

[3] **a)** A.W.Cordes, C.K.Fair, Acta. Cryst (1994) 1621. **b)** F.P.Colonna et al., J. Organomet. Chem. 146 (1978) 235

[4] R.Zimmermann, U.Boesl, D.Lenoir, A.Kettrup, T.L.Grebner, H.J.Neusser, *Int .J. Mass Spectrom. Ion Process.* 145 (1995) 97

[5] **a)** C.Weickhardt, R.Zimmermann, U.Boesl, E.W.Schlag: *Rapid Commun. Mass Spectrom.* 7 (1993) 183 **b)** R.Zimmermann, C.Weickhardt, U.Boesl, D.Lenoir, K.-W.Schramm, A.Kettrup, E.W.Schlag, *Organohalogen Compounds*, Vol. 11 (Dioxin '93 Conference-Proceedings), Federal Environmental Agency, Austria (1993) 91

[6] R.Zimmermann: Thesis 1995, TU München, Herbert Utz Verlag, München (1996), ISBN 3-89675-133-6

[7] A.Amirav, U.Even, J.Jortner, Chem. Phys. Lett. 27 (1980) 21

[8] Y.D.Shin, H.Saigusa, M.Z.Zgierski, F.Zerbetto, E.C.Lim, J.Chem. Phys. 94 (1991) 3511

[9] R.Zimmermann, D.Lenoir, A.Kettrup, H.Nagel, U.Boesl, *26th Symposium (International) on Combbustion, The Combustion Institute, Pittsburgh* (1996) 2859

[10] R.Zimmermann, H.J.Heger, A.Kettrup, U.Boesl, *Rapid Communic. Mass Spectrom.* 11 (1997) 1095

[11] **a)** T.Öberg, J.Bergström, Chemosphere 14 (1985) 1081 **b)** A.Kaune, D.Lenoir. U.Nikolai, A.Kettrup, Chemosphere (1994) 2083 **c)** M. Blumenstock, R. Zimmermann, K.-W. Schramm, A. Kaune, U. Nikolai, D. Lenoir, A. Kettrup, submitted for publication in *J Anal Appl. Pyrolysis*

[12] **a)** R.Zimmermann, E.R.Rohwer, H.J.Heger, U.Boesl, A.Kettrup, German patent DE 19754151.5 (1997), pending **b)** R.Zimmermann, E.R.Rohwer H.J.Heger R.Dorfner, U.Boesl, A.Kettrup, *Proceedings of the 20thInternational Symposium on Capillary Chromatography 1998(CD-ROM)*Editors: P.Sandra, A.J.Rackstraw, I.O.P.M.S Kortrijk, Belgium (1998) KNL04

Overtone Spectroscopy of Jet-Cooled Phenol Studied by Nonresonant Ionization Detected IR Spectroscopy

Shun-ichi Ishiuchi* and Masaaki Fujii**

*The graduate university for advanced studies and **Institute for Molecular Science
Myodaiji, Okazaki 444-8585, JAPAN

Abstract. Vibrational transitions of jet-cooled phenol-h_6 and phenol-d_5 have been measured from 2400 cm^{-1} to 14000 cm^{-1} by nonresonant ionization detected IR spectroscopy. The spectrum shows a well-resolved structure due to the first to the fourth quantum of OH stretching vibrations, CH and CD overtones and various kinds of combination vibrations. It is found that the bandwidth of the OH overtone in phenol-h_6 decreases with increase in the vibrational quantum number, while that in phenol-d_5 decreases. The origin of the bandwidth is discussed in terms of intramolecular vibrational redistribution.

INTRODUCTION

Recently, we have developed a new IR-UV double resonance technique which detects vibrational transitions with high sensitivity. This spectroscopy, named nonresonant ionization detected IR spectroscopy (shortly NID-IR), selectively detects vibrationally excited molecules by nonresonant two-photon ionization due to the UV laser (1, 2). A jet-cooled molecule is excited to a vibrational level in the ground state by an IR laser v_{IR}. The vibrationally excited molecule is selectively ionized by the nonresonant two-photon process due to a UV laser v_{UV}. Because of the ionization detection and weak background signal, this spectroscopy has an advantage in the sensitivity in comparison to the traditional IR absorption spectroscopy and to other IR-UV double resonance spectroscopies (3-5) such as IR dip spectroscopy (6-12).

The key to achieve NID-IR spectroscopy is the selective ionization of vibrationally excited molecules. The simple way to achieve the selective ionization is shown in Fig. 1a. Here, v_{UV} is fixed to energy slightly lower than a half of the ionization potential IP to prevent ionization of a molecule in the ground vibrational level (1). On the other hand, the vibrationally excited molecules can be ionized by the two-photon process due to v_{UV}, because of its initial vibrational energy. Therefore, the ion current is detected only when v_{IR} is resonant to the vibrational level. If a sample molecule has a large ionization cross section for the vertical ionization potential IPv, the selective ionization can be achieved even if v_{UV} is fixed to more than IP/2 (see Fig. 1b). The detailed mechanism of NID-IR spectroscopy has been discussed elsewhere(2).

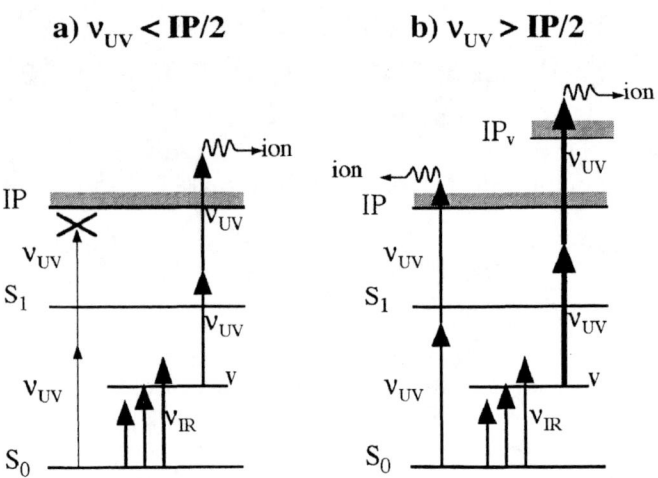

FIGURE 1. Schematic diagram showing the principle of NID-IR spectroscopy when a) $\nu_{UV} < IP/2$ and b) $\nu_{UV} > IP/2$.

In this work, we have applied NID-IR spectroscopy to the overtone transitions of phenol and partially deuterated phenol (C_6D_5-OH) under a supersonic jet condition. The NID-IR spectrum shows well-resolved structures due to the overtones of OH, CH and CD stretching vibrations and various kinds of combination vibrations from 2400 cm^{-1} to 14000 cm^{-1}. Vibrational frequency, anharmonisity, bandwidth of the OH overtones and its relation to the intramolecular vibrational redistribution (IVR) will be discussed.

EXPERIMENTAL

The experimental apparatus for NID-IR spectroscopy has been described elsewhere (1). The second and the third harmonics of Nd^{3+}:YAG lasers (HOYA-Continuum, Lumonics YM1200) were used to pump two dye lasers (Lumonics HD-500). The output of the dye laser pumped by the second harmonics was differentially mixed with the fundamental or the second harmonics of the YAG laser in a LiNbO$_3$ crystal (Inrad) and was converted to a tunable IR laser ν_{IR}. This differential mixing provides the IR laser in the region from 4 μm to 1.2 μm. Below 1.2 μm, the idler beam of OPO laser (Spectra-Physics MOPO-HF) was used. A typical power of the IR laser was 0.1 to 0.5 mJ in the 3 μm region, 1 mJ in the 1.5 μm region, and 10-20 mJ below 1.2 μm region. The laser duration was 6 ns from 4 to 1.2 μm and 2 ns below 1.2 μm. The output of the dye laser pumped by the third harmonics of the YAG laser was frequency-doubled by a KDP crystal and was converted to a UV laser ν_{UV} (~ 1 mJ, 6 ns duration). Both ν_{UV} and ν_{IR} were coaxially introduced into a vacuum chamber and crossed a supersonic jet. The UV laser was delayed 50 ns with respect to the IR laser. A vibrationally excited molecule generated by ν_{IR} was selectively ionized by the nonresonant two-photon transition due to ν_{UV}. The produced cations were pushed into a detector chamber by a repeller at an appropriate voltage (typically 15 V/cm) and were detected by a channel multiplier (Murata Ceratron) through a quadrupole mass filter (EXTREL) . The signal was amplified by a preamplifier (EG&E PARC 115) and was integrated by a digital boxcar (EG&E PARC 4420/4422). The integrated signal was recorded by a personal computer as a function of the IR laser frequency.

The sample vapor at room temperature was seeded in He gas (2 atm), and the mixture was expanded into the vacuum chamber through a solenoid valve. The sample was purchased from Junsei Chemical and was used without further purification.

RESULTS AND DISCUSSION

Figure 2 shows the NID-IR spectrum of jet-cooled phenol. The IR laser ν_{IR} was scanned from 2400 to 11000 cm^{-1} and from 12800 to 14000 cm^{-1}. The spectrum was obtained by monitoring the phenol monomer cation (mass 94) produced by two-photon ionization due to ν_{UV} (34483 cm^{-1}). The NID-IR spectrum shows well-resolved vibrational structures of phenol in S_0. The strongest peak at 3656 cm^{-1} has been assigned to the fundamental OH stretching vibration ν_{OH}, and cluster of peaks at ~3000 cm^{-1} to the fundamental of ν_{CH}. The spectral shape of ν_{CH} is complicated because of the existence of five CH stretching modes and their Fermi resonance. In the region above ν_{OH}, many overtones and combination vibrations are observed. The observation of overtones in a supersonic jet well demonstrates the high sensitivity of the NID-IR spectroscopy.

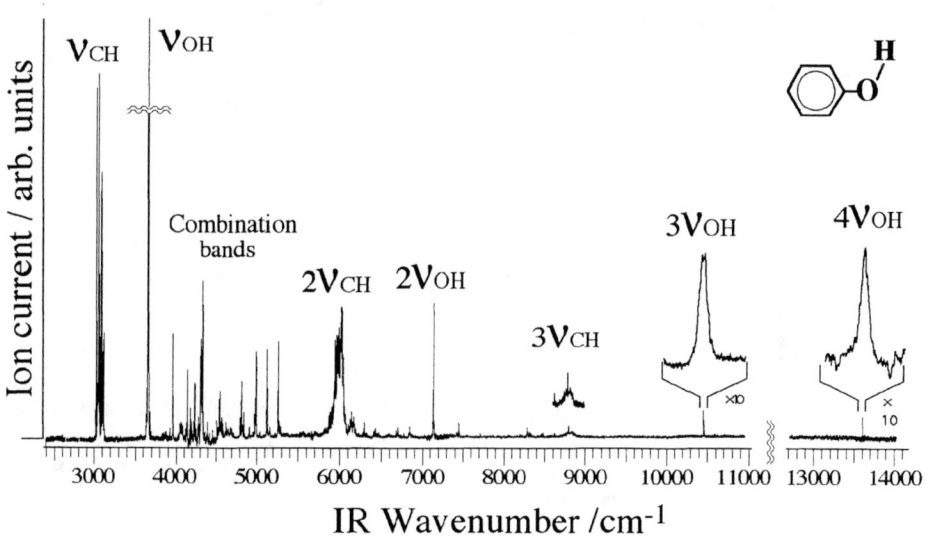

FIGURE 2. NID-IR spectrum of jet-cooled phenol. The spectrum was reported partly in reference 2.

The overtones of OH and CH stretching vibrations are assigned easily because of their vibrational frequency, intensity and band shapes. The clusters of peaks at about 6000 cm^{-1} and 9000 cm^{-1} are assigned to the first (ν=2) and second (ν=3) overtones of ν_{CH}, respectively. Their spectral shapes are equally complicated as the fundamental vibration.

This similarity supports the assignment. The sharp peaks at 7143, 10461 and 13612 cm^{-1} are assigned to the first ($v=2$), second ($v=3$) and third ($v=4$) overtones of OH stretching vibration, respectively. From the analysis, the vibrational frequency ω and anharmonisity $\omega\chi$ for the OH stretching mode are determined to be 3741 cm^{-1} and 84.54 cm^{-1}, respectively. In addition to the overtones of v_{OH} and v_{CH}, many bands appear in the NID-IR spectrum. These bands are assigned to the combination vibrations, such as skeletal or CH bending vibrations built on v_{OH} (2).

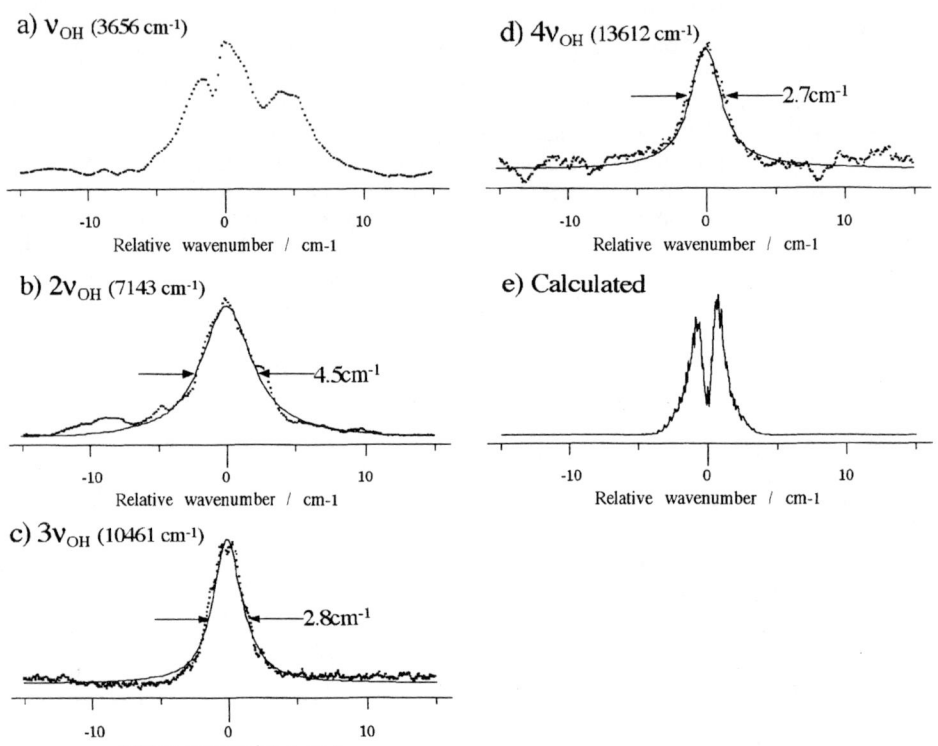

FIGURE 3. Band shape of a) the fundamental vibration (v_{OH}), b) the first overtone ($2v_{OH}$), c) the second overtone ($3v_{OH}$) and d) the third overtone ($4v_{OH}$) of the OH stretching vibration in jet-cooled phenol. The observed shape is shown by a dotted curve, while the best-fit Lorentzian function is shown by a solid curve. Preliminary fitting was reported in reference 2.

Figure 3 shows the band shapes of v_{OH} overtones. The observed shape is shown by a dotted curve. The fundamental vibration of v_{OH} has different shape from the others. This band shape is distorted by the variation of the IR laser power because of accidental IR absorption by atmospheric water. For the others, the IR laser power was almost constant during the measurement. All the overtones of v_{OH} show symmetric band shape close to Lorentian function, which is shown by a solid curve in the figure. The observed band shape is far from the band contour expected by rotational structure. The typical calculated rovibrational band contour is shown in Fig. 3e. Here, we assume that the rotational temperature is 5K and each rotational line has 0.1 cm^{-1} width due to the laser resolution. The direction of OH stretching motion is almost parallel to the b-axis, thus the calculated contour has the rotational structure which has strong P- and R-branches as shown in Fig. 3e. From the comparison, we have concluded that the observed band shape is not determined only by the rotational structure, and is broadened by the intramolecular vibrational redistribution IVR.

The bandwidth of the ν_{OH} overtones becomes narrower in going from the first overtone to the third. The bandwidth of the first overtone $2\nu_{OH}$ is measured to be 4.5 cm^{-1} (FWHM). Those of the second and third overtones (3 ν_{OH} and 4 ν_{OH}) are 2.8 and 2.7 cm^{-1}, respectively. As discussed above, the bandwidth reflects the broadening due to IVR. Therefore the narrower bandwidth suggests the slower IVR rate in higher overtone level. It is opposite to the natural expectation that IVR is accelerated in higher vibrational state because of the rapid increase of the state density. One of the possible explanations for this phenomenon is "door way" in the IVR process (13-15). If the overtone $n\nu_{OH}$ strongly couples with a specific bath mode (door way), the IVR rate is mainly determined by energy gap between $n\nu_{OH}$ and the door way state. Thus, the narrower bandwidth (i. e. slower IVR) in higher overtone can be interpreted as the larger energy gap due to the anharmonisity (15).

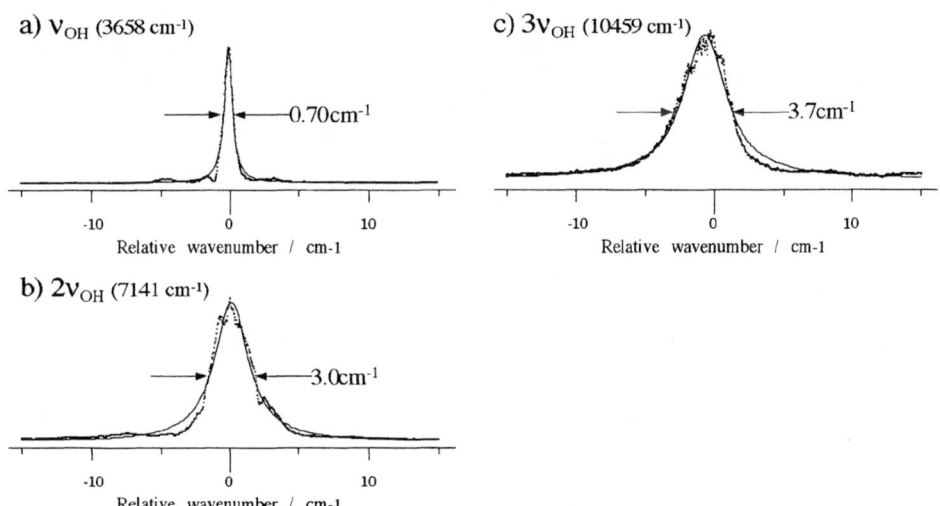

FIGURE 4. Band shape of a) the fundamental vibration (ν_{OH}), b) the first overtone ($2\nu_{OH}$) and c) the second overtone ($3\nu_{OH}$) of the OH stretching vibration in deuterium substituted phenol (C_6D_5-OH). The observed shape is shown by a dotted curve, while the best-fit Lorentzian function is shown by a solid curve.

To confirm the interpretation, the existence of "door way" state must be examined. For this aim, we have been measuring the overtones of various deuterated phenol. If the ν_{OH} overtone has "door way" in IVR, isotope substitution may change the IVR rate drastically, because of the change in the energy gap. Figure 4 shows the band shape of the OH overtones (dotted curve) of jet-cooled C_6D_5-OH measured by the NID-IR spectroscopy. The solid curves show the Lorentzian function fitted to experimental curve. The bandwidth of ν_{OH}, $2\nu_{OH}$ and $3\nu_{OH}$ is measured to be 0.70, 3.0 and 3.7 cm^{-1}, respectively. The bandwidth increases in going from ν_{OH} to $3\nu_{OH}$, therefore the IVR rate is accelerated in higher overtone. This behavior is opposite to that in phenol-h$_6$, therefore it is consistent to the interpretation based on "door way". The theoretical analysis concerned with this interpretation and further experiments such as the deuterium substitution effect on OH are now in progress.

REFERENCES

1. Omi, T., Shitomi, H., Sekiya, N., Takazawa, K. and Fujii, M., Chem. Phys. Lett. **252**, 287-293 (1996).
2. Ishiuchi, S., Shitomi, H., Takazawa, K. and Fujii, M., Chem. Phys. Lett. **283**, 243-250 (1998).
3. Settle, R. D. F. and Rizzo, T. R., J. Chem. Phys. **97**, 2823-2825 (1992).
4. Boyarkin, O. V., Settle, R. D. F. and Rizzo, T. R., Ber. Bunsenges. Phys. Chem. **99**, 504-513 (1995).
5. Boyarkin, O. V. and Rizzo, T. R., J. Chem. Phys. **103**, 1985-1988 (1995).
6. Page, R. H., Shen, Y. R. and Lee, Y. T., J. Chem. Phys. **88**, 5362-5376 (1988).
7. Page, R. H., Shen, Y. R. and Lee, Y. T., J. Chem. Phys. **88**, 4621-4636 (1988).
8. Tanabe, S., Ebata, T., Fujii, M. and Mikami, N., Chem. Phys. Lett. **215**, 347-352 (1993).
9. Ebata, T., Watanabe, T. and Mikami, N., J. Phys. Chem. **99**, 5761-5764 (1995).
10. Ebata, T., Mizuochi, N., Watanabe, T. and Mikami, N., J. Phys. Chem. **100**, 546-550 (1996).
11. Pribble, R. N., Hagemeister, F. C. and Zwier, T. S., J. Chem. Phys. **106**, 2145-2157 (1997).
12. Yoshino, R., Hashimoto, K., Omi, T., Ishiuchi, S. and Fujii, M., J. Phys. Chem., in press.
13. Bray, R. G. and Berry, M. J., J. Chem. Phys. **71**, 4909-4922 (1979).
14. Reddy, K. V., Heller, D. F. and Berry, M. J., J. Chem. Phys. **76**, 2814-2837 (1982).
15. Heller, D. F. and Mukamel. S., J. Chem. Phys. **70**, 463-472 (1979).

RESONANT IONIZATION MASS SPECTROMETRY OF AMMONIA

Titaina Gibert-Legrand, Thierry Gonthiez, Laurent Vivet and Pascal Brault

GREMI université d'Orléans, 45067 Orléans cédex 2, France

Abstract. Our aim is the understanding of the early stages of laser-induced adsorption of NH_3 and its fragments on silicon (100) surface. A quantum effect is expected in the vibrational and rotational state distribution after photodesorption of the surface species. We present the first part of the work which consists to record resonant ionization of ammonia with (2+1) scheme in the 291-297 nm wavelength range. Results show transitions from the ground state to vibrationly electronic excited states $\tilde{C}-7;\tilde{C}'-5;\tilde{C}-6;\tilde{C}'4$, $\tilde{B}-9$. Decomposition during the RIS process occurs leading to the production of N, NH and NH_2 ionic fragments.

INTRODUCTION

The laser desorption of molecules from solid surfaces involves different degrees of freedom ; translationnal energy can be measured by time of flight method (TOF) and vibrational or rotational states can be deduced from the spectroscopic analysis of the products(1). The molecule of interest is NH_3 and the substrate is Si(100).

In order to compare photodesorbed ammonia spectrum and gaseous ammonia spectrum we perform at first the resonant ionization spectroscopy of gaseous NH_3 . Indeed we need a reference spectrum at defined temperature to evaluate further changes in population distribution.The photoionization sheme is (2+1) photons scheme and the wavelength range is between 291nm to 297 nm.

EXPERIMENT

NH_3 is introduced at 300K in a UHV chamber (base pressure 10^{-9} Torr) with a microleakage valve. For the record of the spectra the pressure is regulated to 2.10^{-6}Torr and is limited by the tolerance of the quadrupole mass spectrometer. In front of the entrance of the energy selector, the sample is placed with a 95V reference voltage .The resonant ionization is performd with a Nd-YAG pumped dye laser. The dye is rhodamine 610 frequency doubled. Note that the automatic tracking stage needs a minimal energy to work and that limits the minimal laser power density in figure 3. A spherical lens is used to focus the UV beam typically 0.5cm in front of the mass spectrometer.

In the second stage of the experiment, Si(100) surface exposed to ammonia will be desorbed by a second laser which will be frequency tripled (355nm) or frequency quadrupled (266nm) Nd-YAG laser. Dye laser and photodesorption laser are synchronized with an adjustable delay allowing TOF measure of desorbed species.

Figure 2 shows the fragmentation products. From each level we observe fragmentation. For each fragment the proportion of fragments are typically the same. The amount of the NH_2^+, NH^+ and N^+ fragments corresponds typically to 10% of the NH_3 ions.

DISCUSSION

The selection rules for a two-photon transitions with a linearly polarized laser beam are different for the $\tilde{B}$, $\tilde{C}$ and $\tilde{C}'$ states (3).

For the ground state to the $\tilde{B}$ state transition we consider a $\Delta K = \pm 1$ perpendicular band and $\Delta J = 0; \pm 1; \pm 2$ rules for a two photons absorption (4).
The first point is the interpretation of the (2+1) photon spectra in order to deduce rotational or vibrational temperature. The most convenient line is the $\tilde{B}-9 \leftarrow \tilde{X}$ bands because of the shown resolution. However with the previous rules the identification of the lines are not evident. We investigate now a programme with different parameters as the rotational constants B and C of the both states. The first step will consider that the Hönl London factors are identical for each rotational state.

For the ground state to the $\tilde{C}$ and $\tilde{C}'$ states transition we consider a $\Delta K = 0$ parallel band and a $\Delta J = 0; \pm 1$ one photon transition. This leads to the two photons absorption rules $\Delta K = 0$ and $\Delta J = 0; \pm 1; \pm 2$.
It is not possible to split the lines of the $\tilde{C}'-5 \leftarrow \tilde{X}$. This band contains fewer lines than the $\tilde{B}-9 \leftarrow \tilde{X}$. Let us, one consider the following rotational constants for the ground state B=9.44cm^{-1} and C=6.19cm^{-1}, and for the $\tilde{C}'$ state B'=8.75cm^{-1} and C'=5.5cm^{-1}. The NH_3 molecule is oblate symmetric top then the energy of a rotational level is $F(J;K)=BJ(J+1)-(B-C)K^2$ (5). With a Boltzmann law, we can estimate that the population of the X-(v_2=0; J=2;K=2) level is 0.95% of the $\tilde{X}$(0;0;0) level. These 6 first rotational levels lead to 18 spectral lines to different rotational states of the $\tilde{C}'$ (5;J';K') level. Some of the lines are separated by only 0.02cm^{-1} well below the dye laser linewidth. Our setup does not allow us to resolve the rotational spectrum of this band neither the spectrum of the $\tilde{C}$ level.

In order to determine the degree of saturation of the resonant process, the intensity of principal line of the $\tilde{B}-9 \leftarrow \tilde{X}$ transition has been plotted versus laser power density (figure 3). To keep the identical dye laser condition, the diminution of the laser energy is carried out by putting a transparent attenuator at the entrance of the UV tracking stage. The area of the entire $\tilde{C}'-5 \leftarrow \tilde{X}$ has been plotted versus the laser power density too (not shown). In the Log-Log plot the slope of the curve is 2. This indicates that only the ionization step is saturated at these laser power densities.

RESULTS

The RIS spectrum is shown in figure 1. The spectral resolution of the laser beam is 0.11cm^{-1}. Output energy is recorded and shown on the same figure. Great care has been taken in measuring laser power density in the focal area. Fluctuations of the ammonia signal at different pressures have been recorded in order to calibrate all the signals at a reference pressure of 2×10^{-6}Torr in the chamber. The notation used is $\tilde{B} - 9 \leftarrow \tilde{X}$ e.g. $\tilde{B}$ represents the second excited state of NH_3 and 9 indicates that the quantum number for the ν_2 vibrational mode is 9.

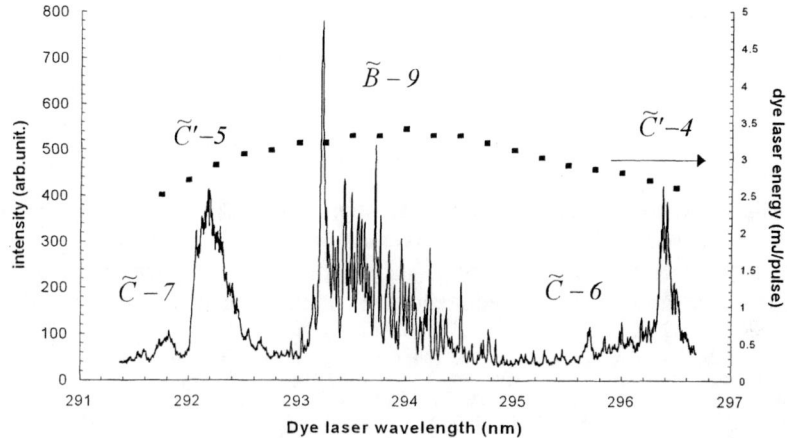

FIGURE 1. Overview of the ammonia (2+1) photon RIS spectrum obtained at 300K. The relevant vibrational states of the electronic states involved are indicated near the lines packets (2).

Such spectra are recorded at masses 16, 15 and 14.

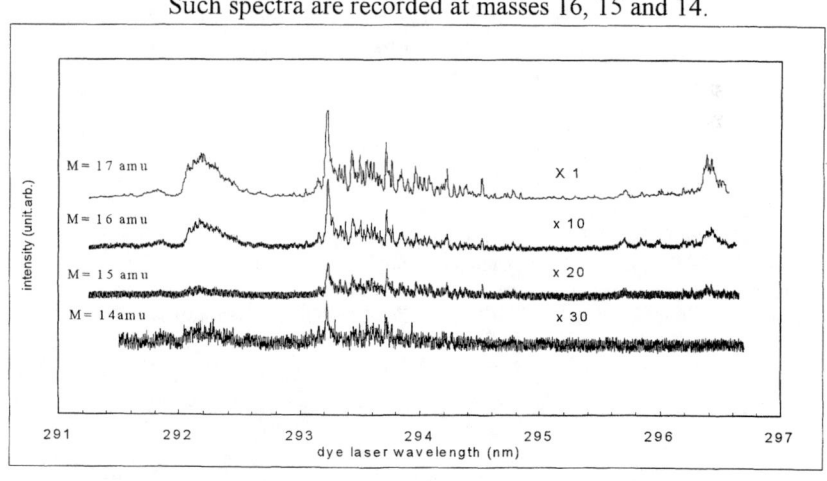

FIGURE 2. RIS spectra at different masses for NH_3 and its fragments.

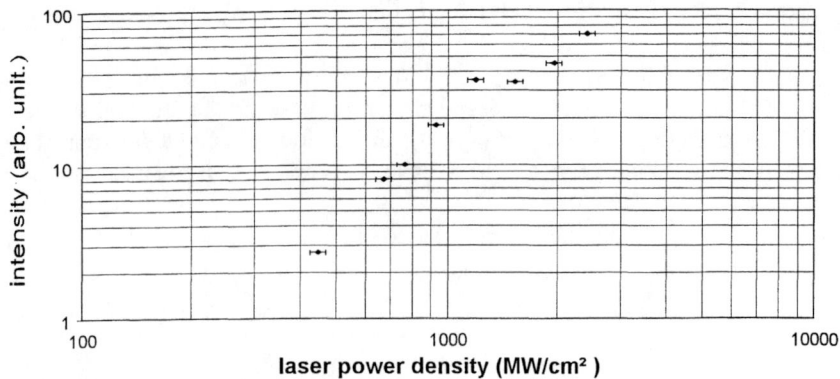

FIGURE 3. Intensity of NH_3^+ signal for different laser power densities. The minimal density is given by the sensitivity of the UV tracking system.

Considering other works concerning the fragmentation processes of ammonia, we suppose that the (2+1) photons energy is enough to reach dissociative potential curves in the ionic state of NH_3. This way of dissociation is most probable way than a dissociation in the excited state which could be different for each electronic transition.

CONCLUSION

The vibrational band are not yet entirely explained because of the complexity of the rotational spectrum.

For an analytical purpose we can choose the $\tilde{C}'-5 \leftarrow \tilde{X}$ transition because it is broader than the rotational lines of the $\tilde{B}-9 \leftarrow \tilde{X}$ band. Note that the setup has a sensitivity which allows easily the measure of one NH_3^+ ions for two laser shots at the wavelength corresponding to $\tilde{C}'-5 \leftarrow \tilde{X}$.

REFERENCES

1. W.Nessler,K.H. Bornscheuer,T.Hertel,E.Hasselbrink, Chemical Physics, **205**,205-219,1996
2. J.H. Glownia,S.J.Riley,S.Colson,G.C.Nieman, J.Chem.Phys. **73**, 4296-4309, 1980
3. M.J. Shultz,J.Wei, J.Chem.Phys.,**92**(10),5951-5958,1990
4. H.Dickinson,D.Rolland,T.P.Softley, Phil.Trans.R.Soc.Lond.A,**355**,1585-1607,(1997)
5. G.Herzberg,Molecular spectra and molecular structure T3 Electronic spectra and electronic structure of polyatomic molecules,Van Nostrand Reinhold Company,1966

Translationally cold Cs$_2$ molecules formation in a magneto-optical trap

A. Fioretti, D. Comparat, C. Drag, A. Crubellier, O. Dulieu, F. Masnou-Seeuws, C. Amiot, and P. Pillet

Laboratoire Aimé Cotton[+],
CNRS II, Bât. 505, Campus d'Orsay, 91405 Orsay cedex, France

Abstract.
We report on the observation of translationally cold Cs$_2$ molecules at temperatures in the 300-70 μK range, produced by the photoassociation of cold Cs atoms in a vapor-cell magneto-optical trap. The detection method is based on pulsed-laser photoionization of Cs$_2$ molecules into Cs$_2^+$ ions, selectively detected through a time-of-flight mass spectrometer. This technique, together with the usual trap-loss method, enables us to perform the spectroscopy of the long-range attractive molecular states below the $6s\,^2S_{1/2} + 6p\,^2P_{3/2}$ dissociation limit.

Laser cooling in the mK-μK range and below, as well as trapping and manipulation of neutral atomic samples are now well established experimental techniques that have opened the way to many exciting research lines, from optical crystals to the realization of dilute quantum-degenerate Bose gases.

In contrast to atoms, direct laser cooling of molecules is very difficult because of the lack of a closed two-level optical pumping scheme for recycling population. More complex cooling schemes and trapping techniques have been proposed [1,2].

A possible scheme for the production of translationally cold ground state molecules is to start from a sample of cold atoms and create bound molecules through laser photoassociation (PA). The short-lived excited molecules should then stabilize by spontaneous emission into a bound vibrational level of the electronic ground state [3].

Many PA experiments in cold alkalis have been successful, opening the way to the observation of long-range dimers and to the determination of long range potential curves [4]. However, in most cases, excited molecules spontaneously decay into continuum levels of the ground electronic state, which lead back to pairs of free (hot) atoms. Those atoms usually escape the trap, giving rise to trap loss, but no translationally cold stable molecules are observed.

We report here on the our recent observation of translationally cold Cs$_2$ molecules in their triplet and singlet ground states, obtained through the intermediate PA of

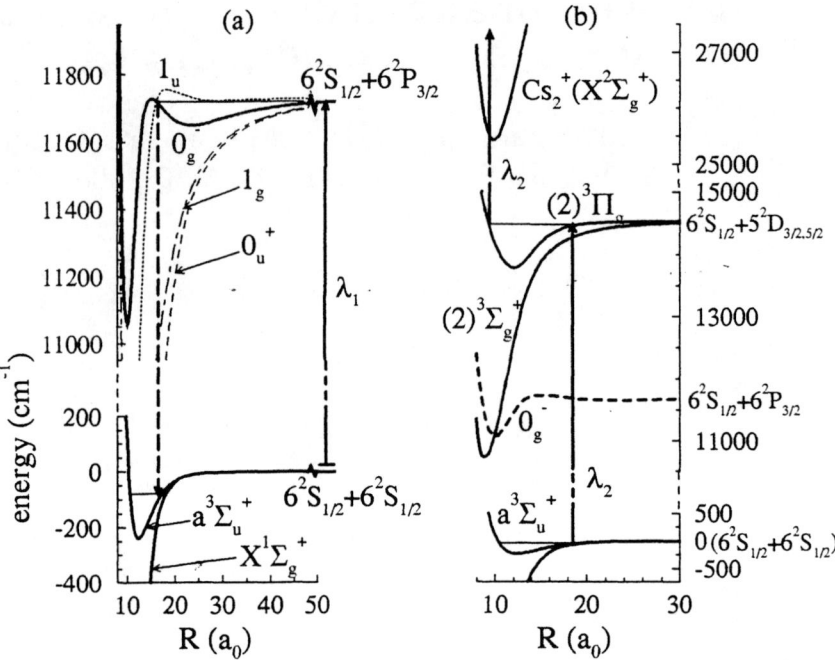

FIGURE 1. (a) Diagram of the Cs$_2$ optical transitions and molecular states relevant for the photoassociation experiment. (b) Two-step ionization scheme for the Cs$_2^+$ detection. All potential curves are deduced from refs. [7,8].

laser cooled cesium atoms in a magneto-optical trap (MOT) [5]. Pairs of cold Cs atoms in their ground state are resonantly excited, using a cw PA laser, to 0_g^- and 1_u long-range rovibrational levels of Cs$_2$ below the dissociation limit $6^2S_{1/2} + 6^2P_{3/2}$ (Fig. 1 (a)). A significant fraction of these photoassociated molecules decay to bound levels of the lowest $^3\Sigma_u^+$ and $^1\Sigma_g^+$ states, yielding long-lived translationally cold molecules. The latter are observed by photoionizing them into Cs$_2^+$ ions (see the scheme in Fig. 1 (b)), which are mass selected through a time of flight, and detected. The very sensitive ion detection, and the use of a cw laser for the PA step, allowed us to perform high-resolution spectroscopy of the excited 0_g^- [6] and 1_u states.

We also report on the first detection in the Cs$_2$ dimer of rovibrational levels belonging to the 0_g^-, 1_g and 0_u^+ series through the measure of the reduction of the fluorescence yield from the MOT. Trapped atoms are indeed expelled from the MOT as a consequence of the PA process.

The principle of the experiment is described in refs. [5,6]. Briefly, we illuminate the trapped, cold Cs atoms with a cw laser to produce the photoassociative transitions:

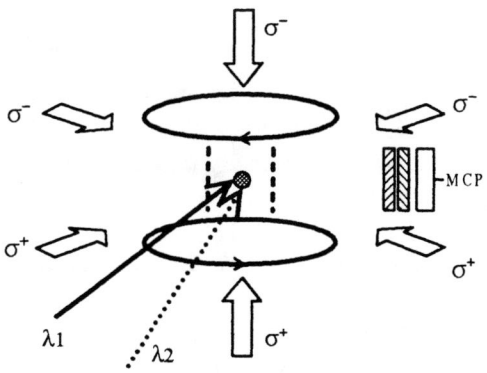

FIGURE 2. (a) Magneto-optical trap scheme with mass-selective ion detection.

$$\text{Cs}\,(6^2S_{1/2}) + \text{Cs}\,(6^2S_{1/2}) + h\nu_1 \rightarrow \text{Cs}_2\,(\Omega_{g,u}(6^2S_{1/2} + 6^2P_{3/2}; v, J))$$

The vapor-cell MOT (Fig 2) can load up to 5×10^7 atoms at a peak density of the order of 10^{11} cm^{-3}, with an estimated temperature of the order of 200 μK [9]. A cooling and a repumping lasers, respectively tuned 13 MHz to the red of the $6^2S_{1/2}(F = 4) \rightarrow 6^2P_{3/2}(F' = 5)$, and resonant with the $6^2S_{1/2}(F = 3) \rightarrow 6^2P_{3/2}(F'' = 4)$ atomic transitions, keep atoms cycling between the $6^2S_{1/2}(F = 4)$ ground state and the $^2P_{3/2}(F' = 5)$ excited state. The cold atoms are illuminated by the beam of a laser (Ti-Sapphire Ring Laser, pumped by a 20 W Ar$^+$ laser), focused on a diameter of roughly 500 μm, leading to a maximum available intensity in the MOT zone up to 1 kW cm^{-2}. Its frequency ν_1 can be tuned to any value in the range 750-900 nm, and continuously scanned over a 30 GHz range. Its linewidth is of the order of 1 MHz. The frequency scale of the spectra has been calibrated by using Cs saturated absorption lines, a Fabry Perot interferometer, and the absorption lines of iodine.

Cs$^+$ and Cs$_2^+$ ions are produced by a pulsed dye laser (7 ns pulse duration, 1 mJ pulse energy, focused to a 1 mm^2 spot) pumped by the second or the third harmonic of a Nd-YAG laser, running at a 10 Hz repetition rate, and tunable over the range $\lambda_2 \sim$500-750 nm. After the laser pulse, a pulsed high voltage field (4 kV, 0.5 μs) is applied at the trap position by means of a pair of grids spaced 15 mm apart to accelerate the ions. The ions are expelled from the interaction region in a 6 cm free field zone constituting a time-of-flight mass spectrometer, they are detected by a pair of micro-channel plates, and independently recorded with gate integrators. The relative magnitude of the Cs$^+$ and Cs$_2^+$ depends strongly on the pulsed laser wavelength and intensity. The Cs$_2^+$ ion signal depends linearly on intensity while the Cs$^+$ one is quadratic, being mostly due to two-photon ionization of atomic cesium in the $6^2P_{3/2}(F' = 5)$ excited state.

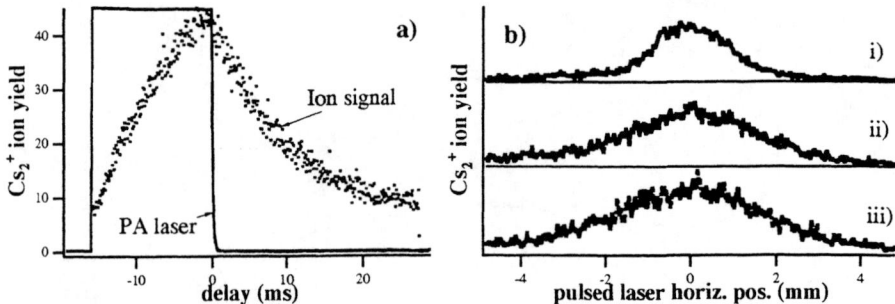

FIGURE 3. (a) Ion signal as a function of the delay between the ionizing paser pulse and the switch-off of the cw PA laser. (b) Spatial analysis of the molecular density by varying the ionizing laser horizontal position: (i) at the MOT position, (ii) 1.3 mm below and (iii) 2.1 mm below.

Fig. 3 (a) shows that the Cs_2^+ ions we detect are not produced by direct photoionization of the excited rovibrational molecular levels, but from photoionization of long-lived Cs_2 molecules. The ion signal exibits rising and falling times, with respect to the application of the PA laser, in the millisecond range. This timescale is 5 orders of magnitude larger than molecular excited-state lifetimes, and is related only to the residence time of the falling molecules in the ionizing volume. We also observed Cs_2^+ ion without applying the PA laser, meaning that PA is already produced in the MOT by the trapping and repumping lasers alone. Fig. 3 (b) depicts the spatial analysis of the falling molecular cloud, which leads to a measure of the temperature of the molecules. Temperatures af about 300 μK are measured in normal operating conditions i.e. close to that estimated for the trapped atomic sample [9]. If a further sub-Doppler cooling phase [10] is applied just before the PA pulse, temperatures as low as 70 μK are measured, with 30 % uncertainty.

Fig. 4 shows a typical vibrationally resolved photoassociation spectrum obtained from both ion and fluorescence detection. Lines are detected over a 80 cm^{-1} range of detunings in the ion spectrum, while the signal in the fluorescence yield is greatly reduced beyond 30 cm^{-1}. Two series of lines with different widths are visible in the ion spectrum and three in the fluorescence spectrum, one of which coincides with the longer one in the ionic signal. Four attractive states (0_u^+, 1_g, 0_g^-, 1_u) are expected to be populated when the PA laser is detuned to the red of the D_2 line (Fig. 1 (a). According to their long-range behaviour and their calculated hyperfine structure [5], the two series in the ion signal are attributed to the 0_g^- (thin lines) and to the 1_u (broad lines) states. Both of them are long-range states [11] with shallow secondary wells (resp. $\simeq 78\ cm^{-1}$ and $\simeq 7\ cm^{-1}$) located at $\simeq 24\,a_0$ and $\simeq 30\,a_0$ respectively. Thus they are unlikely to be detected by conventional molecular spectroscopy because of the Franck-Condon principle. We have thus performed high-resolution spectroscopy of the 0_g^- [6], and of the 1_u states. In contrast with the other alkalis, the double-well of these states in Cs_2 induces a Condon point at intermediate distances and provides an efficient spontaneous decay channel

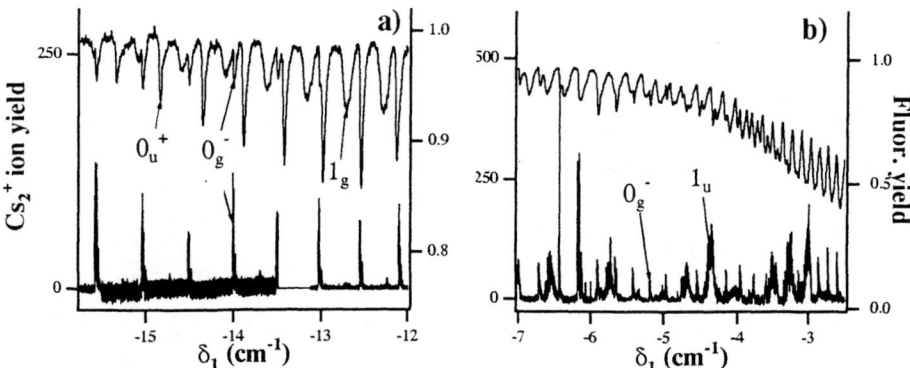

FIGURE 4. Cs_2^+ ion signal (lower signal) and trap fluorescence yield (upper signal) versus the detuning δ_1 of the PA laser. The origin is at the $6^2S_{1/2}(F = 4) \rightarrow 6^2P_{3/2}(F' = 5)$ atomic transition. The PA laser step is 15 MHz and ion counts are averaged on 10 laser pulses. Notice the different vertical scales for Figs. (a) and (b).

towards rovibrational levels of the molecular singlet and triplet ground states. On the contrary, the 1_g and 0_u^+ single-well states are observed only in the fluorescence signal, meaning that they are not decaying efficiently into ground state molecules.

We cross-checked our interpretation of the ionization process by keeping λ_1 fixed and scanning the wavelength λ_2 of the pulsed laser in most of the 500-750 nm range. While we observe a very weak Cs_2^+ ion signal for most wavelengths, we get two windows of a few hundred wavenumbers around $\lambda_2 \sim 716$ and 554 nm for which the signal is much larger, for any fixed λ_1. These ionization regions correspond to a two-step ionization via the intermediate vibrational levels $\Omega_{g,u}(6^2S_{1/2} + 5^2D_{3/2,5/2})$ and $\Omega_{g,u}(6^2S_{1/2} + 7^2S_{1/2})$ (Fig. 1 (b)). Different spectra are obtained tuning the PA laser on different lines. Fig. 5 shows two spectra corresponding to photoionization of $^3\Sigma_u^+$ (a) and $^1\Sigma_g^+$ (b) ground-state molecules, which are the decay product of respectively 0_g^- and 1_u excited states. Those spectra present very complex structures that are expected to be due to several bands of transitions between the molecular ground states populated by spontaneous emission and the molecular states belonging to the $6^2S_{1/2} + 5^2D_{3/2,5/2}$ multiplet.

In conclusion, we have reported for the first time the observation of translationally cold Cs_2 molecules, produced through photoassociation of Cs cold atoms, and detected using a pulsed laser ionization into Cs_2^+. We performed also for the first time PA spectroscopy of Cs_2 through the trap loss detection. The two spectra differs noticeably, the former showing only the 0_g^- and 1_u series, and the latter showing the 1_g, 0_u^+, and 0_g^- series. This result is interpreted as a consequence of the Condon point at intermediate distance provided by the double-well shape of respectively the 0_g^- and the 1_u potentials, which are responsible for the existence of a rather efficient channel in spontaneous emission for the creation of ground state molecules. PA of Cs cold atoms is thus a favorable method for the formation of translation-

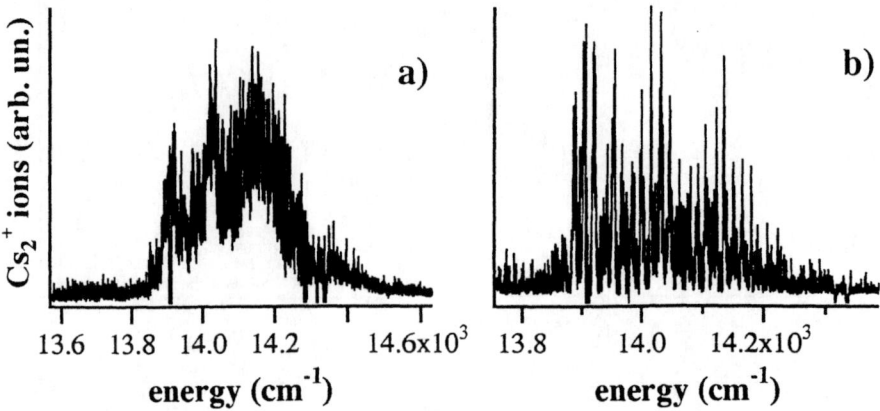

FIGURE 5. Cs_2^+ ion signal versus the pulsed laser frequency ν_2. In (a) the PA laser frequency ν_1 is kept fixed at -12 GHz from the D_2 atomic transition, exciting mostly a high lying 0_g^- states. In (b) it is tuned to the 1_u ($v = 3$) resonance line.

ally cold molecules. Our PA experiment should open a way to the physics of cold molecules. In the future, dipolar or magnetic traps could be developed to store these molecules.

We thank M. Allegrini, A. Bleton, T.F. Gallagher, P. Lett, J. Pinard, E. Tiemann, and J. Vergès for contributions and discussions. A. Fioretti is recipient of an European grant, TMR program, contract n. ERBFMBICT961218.
+ Laboratoire Aimé Cotton is associated to Université Paris-Sud.

REFERENCES

1. Y. B. Band and P. S. Julienne, Phys. Rev. A **51**, R4317 (1995).
2. J.M. Doyle, B. Friedrich, J. Kim, and D. Patterson, Phys. Rev. A **52**, R2515 (1995).
3. H. R. Thorsheim, J. Weiner, and P.S. Julienne, Phys. Rev. Lett. **58**, 2420 (1987)
4. P. D. Lett et al., Phys. Rev. Lett. **71**, 2200 (1993); J. D. Miller et al., Phys. Rev. Lett. **71**, 2204 (1993); E. R. I. Abraham et al., J. Chem. Phys. **103**, 7773 (1995); H. Wang et al., Phys. Rev. A **53**, R1216 (1996).
5. A. Fioretti, D. Comparat, A. Crubellier, O. Dulieu, F. Masnou-Seeuws, and P. Pillet, Phys. Rev. Lett. **80**, 4402 (1998).
6. A. Fioretti, D. Comparat, C. Drag, C. Amiot, O. Dulieu, F. Masnou-Seeuws, and P. Pillet, submitted to Europhys. J. D, (1998).
7. M. Marinescu and A. Dalgarno, Phys. Rev. A, **52**, 311 (1995).
8. N.Spiess, Ph.D thesis, Universität Kaiserslautern, (1989), unpublished.
9. S. Grego, M. Colla, A. Fioretti, J.H. Müller, P. Verkerk, and E. Arimondo, Opt. Comm., **132**, 519 (1996).
10. C. Monroe, W. Swann, H. Robinson, and C. Wieman, Phys. Rev. Lett. **65** (1990).
11. W. C. Stwalley, Y. H. Uang, and G. Pichler, Phys. Rev. Lett., **41**, 1164 (1978).

SESSION VII
FEMTO-SECOND AND HIGH INTENSITY RIS

The Dissociation Dynamics of Diatomic Molecules in Intense Laser Fields

Jan H Posthumus and Keith Codling

JJ Thomson Physical Laboratory
Whiteknights, Reading, RG6 6AF, UK

Abstract. The kinetic energies and angular distributions of the fragment ions resulting from the multielectron dissociative ionisation of diatomic molecules in intense laser fields have been explained, in the past, in terms of laser-induced stabilisation and laser-induced alignment. We have developed a classical field ionisation model that is able to explain recent results in I_2 without recourse to either mechanism. However, in the lighter molecules laser-induced alignment undoubtedly occurs and an explanation of the kinetic energies requires, among other things, a greater understanding of the process of laser-induced dissociation.

MULTIELECTRON DISSOCIATIVE IONISATION

In 1987, the Reading group published some preliminary results on the multiphoton ionisation of N_2 (1) using a laser of pulse length 600fs, wavelength 600 nm and focused intensity of about 10^{15} W cm^{-2}. The highest channel reached was the (2,3) channel (2):

$$N_2 + nh\nu \rightarrow [N_2^{5+}] \rightarrow N^{2+} + N^{3+} + 38eV$$

To achieve such a high state of ionisation required the correlated action of many 2eV photons and it was therefore suggested that the multielectron dissociative ionisation (MEDI) process might best be described in terms of classical field ionisation. At the above intensity, the field across the molecule, approaching 10V Å^{-1}, is more than enough to field ionise a valence electron with a binding energy of 15eV. The measured dissociation energies of the various (Q_1, Q_2) channels allowed us to reject two extreme hypothesis, 'instantaneous' removal of all 5 electrons at the neutral molecule internuclear separation, $R_e = 1.1$Å, and 'slow' removal; in the latter case the N^+, N^{2+} and N^{3+} ions would have had quite similar energies. In fact, the steady increase in dissociation energy was taken as an indication that the molecule ionises sequentially as the laser intensity increases to its peak value (3,4).

Moreover, because of the molecule's elongated shape, the potential difference created by the laser E-field is larger along its axis than at right angles to it. Consequently, electrons are ejected preferentially along the E-field and since the dissociation process is fast compared to a typical rotation time, the ions are also ejected in this direction. The first experiments on N_2 showed strong peaking along the E-field of the linearly polarised laser beam, perfectly consistent with the model.

LASER-INDUCED DISSOCIATION

When a valence electron is removed from a molecule, the molecular bond tends to become weaker. As further electrons are sequentially removed, the molecule has no bound states and Coulomb explosion results. However, when a molecule is subjected to an intense laser field, the molecular ion will dissociate even when it might otherwise be expected to be quite strongly bound. In this case the laser E-field distorts the bond-forming electron cloud to such an extent that the bond breaks. In H_2, for example, the lower-than-expected H^+ energies were explained in terms of 'bond-softening' (5), where one considers the H_2^+ ion 'dressed' by a number of photons of the laser field. A similar process can be expected to occur in other molecules.

One particular problem in analysing the outcome of a MEDI process is that the ionisation and dissociation timescales overlap. This has been overcome, to some extent, by using short laser pulses (<100fs) and a heavy molecule such as I_2, whose vibrational period exceeds 150fs. Under appropriate circumstances one can then observe the charge-asymmetric (0,2) channel, for example. Perhaps more important in the context of trace analysis of more complex molecules, the use of femtosecond pulses means that one can observe the parent molecular ion with considerable strength (6), since the relative importance of laser-induced dissociation is reduced.

LASER-INDUCED STABILISATION

A number of experiments in the early '90's (7,8,9) indicated that the kinetic energy release, E_c, of all (Q_1,Q_2) channels were, to a good approximation, a constant fraction, C_m, of the Coulomb energy at $R = R_e$. That is

$$E_c(eV) = C_m \frac{14.4 Q_1 Q_2}{R_e} = \frac{14.4 Q_1 Q_2}{R_c} \tag{1}$$

The simplest explanation of this behaviour was that the transient molecular ions were stabilised by the laser E-field at a critical distance, $R_c(= R_e/C_m)$. For the lighter molecules, N_2, CO and O_2 $C_m \approx 0.45$ and for the heavier halogens Cl_2 and I_2, $C_m \approx 0.70$.

Laser-induced stabilisation could also be invoked to explain another surprising fact, that the kinetic energy releases were independent of laser pulse rise time (7,8). This laser-induced stabilisation was thought to be similar to the vibrational trapping in H_2^+ (10), but it was hard to imagine a mechanism that could stabilise the higher channels, such as the (5,5) channel in I_2.

The Reading group approached the question of stabilisation on two fronts, one experimental the other theoretical. Experimentally, an interferometric technique was developed (11) to modify the laser pulse rise time without changing any other laser characteristics. Experiments were performed on I_2 using a wavelength of 750 nm, a pulse width of about 70fs and a focused intensity of about 10^{15} W cm^{-2}. The interferometric technique allowed a change of pulse width from 55 to 100fs and ion time-of-flight (TOF) spectra were obtained for both. The kinetic energies of the I_2 fragments were found to decrease with increasing rise time, as predicted by the field ionisation model (see below) and in contradiction to the ideas of laser-induced stabilisation.

Experiments on N_2 using pulse durations of 55fs and 400fs and employing covariance mapping also showed conclusively that the longer pulses yielded ion fragments with lower kinetic energies (12).

THE FIELD IONISATION, COULOMB EXPLOSION MODEL

We have developed a classical field ionisation, Coulomb explosion model to describe the dissociative ionisation process (12). We assume that the molecule is aligned with the laser E-field. Figure 1 illustrates the procedure for finding the appearance intensity for the (1,1) channel of I_2. At an internuclear separation R = 5 a.u. the energy level is well above the central barrier and an intensity of 5.3×10^{13} W cm^{-2} is required for over-the-barrier ionisation. For a separation R≈7.5 a.u. (not shown) the inner barrier rises to localise the electron. Between R = 7.5 and 9.8 a.u., the electron in the left well is Stark-shifted and at R = 9.5 a.u. an intensity of only 1.2×10^{13} W cm^{-2} is required for over-the-barrier ionisation. Above R = 10 a.u., however, the central barrier inhibits ionisation and at R = 14 a.u. the appearance intensity rises to 3×10^{13} W cm^{-2}.

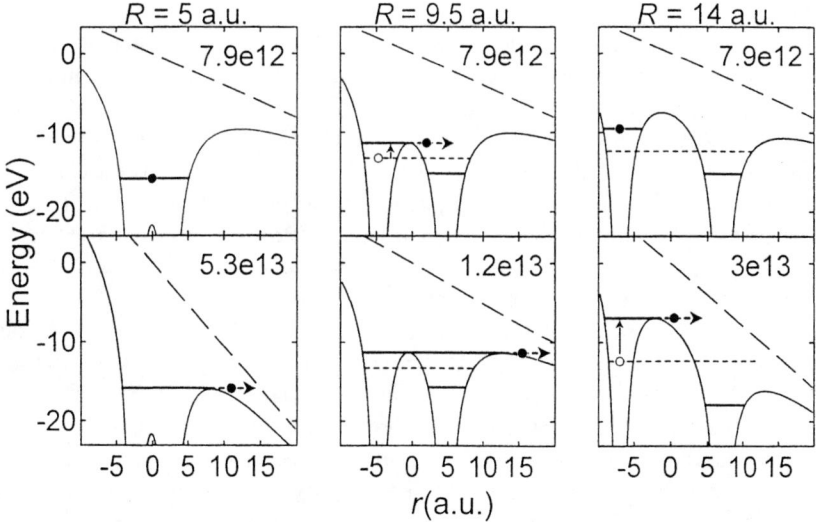

FIGURE 1. Double well potentials for I_2^+ at three internuclear separations and a number of laser intensities.

The complete appearance intensity curve for the (1,1) channel is shown in the lowest solid curve of Fig. 2, along with similar curves for other (Q_1,Q_2) channels. One notes that the ion separation that maximises the field ionisation process is *virtually the same* for all channels. At this point one can calculate classical trajectories for molecules situated in various parts of the laser focus. The dashed (-dot) curves in Fig. 2 trace the temporal evolution of laser intensity versus internuclear separation for molecules in 4 regions of the focus and 2 pulse lengths, 150 and 400fs.

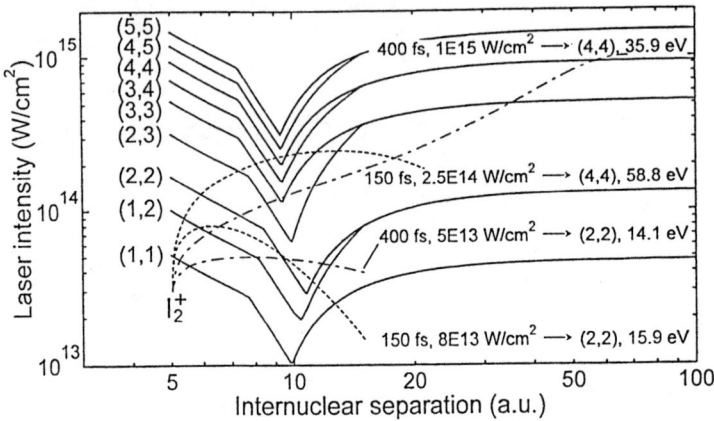

FIGURE 2. Threshold intensities of the (Q_1,Q_2) channels of I_2 (solid curves) and trajectories (dashed curves).

The process of over-the-barrier ionisation occurs each time a specific trajectory first crosses an appearance intensity curve. One sees that with a 150fs pulse of intensity 8×10^{13} W cm^{-2}, the (2,2) channel is reached at R $\approx$ 9.5 a.u., giving a dissociation energy of 15.9 eV. With a 400fs pulse of intensity 5×10^{13} W cm^{-2}, the trajectory results in an energy release of 14.1 eV for the (2,2) channel. That is, the energy release appears to be *independent* of pulse rise time. Perhaps more importantly, the concept of laser-induced stabilisation is not required to explain the lack of variation of dissociation energy with rise time; one is simply observing rapid, sequential ionisation in the region of R_c. At longer pulse lengths and higher channels post dissociative ionisation occurs (12).

We have recently performed a double-pulse experiment to check whether this critical distance has any validity (12). The (0,2) channel of I_2 lends itself perfectly to this task. A laser of wavelength 750 nm and pulse width 55fs creates the slowly dissociating (0,2) channel. The second pulse, which causes the (0,2) to (1,2) transition, was delayed with respect to the first. Figure 3(a) shows the effect of the second pulse on the TOF spectra. For delays ranging from 100 to 480fs one observes an extra peak (arrowed) due to creation of the (1,2) channel. The curves in Fig. 3(b) explain why the peak moves to lower energies as the delay is lengthened. Calculations indicated that the second pulse should be most effective at a delay of about 250fs; this was found experimentally to be the case.

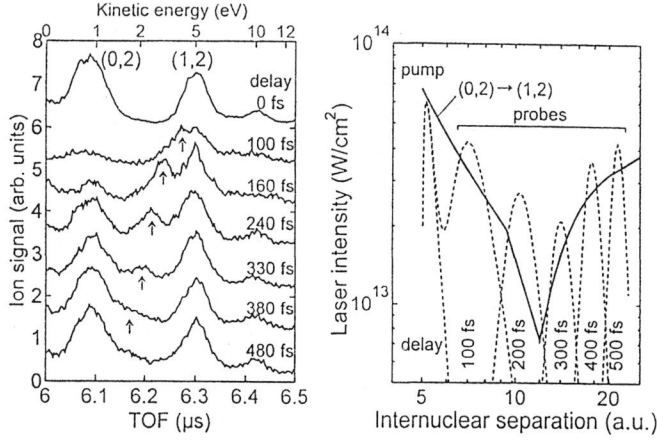

FIGURE 3. I^{2+} fragments in the I$_2$ TOF spectra, the arrow indicating the (0,2) to (1,2) transition. The curves on the right explain why the TOF of the arrowed 'backward' peak decreases with increasing delay.

LASER-INDUCED ALIGNMENT

We have seen that, for a specific laser E-field, enhanced ionisation is most effective when molecules are aligned with the field. In early publications it was suggested that the strong peaking of fragment ions along the E-field might be explained in purely geometrical terms (1). It was then pointed out that experiments on the angular distribution of fragment ion from I$_2$ using sub-100fs pulses could not be explained in this way, but required laser-induced alignment associated with the molecules' polarisability (13). Double-pulse techniques appeared to confirm that laser-induced re-orientation does occur in both CO and I$_2$ (14,15). However, the interpretation of these various experiments is questionable, since they preceded our understanding of enhanced ionisation.

We have recently performed experiments (16) on I$_2$, N$_2$, and H$_2$ and in the case of I$_2$ in particular, have attempted to answer the question as to the extent to which the angle-dependent enhanced ionisation phenomenon (a pure geometrical effect) is responsible for the highly anisotropic angular distribution. We have calculated threshold intensities for a range of angles, θ, between laser E-field and molecular axis. In this case the maximum Stark shift is now the dot-product, ½ E.R. Figure 4 shows the ionisation threshold intensity versus internuclear separation for a range of θ values, for the I$^+$ + I$^+$ channel. The thresholds vary by a factor of 5 for orthogonal polarisations.

When a molecular ion dissociates, it is the laser intensity that pertains when the ion reaches R$_c$ that determines the final charge state. We see, for example, that when a molecule lies at an angle of 75° to the E-field, the required intensity is 3×10^{13} W cm^{-2}, whereas at 0° it is only 10^{13} W cm^{-2}. Because of the form of the variation of intensity in the laser focus, the

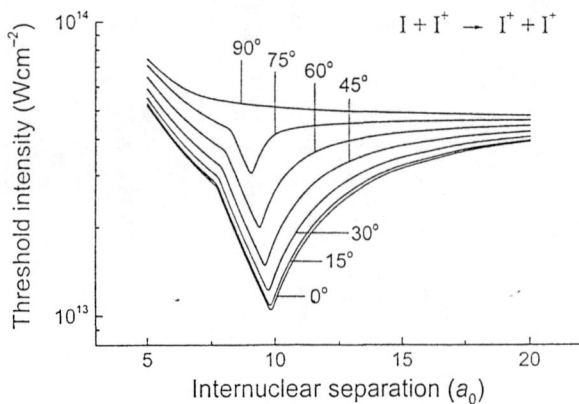

FIGURE 4. Threshold intensities as a function ion separation for a series of angles between laser E-field and molecular axis.

focal volume involved for molecules at 75° is considerably less than for fully aligned molecules. Assuming random orientation of the I_2 molecules and no laser-induced rotation, the angular distributions will depend critically upon the shell volumes as defined by the threshold intensities. Figure 5(a) shows the relative size of shells as a function of angle for the (1,2) channel of I_2 for a number of laser intensities. The experimental results (16) for the angular distributions of this channel are shown in Fig. 5(b). The overall agreement between the two suggests that the angular anisotropy in I_2 is chiefly the result of geometric rather than dynamic alignment.

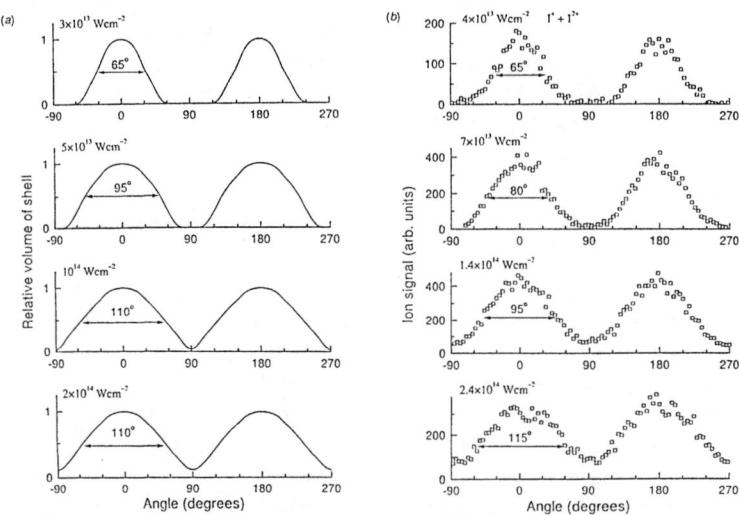

FIGURE 5. (a) Relative volumes of the $I^+ + I^{2+}$ shell as a function of angle between laser E-field and molecular axis; (b) experimental angular distribution of fragments.

We have also measured the angular distribution of H_2 and N_2 fragment ions as a function of laser intensity. These lighter molecules are more susceptible to re-orientation. Perhaps more importantly, due to their high ionisation potentials they can be subjected to higher laser E-fields and torques before they dissociate. In the case of H_2 at 10^{14} W cm^{-2} the angular width is reduced by over a factor of 3 compared to I_2. Moreover, no H^+ ions are observed at 90° to the E-field, whatever the intensity; this behaviour is quite different to that of I_2, see Fig. 5(b).

We have recently performed a double-pulse experiment (17) to confirm that laser-induced alignment occurs in H_2. The first 50fs pulse arrives at the focus with its E-field perpendicular to the axis of the TOF analyser. A second pulse with its E-field along the axis arrives typically 2ps later, but other delays have been used. This second pulse is focused more sharply. Figures 6(a) and (b) show the changes in the TOF spectrum of H_2 as the pump beam moves through the probe beam, at the focus. We see two forward and two backward H^+ peaks from the probe beam and a central, low-energy peak from the pump (and probe) beam. As expected, the ions from the probe beam disappear completely when the two beams overlap; the pump beam has rotated all H_2 molecules so that they miss the detector.

This experiment does not, of course, tell us at what stage the dynamic alignment occurs, but we suspect that dissociation and re-orientation occur simultaneously during the bond-softening process. In the case of I_2 the TOF spectra are virtually unchanged as a function of pulse-probe overlap, confirming that no significant re-alignment has taken place.

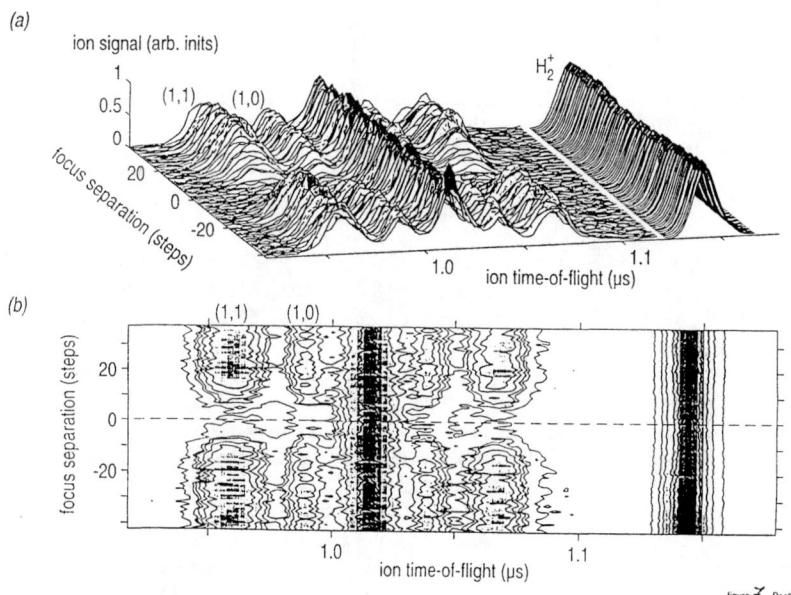

FIGURE 6. (a) H_2 TOF spectra as a function of the spatial overlap of pump and probe; (b) contour maps of the data.

COHERENT CONTROL OF DISSOCIATION DYNAMICS

We have described briefly how a Michelson interferometer was used to modify the rise time by changing from constructive to destructive interference, a simple example of coherent control. We have recently performed a more complex experiment using the fundamental and second harmonmic at 750 and 375 nm and a pulse length of 85fs. When the phase between the two is changed by π, the superposed E-field maximum reverses direction (18). One might therefore expect the direction of emission of H^+ ions to be reversed and indeed such a reversal is observed experimentally; an enhancement of the forward peak became an enhancement in the backward peak. The ions were found to be ejected counter-intuitively but this behaviour was explained in the context of the field ionisation model (19).

ACKNOWLEDGEMENTS

We are pleased to acknowledge the Engineering and Physical Sciences Research Council (UK) for their financial support. Special thanks go to Drs P.F. Taday and A.J. Langley of Rutherford Appleton Laboratory for their expert assistance with the experiments described here.

REFERENCES

1. Frasinski, L.J., Codling, K., Hatherly, P.A., et al, *Phys.Rev.Lett.* **58**, 2424-2427 (1987).
2. Frasinski, L.J., Codling, K., and Hatherly P.A., *Phys.Lett.A* **142**, 499-503 (1989).
3. Codling, K., Frasinski, L.J., Hatherly, P.A., and Barr, J.R.M., *J.Phys.B* **20**, L525-L531 (1987).
4. Codling, K., Frasinski, L.J., and Hatherly, P.A., *J.Phys.*B **22**, L321-L327 (1989).
5. Bucksbaum, P.H., Zavriyev, A., Muller, H.G., and Schumacher, D.W., Phys.Rev.Lett. **64**, 1883-1886 (1990).
6. Singhal, R.P., Ledingham, K.W.D., Kosmidis, C., et al *Chem.Phys.Lett.***253**, 81-86 (1996).
7. Cornaggia, C., Lavancier, J., Normand, D., et al, *Phys.Rev. A* **44**, 4499-4505 (1991).
8. Schmidt, M., Normand, D., and Cornaggia, C., *Phys.Rev. A* **50**, 5037-5045 (1994).
9. Hatherly P.A., Stankiewicz, M., Codling, K., et al, *J.Phys. B* **27**, 2993-3003 (1994).
10. Zavriyev, A., Bucksbaum, P.H., Squier, J., and Saline, F., *Phys.Rev.Lett.***70**, 1077-1080 (1993).
11. Giles, A.J., Posthumus, J.H., Thompson, M.R., et al, *Opt.Commun.***118**, 537-541 (1995).
12. Posthumus, J.H., Giles, A.J., Thompson, M.R., and Codling, K., J.Phys. B **29**, 5811-5829 (1996).
13. Strickland, D.T., Beaudoin, Y., Dietrich, P., and Corkum, P.B., *Phys.Rev.Lett.*, **68**, 2755-2758 (1992).
14. Normand, D., Lompré, L.A., and Cornaggia, C., *J.Phys B,* **25**, L497-L503 (1992).
15. Dietrich P., Strickland, D.T., Laberge, M., and Corkum, P.H., *Phys.Rev. A***, **47**, 2305-2311 (1993).
16. Posthumus, J.H., Plumridge, J., Thomas, M.K., et al, *J.Phys. B* , **31**, L553-L562 (1998).
17. Posthumus, J.H., Plumridge, J., Codling, K. et al, *Laser Phys.* submitted.
18. Thompson, M.R., Thomas, M.K., Taday, P.F., et al, *J.Phys. B,* **30**, 5755-5772 (1997).
19. Posthumus, J.H., Thompson, M.R., Giles, A.J., and Codling, K., *Phys.Rev. A*, **54**, 955-957 (1996).

Femtosecond Dynamics of Photoinduced Chemical Reactions Studied by Intense-Field Dissociative Ionization

W. Fuß, K. L. Kompa, W. E. Schmid, and S. A. Trushin

Max-Planck-Institut für Quantenoptik, Postfach 1513, D-85740 Garching, Germany

Abstract. This work uses intense-field dissociative ionization to probe the chemical dynamics and profits from the fact that different positions on the potential energy surface have different probabilities of ionization and subsequent ion dissociation. Therefore monitoring the parent and various fragment ions as function of the pump-probe delay allows the molecule to be followed on the potential energy surface all along the primary photochemical reaction path. We have applied this approach to study the femtosecond dynamics of a number of photoinduced isomerization and dissociation reactions in the gas phase including pericyclic hydrogen migration of 1,3,5-cycloheptatriene and UV photodissociation of the transition metal carbonyls $Mo(CO)_6$, $W(CO)_6$, $Fe(CO)_5$.

INTRODUCTION

Femtosecond time-resolved multiphoton ionization combined with mass selective detection has proved to be a powerful tool in time-resolved studies of chemical reaction dynamics (1). Its application is straightforward in two cases: 1. the decay of the initially excited electronic state will be reflected in a decaying signal of the parent ion; 2. the formation of products in a neutral fragmentation reaction may be monitored by observing the appropriate ionic fragments. However monitoring isomerization and internal conversion does not seem promising by this technique, since the product and educt have the same mass. Recently we have demonstrated, however, that sometimes isomers can easily be distinguished from each other (e.g., hexatriene versus cyclohexadiene), if the molecules are ionized by non-resonant radiation at high intensity under conditions of barrier suppression (2,3). In this case ionization is assumed to be caused by the electric field of a long-wavelength intense laser. In a classical model an intense-laser field can suppress the ionization barrier and cause efficient ionization of an atom or molecule. This model predicts that the threshold intensity for ionization of atoms by barrier suppression scales as the 4-th power of the ionization potential (IP). If the ionization efficiency mainly depends on IP, we can expect that every electronically excited state should give an enhanced ion signal independent of its electronic and vibrational nature. Since IP will change along the reaction path (because the ionic and neutral surfaces are not parallel), we can infer that the ionization probability depends on the position on a potential surface, and such a dependence is sufficient to allow the measurement of time constants. Ionization by an intense-laser field is often followed by extensive fragmentation which provides additional information on transient intermediate states and/or species. This can be called intense-laser field dissociative ionization (IFDI). This use of ionization to probe the chemical dynamics profits from the fact that different positions on the potential energy surface (PES) have different probabilities of ionization and subsequent ion dissociation. The most important reason for the selectivity is that excess vibrational energy is transferred from the neutral to the ion, resulting in fragmentation of the latter. One source of the excess energy is conversion of electronic to vibrational energy when a molecule runs down a neutral PES or when during internal conversion or chemical reaction it crosses from an initially excited state to a lower state, since total energy is conserved. If the vibrational energy (which is transferred to the ion) exceeds the threshold for fragmentation, the parent ion will not be observed anymore, but a fragment will be observed instead. Thus when neutral molecules run down potential energy surfaces, more and more electronic energy is converted to vibration, and the ions observed will be lighter and lighter. Hence, if the fragmentation thresholds are suitable, ionic fragmentation can serve to determine roughly the vibrational energy content and thereby help to distinguish different electronic states. Another source of the excess energy is that at some locations on the potential energy surface, the ion is generated with excess electronic energy. (The minimum number of probe photons required is then, of course, larger.) Besides the excess electronic and vibrational energy, fragmentation is also caused by photodissociation of the ion.

A special feature of field ionization at high laser intensities is multielectron dissociative ionization, i.e., multiple ionization followed by dissociation through Coulomb explosion of the charged fragments. Recently it has been found that multiple ionization is dramatically enhanced in various molecular and cluster systems at some critical internuclear distances, larger than the equilibrium values. Corkum and coworkers proposed to use the sensitivity of IFDI to the molecular configuration as a probe in time-resolved studies of a dissociative system (4). Castleman and coworkers used it to "arrest intermediate states in a chemical reaction on a femtosecond time scale" (5). Therefore monitoring the parent and various fragment ions as function of the pump-probe delay allows the molecule to be followed on the PES all along the primary photochemical reaction path. This approach was successfully applied to study ultrafast electrocyclic ring opening of 1,3-cyclohexadiene (2,3,6), and UV photodissociation of $Cr(CO)_6$ (7). In this present paper, we present new results on pericyclic hydrogen migration in 1,3,5-cycloheptatriene and UV photodissociation of transition metal carbonyls $Mo(CO)_6$, $W(CO)_6$, $Fe(CO)_5$.

The experimental system, based on a Ti-sapphire laser has been described in detail elsewhere (2,3,6,7). The species under study were excited by ca. 140 fs pulses at 267 nm. The subsequent motion of molecules on the PES was monitored by ionization by 110 fs pulses at 800 nm with variable delay. Measurements were performed in the gas phase with mass selective detection of the ions.

HYDROGEN MIGRATION IN CYCLOHEPTATRIENE

H migration in cycloheptatriene is a photochemically allowed [1,7]-sigmatropic pericyclic reaction. The product is distinguishable from the educt only in substituted derivatives of this molecule. We have investigated 1,3,5-cycloheptatriene C_7H_8 as well as several 7-alkyl derivatives and fully deuterated species by the same transient ionization method as we have used for cyclohexadiene (2,3,6). Again we found ultrashort lifetimes for both the bright state ($\tau_{1A''} = 60$ fs) and the dark state ($\tau_{2A'} = 70$ fs). Methylation at C^7 practically does not change these times, whereas perdeuteration substantially lengthens them ($\tau_{1A''} = 95$ fs, $\tau_{2A'} = 150$ fs for $CD_3C_7D_7$). Our deuteration results strongly support the idea that internal conversion is connected with the photochemical H (D) shift and that the two processes have the initial part of their pathway in common (8,9).

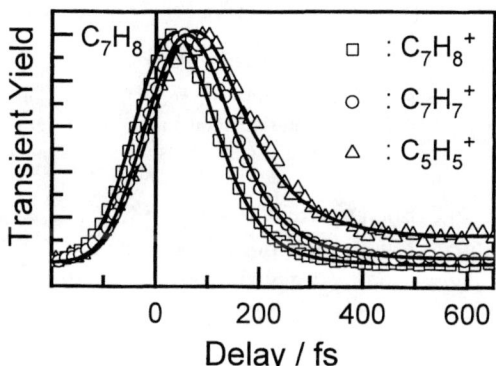

FIGURE 1. Relative ion yield for parent and fragment ions vs the time delay between pump and probe pulses. The symbols show the experimental data and the lines present results of simulation within a rate equation model with 4 levels (windows of observation) depicted in Fig.2. The parent ion signal has a tail decaying with time constant τ_{23}, while the fragments are noticably shifted and have tails decaying with time constant τ_{34}. The transient parent ion signal is only generated from the 1A'' surface, while the fragments contain substantial contributions from vibrationally hot lower excited and ground states.

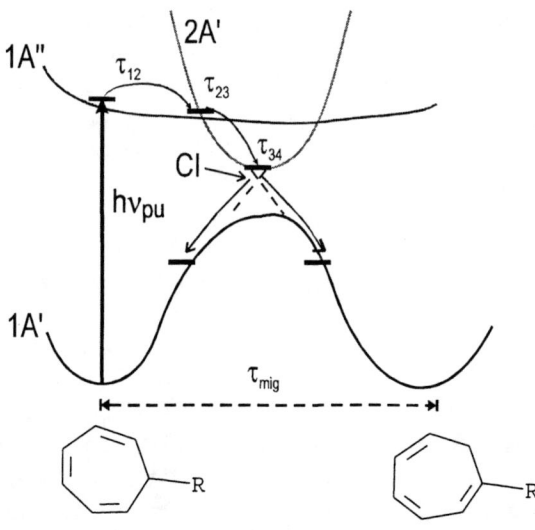

FIGURE 2. Schematic representation of the early-time dynamics of hydrogen migration in cycloheptatrine C_7H_8 and its derivatives $CH_3C_7H_7$ and $CD_3C_7D_7$. Whereas the potential surfaces for this system have not been calculated, the experiments (in our interpretation) indicate that it follows the general scheme suggested in refs. (2,3,6,8) for photochemical pericyclic reactions, with a collection funnel and a continuous path downwards via an easily accessible conical intersection.

time, fs	C_7H_8	C_8H_{10}	C_8D_{10}
τ_{12}	10	5	5
τ_{23}	60	66	95
τ_{34}	70	95	150
τ_{mig}	160	190	270

UV-DECOMPOSITION OF TRANSITION METAL CARBONYLS

Vibrational Coherence in Decomposition of $M(CO)_6$, M= Cr, Mo, W

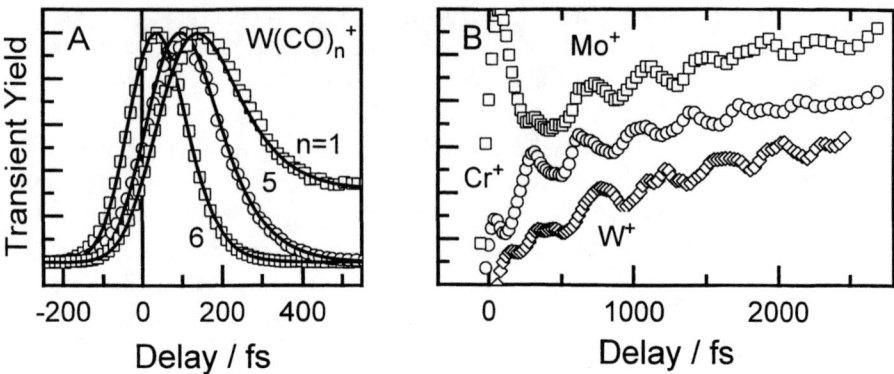

FIGURE 3. Pump-probe delay kinetics of the transient yield of the parent and some fragment ions in decomposition of $M(CO)_6$. A: initial kinetics of $W(CO)_6$; B: coherent oscillation and secondary growth of the atomic ion signals. The symbols are the experimental data and the lines are results of simulation.

The initial dynamics of the single-photon dissociation of $M(CO)_6$ into $M(CO)_5$ + CO in the gas phase is very similar for all three group 6 transition metals Cr, Mo, and W (7,10). After excitation at 267 nm, the parent and all fragment ions produced by the delayed probe pulse at 800 nm show different temporal behavior at delay times < 200 fs as shown in Fig. 3a. This indicates that the molecules have already gone through several different states or configurations during this short time interval. With longer delays the heavy ions disappear, but the light fragment ions $M(CO)_n^+$, n = 4,...0, (for Cr 3,...0) persist and exhibit coherent oscillations. The period t_{BPR} and phase are independent of n, indicating a common precursor of the signals. We assign the oscillation to Berry pseudorotation (BPR) of the neutral precursor $M(CO)_5$ (7). Our observations support the pathway of $M(CO)_6$ photodecomposition proposed by Burdett et al. (11) and depicted in Fig. 4. The ground-state product $M(CO)_5$ (C_{4v}, 1^1A_1) is produced within 100-150 fs and vibrates along the BPR coordinate (C_{ax}-M-C_{eq} bending). These oscillations are damped with a time constant of ca. 1 ps by thermal dissociation of vibrationally hot $M(CO)_5$ to produce $M(CO)_4$ which also gives rise to a slight secondary growth of small ionic fragments (Fig. 3b). On ionization of vibrationally hot $M(CO)_5$ followed by thermal dissociation and photodissociation, the $M(CO)_5^+$ ion transfers these oscillations to $M(CO)_n^+$, n = 4...0. A simulation of the initial kinetics by a rate equation model with 5 levels (windows of observation) detailed in ref. (7) allows the time constants presented in the caption of Fig. 4 to be derived. The shapes of the parent ion signals are in all cases well fitted to a single exponential with a time constant τ_{12} convoluted with the pump and probe pulses. The longest time constant τ_{45} corresponds to the decay of the light fragments, and the other time constants are responsible for the temporal shift of the fragments without significant broadening.

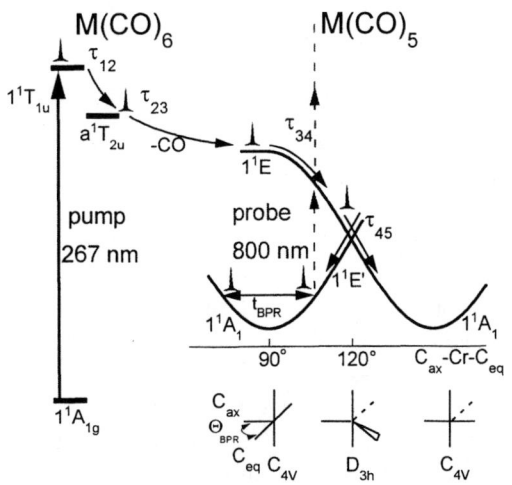

FIGURE 4. Schematic representation of the early-time dynamics of UV decomposition of $M(CO)_6$ M=Cr, Mo, W. The Jahn-Teller effect induces a conical intersection between the $1^1E'$ and 1^1A_1 states of $M(CO)_5$ in a trigonal-bipyramidal geometry, thus providing a path for ultrafast internal conversion from an electronically excited state to the ground state and for the excitation of coherent oscillations along the C_{ax}-M-C_{eq} bending coordinate with a period of t_{BPR}.

time, fs	Cr	Mo	W
τ_{12}	23	30	46
τ_{23}	10	20	23
τ_{34}	20	30	30
τ_{45}	40	46	70
t_{BPR}	350	396	418

Where is Intersystem Crossing in $Fe(CO)_5$ UV Decomposition?

Kinetics of transient ions in 267 nm single-photon decomposition of $Fe(CO)_5$ probed by IFDI at 800 nm is formally rather similar to $M(CO)_6$ (compare Figs. 3 and 5): The parent and heavy fragment ions $Fe(CO)_n^+$, n=5,4,3 show an initial spike only and disappear after 300 fs; the lighter fragments $Fe(CO)_n^+$, n=2,1,0 decrease after the initial spike and then show a rising feature. After about 10 ps the ion yields do not change anymore. However there is a striking difference between $M(CO)_6$ and $Fe(CO)_5$: in the latter case the fragments show no oscillation, and the secondary growth is much more pronounced and noticeably slower (3.3 ps vs 1 ps). The main features of the kinetics

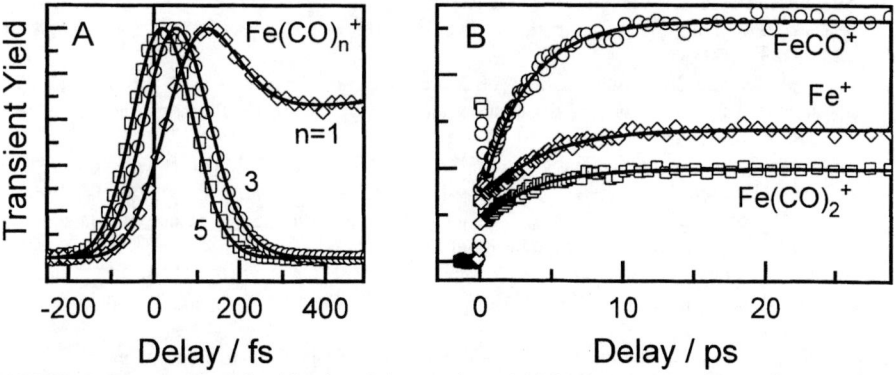

FIGURE 5. Pump-probe delay kinetics of the transient yield for the parent and some fragment ions in decomposition of $Fe(CO)_5$. A: initial kinetics; B: secondary growth of $Fe(CO)_n$, n=3,2,1. The symbols show the experimental data and the lines present results of simulation.

$$\text{Fe(CO)}_5(1^1A'_1) \xrightarrow{h\nu} \text{Fe(CO)}_5(6^1E') \xrightarrow{21\,fs} \text{IS1} \xrightarrow{15\,fs} \text{IS2}$$

$$\xrightarrow{30\,fs} \text{IS3} \xrightarrow{47\,fs} \text{P1} \xrightarrow{3300\,fs} \text{P2}$$

for all ions can be reproduced by a rate-equation model with the 6 levels indicated in the above scheme taken from ref. (10). After excitation to a metal-to-ligand charge transfer state $^1E'$, the molecule passes through three short-lived (<100 fs) intermediate states (IS) and then comes to a long-lived (3.3 ps) product state P1 which decays to a final product state P2. A previous calculation suggests that photodissociation of $Fe(CO)_5$ to $Fe(CO)_4$ proceeds via intersystem crossing (ISC) between $^1E'$ and $^3E'$ states of $Fe(CO)_5$ (12). The $^3E'$ state of $Fe(CO)_5$ then dissociates to $Fe(CO)_4$ in its triplet ground state $^3B_2$. According to the calculation, the ISC should occur as an early step. However this is in contradiction with our observation of sub-100-fs lifetimes of the four early steps and with the experiment on two-photon (400 nm) decomposition (13), where the slowest time constant was found to be 230 fs. ISC is expected to take longer. We suggest that after internal conversion to a repulsive singlet ligand field state (IS1 in the above scheme), the molecule dissociates, forming some electronically excited state (IS2) of $Fe(CO)_4$, which after two internal conversions via IS3 arrives at the lowest singlet state $^1A_1$ of $Fe(CO)_4$ (P1). Then it undergoes within 3.3 ps ISC to the ground triplet state $^3B_2$ in parallel with a thermal dissociation to produce the end product P2, $Fe(CO)_3$ in its ground state $^3A_2$. All the product states in question are so close in energy (<1eV) that there is no chance to tune them through resonance with the probe laser. This is probably why no coherent oscillation was observed in the $Fe(CO)_n$ system.

REFERENCES

1. Zewail, A. H. *Femtochemistry: Ultrafast Dynamics of the Chemical Bond*; World Scientific: Singapore, 1994; Vols I and II and references therein.
2. Trushin, S.A., Fuss, W., Schikarski, T., Schmid, W.E., and Kompa, K.L., *J. Chem. Phys.* **106**, 9386-9389 (1997).
3. Fuss, W., Kompa, K.L., Schikarski, T., Schmid, W.E., and Trushin S.A., " Probing of ultrafast photoinduced isomerization and dissociation reactions by intense-field dissociative ionization" in *SPIE Proc.* **3271**, 114-130 (1998).
4. Seideman, T, Ivanov, M.Yu., and Corkum, P.B., *Chem.Phys.Lett.* **252**, 181-188 (1996).
5. Folmer, D.E., Poth, L., Wisnievski, E.S., Castleman, Jr, A.W.,*Chem.Phys.Lett.* **287**, 1-7 (1998).
6. Fuss, W., Kompa, K.L., Schikarski, T., Schmid, W.E. , Trushin, S.A., Celani, P., Garavelli, M., and Olivucci, M., *J. Chem. Phys.*, to be submitted.
7. Trushin, S.A, Fuss, W., Schmid, W.E., Kompa, K.L., *J. Phys. Chem.* **102**, 4129-4137 (1998).
8. Fuss, W., Lochbrunner, S., Müller, A.M., Schikarski, T., Schmid, W.E., and Trushin S.A., *Chem. Phys*, 232, 161-173 (1998).
9. Diemer, S., Fuss, W., Kompa, K.L., Schmid, W.E., and Trushin S.A.,in preparation.
10. Trushin, S.A., Fuss, W., Schmid, W.E., and Kompa, K.L.,in preparation.
11. Burdett, J. K., Grzybowski, J. M., Perutz, R. N., Poliakoff, M., Turner, J. J., and Turner, R. F., *Inorg. Chem.* **17**, 147-154 (1978)
12. Veillard, A., Daniel, C., and Strich, A., *Pure & Appl. Chem.* 60, 215-221 (1988)
13. Banares, L., Baumert, T., Bergt, M., Kiefer, B., Gerber, G., *J. Chem. Phys.* **108**, 5799-811 (1998)

SESSION VIII
APPLICATIONS II

Lineshapes and Optical Selectivity in Double-Resonance RIMS Measurements

W. Nörtershäuser[†], K. Blaum[†], B. A. Bushaw[*], P. Müller[†],
N. Trautmann[‡], K. Wendt[†]

[†] Institut für Physik, [‡] Institut für Kernchemie,
Johannes Gutenberg-Universität, D-55099 Mainz, Germany
[*] Pacific Northwest National Laboratory, Richland, WA 99352, USA

Abstract: Lineshapes for double-resonance three-photon ionization have been evaluated using the density matrix formalism. Experimental conditions were incorporated by considering realistic distributions for atomic velocity and angular divergence, as well as spatial laser intensity profiles. Results for the

$$4s^2\ {}^1S_0 \rightarrow 4s4p\ {}^1P_1 \rightarrow 4s4d\ {}^1D_2 \rightarrow Ca^+$$

ionization scheme of calcium are discussed and compared with experimental measurements. The calculations accurately reproduce isotopic selectivities achieved in the experiment. Moreover, these measurements have been carried out with dynamic range of $>10^6$, revealing many subtle effects of simultaneously present coherent and incoherent excitation processes.

INTRODUCTION

Ultratrace determination of long-lived radioisotopes in the presence of stable neighboring isotopes requires both extremely high isotopic selectivity and detection efficiency. Both can be provided by RIMS with cw lasers: Combining the optical selectivity of resonant excitation with the selectivity of the mass separation step can yield the overall requisite isotopic abundance sensitivity, and elimination of duty-cycle constraints can lead to extremely efficient detection. Different applications of this approach have been demonstrated for a variety of problems (see e.g. (1,2)). The optical selectivity that can be realized in the excitation process depends not only on isotope shifts and hyperfine structure, but also crucially on the lineshape function of the resonance signal. To realize both extremely high isotopic selectivity and ionization efficiency, multi-step resonance excitation with narrowband cw lasers is usually required. However, in this experimental approach, lineshapes and intensities are strongly affected by coherences between atomic states introduced by the light fields. These must be considered in the theoretical description of the resonance ionization process. A reasonable prediction of the lineshapes in multi-step RIS, combined with the knowledge of isotope shifts and hyperfine structure constants, enables calculation of realistic isotopic selectivities achievable with the excitation scheme under investigation. In this paper we compare lineshapes calculated with the density matrix formalism to high- dynamic-range experimental measurements.

THEORY

Theoretical modeling of lineshapes in high-resolution multiple-resonance ionization spectroscopy requires the description of the excitation process with the density matrix (DM) formalism. The interaction of the atoms with the extremely narrowband radiation of single-frequency cw lasers establishes coherences between the different states of the excitation ladder which are considered explicitly by the off-diagonal elements of the DM. These coherences have a pronounced influence on the lineshapes of the observed ionization signal and are not predicted by simple rate equation approximations. While a detailed description of the DM formalism can be found in a number of textbooks, e.g. in (3), we provide here only a brief description together with the working equations necessary for the calculations presented.

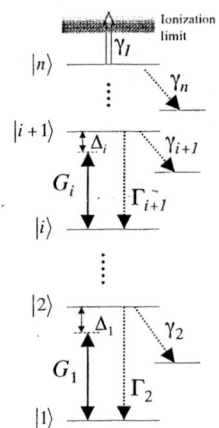

FIGURE 1. Schematic representation of a generalized n-state excitation ladder

A schematic of the excitation path in these experiments, including the parameters necessary for the DM description, is given in Figure 1. The atomic parameters Γ_j are half the Einstein A_{jk} ($k = j-1$) coefficients for spontaneous decay within the excitation ladder and γ_j are half the corresponding coefficients for decay out of the ladder system. A detuning Δ_j from the exact resonance condition is allowed for each radiation field. The coupling between the two states involved in the transition $j \leftrightarrow j+1$ is described by the parameter G_j which is half the corresponding Rabi-frequency and is related to the Einstein A_{kj} coefficient, the laser intensity I_j, and the wavelength λ_j of the individual transition according to

$$G_j = \sqrt{\frac{3 A_{j+1,j} I_j \lambda_j^3}{8\pi h c}} \quad (1)$$

With these parameters, the equations of motion describing the time evolution of the DM elements in a generalized multi-step ladder system are

$$\left.\frac{d}{dt}\rho_{jk}\right|_{j \geq k} = \left(i \sum_{\ell=k}^{j-1} \Delta_\ell - \Gamma_j - \gamma_j - \Gamma_k - \gamma_k \right) \rho_{jk} + 2 \Gamma_{k+1} \rho_{k+1,k+1} \delta_{jk}$$
$$- i G_{k-1} \rho_{j,k-1} - G_k^* \rho_{j,k+1} + G_{j-1}^* \rho_{j-1,k} + i G_j \rho_{j+1,k} \quad (2)$$

$$\left.\frac{d}{dt}\rho_{jk}\right|_{j<k} = \frac{d}{dt}\rho_{kj}^*.$$

The rate of ionization $\gamma_I = \sigma_{PI} I/h\nu$, with corresponding cross section σ_{PI}, is included in the γ_j coefficient of the uppermost level $j=n$. The number of ionized atoms ρ_I will evolve according to

$$\dot\rho_I = 2 \gamma_I \rho_{nn}. \quad (3)$$

To calculate the ionization efficiency for atoms passing through the laser beams, we start with all population in the ground state ($\rho_{ij} = \delta_{i1} \delta_{j1}$) at time $t = 0$ and numerically integrate the equations over the interaction time. During this integration the realistic (~ Gaussian) spatial intensity profiles of the lasers are taken into account and consequently the interaction strengths G_j and the ionization rate γ_j become time-dependent. In addition, atoms in a real atomic beam have a velocity distribution and residual angular divergence, which leads distributions of both Doppler-shifted laser detunings and interaction times. For an accurate prediction of experimental results, subsequent integrations must be performed over these additional degrees of freedom. The normalized flux distribution for a collimated atomic beam derived from an effusive thermal source is given by

$$\Phi(v) = 2v^3 \alpha^{-4} \exp(-v^2/\alpha^2) dv \qquad (4)$$

where $\alpha = \sqrt{2kT/m}$ is the most probable velocity. The angular distribution was assumed to be a square-wave function, which is appropriate for the small angles (2°-5°), limited by mechanical apertures, used in our measurements. A more detailed description of the numerical integration will be given in a forthcoming publication (4).

EXPERIMENTAL

To test the theoretical predictions, double-resonance ionization experiments were performed for the

$$4s^2\ {}^1S_0 \xrightarrow{422.7\ nm} 4s4p\ {}^1P_1 \xrightarrow{732.8\ nm} 4s4d\ {}^1D_2 \xrightarrow{514.5\ nm} Ca^+ \qquad (5)$$

ionization ladder of calcium. This excitation scheme can be treated as an essentially closed system, except for the ionization step. Branching ratios for decay to other states, not included in the excitation ladder, are estimated to be «1% for both bound excited states.

The wavelengths for the resonant excitations were provided by a Coherent 699-21 dye laser with stilbene 3 (422.7 nm) and a Coherent 899-21 titanium-sapphire laser (732.8 nm), both pumped by argon ion lasers. The 514.5 nm ionization wavelength has photon energy sufficient to ionize out of the $4s4d\ {}^1D_2$ state without affecting the intermediate $4s4p\ {}^1P_1$ state. Short-term jitter of the tunable lasers matched the 500 kHz specification for these commercial systems. Frequency-drift stabilization and frequency scanning were accomplished by computer-controlled fringe-offset-locking relative to a single mode He:Ne laser in a 150 MHz confocal Fabry-Perot interferometer (5).

The rest of the experimental setup consists of a graphite crucible for sample atomization (6) and a commercial quadrupole mass spectrometer (QMS) for mass analysis and ion detection. A detailed description can be found in (7). The atomic beam intersects the three laser beams at right angles within the QMS ionizer. The 3 mm entrance aperture of the ionizer defines the residual divergence of the atomic beam. This divergence can be adjusted by changing the distance of the crucible from the aperture. Measurements reported here were performed with a residual divergence of either 1.9° (far-coupled) or 5.2° (close-coupled). The latter configuration is preferable for analytical measurements since it maximizes transport efficiency and hence sensitivity. The two resonant laser beams are counterpropagating, resulting in further reduction of Doppler width due to the partial cancellation of photon momentum.

RESULTS AND DISCUSSION

For comparison between theory and experiment, a number of one- and two-dimensional laser scans were performed and the experimental parameters were used to calculate the theoretically expected lineshapes. Figure 2 shows results of one-dimensional frequency scans where the frequency of one laser is held constant while the other is scanned. In these figures points represent experimentally recorded data while the solid curve is the result of the corresponding DM calculation. Experimental data is corrected for dead-time effects in the counting electronics and a slow exponential decrease in atomization rate during the measurement. The only adjustment of the theoretical curve was normalization to the experimental peak maximum. Experimental parameters used in the DM calculation are given explicitly in the figure caption. In Figure 2(a) the second-step laser was held on resonance ($\Delta v_{733}= 0$ MHz) while the first-step laser was tuned across the resonance. The wings of the peak are very steep and the function cannot be described by a Voigt profile, which can be understood by realizing that detuning of the first-step laser decouples the transition not only from the two-photon resonance but also simultaneously from the intermediate level. The result of the DM integration reproduces almost perfectly the experimental data over the full dynamic range of more than five orders of magnitude. Figure 2(b) shows the corresponding case with the second-step laser fixed 500 MHz below resonance. The lineshape exhibits two distinct peaks which are separated by the detuning of the second-step laser. The larger peak ($\Delta v_1= 500$ MHz) corresponds to the coherent two-photon transition ($\Delta v_{423}+ \Delta v_{733} = 0$), while the peak at $\Delta v_{423} = 0$ MHz represents an incoherent two-step transition: The resonant laser leads to a

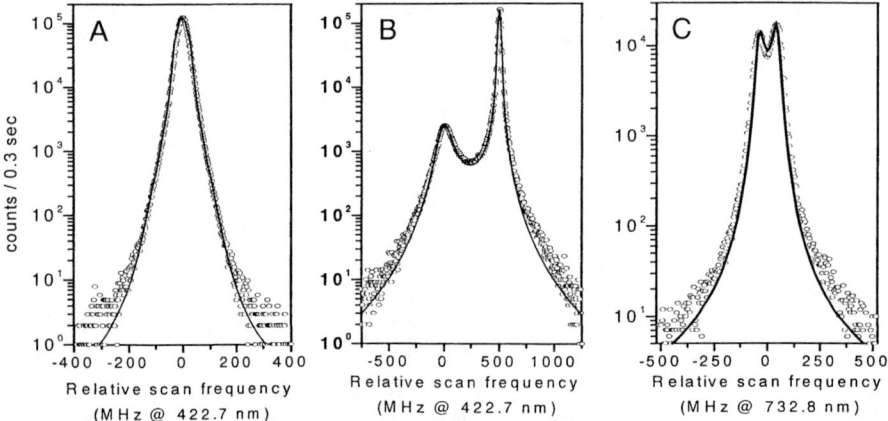

FIGURE 2. Experimental (points) and theoretical (curve) resonance lineshapes of ^{40}Ca in double-resonance ionization for different laser detunings and laser intensities. Mass spectrometer is tuned to mass 40. Experimental parameters considered in theory are: residual full beam divergence (ϵ), detuning of the two lasers (Δv_X), diameter of the Gaussian laser beams $\sigma_{423}=1.7$ mm, $\sigma_{733}=2.1$ mm (FW@1/e), and temperature of the graphite crucible (T = 1200 K). Values used are A) $\epsilon = 5.2°$, $\Delta v_{733}= 0$ MHz, $P_{423}= 0.3$ mW, $P_{733}= 20$ mW; B) $\epsilon = 1.9°$, $\Delta v_{733}= -500$ MHz, $P_{423}= 30$ mW, $P_{733}= 300$ mW; C) $\epsilon = 1.9°$, $\Delta v_{423}= -7$ MHz, $P_{423}= 30$ mW, $P_{733}= 3$ mW.

population of the intermediate state, on which the second-step laser acts within the Lorentzian wings of the homogeneous linewidth. This "single-photon resonance" is about two orders of magnitude smaller than the two-photon resonance. Nevertheless, the theory reproduces its relative intensity and shape in excellent agreement with the experimental data. Note that comparing Figure 2(a) and 2(b) is relevant to determining isotopic selectivity: 2(a) represents the on-resonance response of a desired target isotope while 2(b) corresponds to the spectrum of an interfering isotope when the frequency of the fixed laser is detuned to resonance for the target. As a final example for one-dimensional frequency scans, Figure 2(c) shows the case of an intense laser in the first step, fixed 7 MHz below resonance, and a weak scanning laser in the second step ("pump and probe" arrangement). Consequently, the resonance exhibits a slightly asymmetric Autler-Townes (also called AC-Stark) splitting, and is again well reproduced by theory.

Two-dimensional laser scans are instructive for obtaining a better overview of the double-resonance structure and evaluating selectivity for multiple isotopes. In Figure 3(a) the result of such a measurement, with relatively high power (30 mW) in the first and low power (3 mW) in the second transition, is shown as a logarithmic contour plot of the signal intensity. This measurement was carried out in the far-coupled configuration. All features predicted by theory and discussed thus far can be clearly observed. The strong diagonal ridge corresponds to the coherent two-photon transition. The weaker perpendicular ridges are the incoherent "single-photon resonances" and the peak center shows the Autler-Townes splitting due to the high laser intensity in the first step. Additionally, a weak pair of ridges at an angle of about 20° to the diagonal two-photon ridge and passing through each of the Autler-Townes maxima are observable. The origin of this off-resonant structure lies in the Doppler distribution of the first transition. When the detuning of the first step laser is within a few Doppler widths of the first resonance, there is still significant population of atoms with resonance frequency Doppler shifted to the laser frequency. These atoms are effectively transferred into the first excited state. Then, if the second laser is simultaneously tuned to their Doppler-shifted resonance frequency in the second step, which occurs for counterpropagating beams when

$$\Delta v_2 = -\Delta v_1 \frac{v_2}{v_1}, \qquad (6)$$

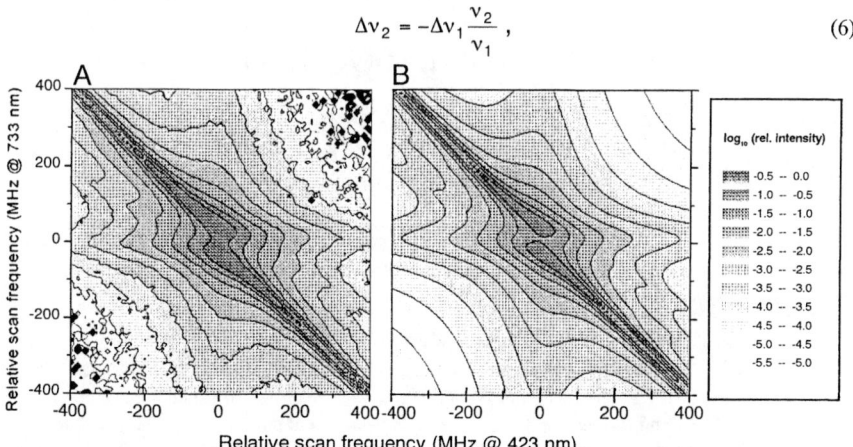

FIGURE 3. Logarithmic contour plot of the two-dimensional lineshape of ^{40}Ca in experiment (A) and theory (B).

and a resonance signal is observed for the corresponding Doppler class. Due to the high intensity in the first transition, AC-Stark splitting leads to the production of two ridges; at lower intensities, only a single Doppler-shifted ridge is observed passing through the center of the (now unsplit) main peak.

The first calculation of this two-dimensional spectrum with a residual divergence of 1.9° showed good agreement with the experimental result but a too little intensity in these Doppler ridges. We assume that the background pressure of calcium in the oven chamber acts as a secondary beam source. This chamber is connected to the main vacuum chamber, containing the QMS and ionization region, by a small hole. This hole is located much closer to the entrance aperture of the QMS ionizer than the graphite crucible; hence the residual divergence of this beam is about ten times larger ($\varepsilon = 20°$). Indeed, a second calculation with a 1% admixture of this "secondary source" results in excellent agreement for the intensity of these ridges as is shown in Figure 3(b).

CONCLUSION AND OUTLOOK

A system of DM equations for a generalized multi-step ladder excitation is given and has been used for the special case of double-resonance ionization of calcium. It has been shown that the numerical integration of the DM equations leads to a very accurate prediction of lineshapes and hence, combined with isotope shift and hyperfine structure information, a very good estimation of the selectivity which can be achieved with such an excitation scheme. Experimental investigations covered a wide range of laser intensities and different detunings of the resonant lasers. Excellent agreement between theory and experimental results has been observed in all cases; also for two-dimensional laser scans which reveals the complexity of lineshapes in multi-resonant excitation.

Triple-resonance calculations have already been performed and corresponding experiments on calcium and strontium are under development at the University of Mainz and Pacific Northwest National Laboratory, respectively.

ACKNOWLEDGEMENT

This work was supported by the Office of Basic Energy Sciences of the U.S. Department of Energy under Contract DE-AC06-76RLO 1830, the Deutsche Forschungsgemeinschaft under Contract Tr336/1-3, and the DAAD.

REFERENCES

1. Bushaw, B. A. and Munley, J. T., *Inst. Phys. Conf. Ser.* **114**, 387-392 (1991).
2. Wendt, K., et al., *AIP Conf. Proc.* **388**, 361-364 (1997).
3. Shore, B. W.: *The Theory of Coherent Atomic Excitation*, John Wiley & Sons, New York (1990).
4. Bushaw, B. A., Nörtershäuser, W., and Wendt, K., *Spectrochim. Acta B* (to be published).
5. Bushaw, B. A., Cannon, B. D., Gerke, G. K., and Whitaker, T. J.*Opt.Lett.* **11**, 422-424 (1986).
6. Bushaw, B. A., and Gerke, G. K., *Inst. Phys. Conf. Ser.* **94**, 277-280 (1989).
7. Nörtershäuser, W., Trautman, N., Wendt, K. Bushaw, B.A., *Spectrochim. Acta B*, **53**, 709-721 (1998).

Diode-Laser-Based RIMS Measurements of Strontium-90

B. A. Bushaw and B. D. Cannon

Pacific Northwest National Laboratory, Richland, WA 99352, USA

Abstract: Double- and triple-resonance excitation schemes for the ionization of strontium are presented. Use of single-mode diode lasers for the resonance excitations provides a high degree of optical isotopic selectivity: with double-resonance, selectivity of $>10^4$ for ^{90}Sr against the stable Sr isotopes has been demonstrated. Measurement of lineshapes and stable isotope shifts in the triple-resonance process indicate that optical selectivity should increase to $\sim 10^9$. When combined with mass spectrometer selectivity this is sufficient for measurement of ^{90}Sr at background environmental levels. Additionally, autoionizing resonances have been investigated for improving ionization efficiency with lower power lasers.

INTRODUCTION

The measurement of the radiotoxic isotope ^{90}Sr by RIMS methods has received considerable interest at this conference over the last few years. Double-resonance excitation with cw dye lasers (1) demonstrated detection limits in the low attogram range, however, unfavorable isotope shifts limited optical isotopic selectivity to 10^2. Measurement by fast beam collinear spectroscopy (2) has demonstrated overall isotopic selectivity of $\sim 10^{11}$ and detection limits of ~ 1 fg, and has now been demonstrated with a number of real environmental samples (3). Unfortunately, both of these approaches require large complex instrumentation which is not amenable to routine analytical measurements.

More recently, we demonstrated a new approach for double-resonance excitation of Sr based on solid-state diode lasers (4). The excitation scheme is shown as path A in Figure 1. Two single-mode diode lasers operating near 689 and 688 nm populate the 5s6s $^3S_1$ state of Sr, followed by photoionization at 488 nm with an Ar ion laser. This work measured the isotope shifts and hyperfine structure for all stable isotopes as well as ^{90}Sr in both resonance transitions. Because of favorable isotope shifts and lineshapes, optical selectivities of $>10^4$ against all stable Sr isotopes were achieved. Partial saturation of the ionization step using 488 nm radiation was possible because of a fortuitous near coincidence with the 4d6p $^3P_1$ autoionizing state. Combining optical isotopic selectivity with that of a quadrupole mass spectrometer, demonstrated analytical figures of merit with graphite furnace atomization were overall isotopic selectivity of 1.5×10^{10} and detection limits of 0.8×10^{-15} g.

While this performance is sufficient for a number of real applications, measurements capable of approaching the northern hemisphere background levels of ^{90}Sr ($<10^{-12}$ relative to ^{88}Sr) require ~ 2 orders of magnitude increase in selectivity and a similar lowering of detection limits is desirable for smaller sample sizes. The double-resonance excitation scheme still has several drawbacks which include: (i) The first transition has a natural linewidth of 7.5 kHz because it is formally spin forbidden. While the ~ 10 mW of power

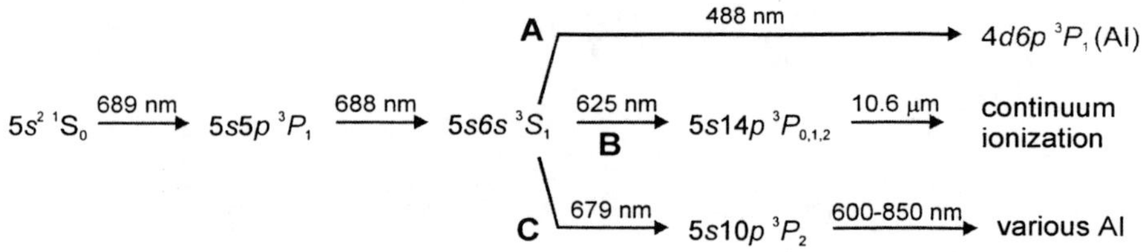

FIGURE 1. Double- and triple-resonance excitation schemes for isotopically selective ionization of strontium using diode lasers. AI indicates autoionizing states.

available from the diode laser can effectively excite atoms within the resonance width, only a small fraction of the residual inhomogeneous Doppler distribution is truly in resonance and resulting excitation efficiency is low. (ii) The second excited state, $5s6s\ ^3S_1$ with a lifetime of ~13 ns, decays back into all three $J = 0,1,2$ components of $5s5p\ ^3P$ and population is rapidly lost into the "dark" $J = 0,2$ components. It would be desirable to have a subsequent excitation to a higher-lying bound state where excitation rates could be more competitive with decay into the "dark" states. (iii) While diode lasers have significantly reduced instrumental complexity and cost, a large argon ion laser is still used for the ionization step. It would be desirable to find smaller, less expensive lasers that can either provide much higher photon fluxes (e.g. CO_2 laser) or can be tuned (e.g., high-power diode lasers) to highly favorable autoionizing resonances. Preliminary studies that address possible improvements are discussed in the following sections.

FREQUENCY-MODULATED EXCITATION OF $5s^2\ ^1S_0 - 5s5p\ ^3P_1$

As noted above, efficiency in the first resonance transition is limited by the ratio of homogeneous linewidth to inhomogeneous Doppler width. This may be partially mitigated by power broadening, however, power broadening of the natural width increases only with the square-root of the laser intensity and, if large, will significantly reduce isotopic selectivity. In the ideal case of narrowband excitation of a two-level system with only the natural width, the excitation rate is independent of the upper state lifetime and given by $\Phi\lambda^2/2\pi$, where Φ is the laser intensity in photons/cm^2·s and λ is the transition wavelength. Thus, for a 10 mW laser operating at 689 nm and 2 mm beam diameter, one predicts an (ideal) excitation rate of ~10^9 s^{-1}, which is much faster than a typical beam crossing time of 3 μs for a thermal atom (T=1300K, M=90). Obviously, excitation of truly resonant atoms will be well saturated but, under the given conditions, the power-broadened homogeneous width is still only ~1 MHz. When compared to an experimental Doppler width of ~30 MHz, the excitation efficiency can not exceed a few percent.

Our approach to improving efficiency in this transition has been to use frequency modulation (FM) of the diode laser, effectively performing a rapid sweep of the laser frequency through the Doppler profile over a period corresponding to the average beam crossing time. When compared to simple saturation broadening this has two advantages: (1) Increase in effective linewidth only requires corresponding linear increases in laser intensity to maintain the same (on resonance) excitation rate and (2) the far tails of the lineshape still correspond to the ~1 MHz power-broadened width. Beyond the simplistic considerations outlined in the previous paragraph, we have used phase-randomized

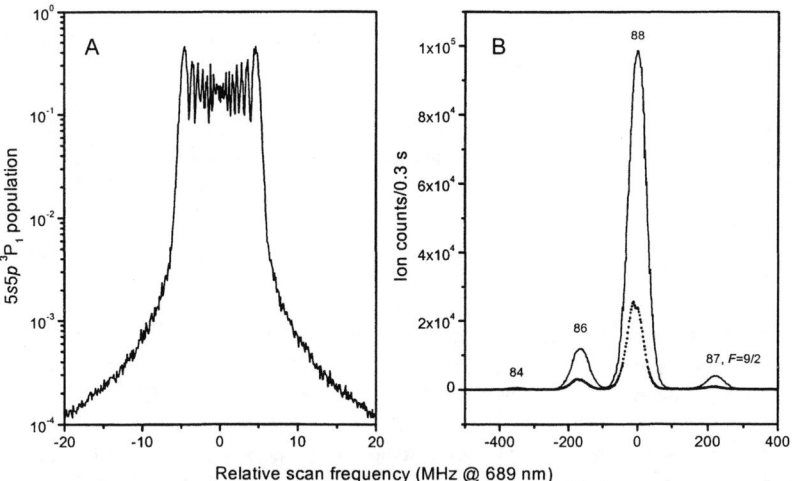

FIGURE 2. Increasing efficiency with FM excitation. A) Phase-randomized density matrix simulation for 10 MHz modulation width at 300 kHz rate and no Doppler broadening. B) Experimentally observed enhancement: upper trace (solid) with 20 MHz modulation width at 300 kHz, lower trace (points) without modulation. Mass spectrometer is set to allow transmission of all masses.

density matrix calculations to model this process. Figure 2A shows the response calculated in the absence of Doppler broadening for a sinusoidal modulation with 10 MHz peak-to-peak width at a 300 kHz rate. The resulting effective lineshape reproduces the well known sum of Bessel functions characteristic in FM methods, convoluted with a narrow (power-broadened) Lorentzian that gives the small tails. When applied to a population with 30 MHz Doppler width and 20 MHz modulation width, the density matrix calculations predict a 6-fold improvement in excitation efficiency, with negligible increase in apparent linewidth. Figure 2B shows the experimental verification of this using direct photoionization with a UV Ar ion laser to monitor 5s5p $^3P_1$ population. The upper trace is recorded with frequency modulation and the lower trace without. The observed enhancement is a factor of 4, approaching the theoretical prediction. The remaining discrepancy can be (partially) attributed to frequency jitter in the diode laser system (~1.5 MHz) that may already effectively act as a small FM component, even without externally driven FM. Consistent with the theoretical predictions, the observed linewidth (FWHM) increase only from 38.5 to 41.3 MHz when the FM is applied. Although the improvement in excitation efficiency is not as dramatic as initially expected from the ratio of modulated to power broadened widths, a factor of 4 improvement is still significant and may be implemented very easily with diode lasers simply by modulating the injection current.

TRIPLE-RESONANCE EXCITATION

In an effort to develop excitation schemes with increased selectivity and more favorable ionization steps, we have investigated triple-resonance schemes as illustrated by pathways B and C in Figure 1. These begin the same as the previously studied double diode laser excitation to the 5s6s $^3S_1$ state. Rather than direct photoionization, an additional resonance excitation to a 5snp $^3P_{0,1,2}$ Rydberg state is performed, followed by photoionization with a separate laser. We have examined two different possibilities: a "brute force" method using a CO_2 laser, and one using a tunable laser to reach strong autoionizing (AI) resonances.

The CO_2 laser approach requires that the Rydberg state lie within ~1000 cm^{-1} of the ionization limit, which, for Sr, means that the lowest lying accessible states are of the 5s14p ^3P configuration and implies a wavelength of ~625 nm for the third-step resonance laser. In current investigations, we used a single-frequency dye laser with rhodamine-B to reach this wavelength but it could be reached with diode lasers. After the initial excitation to the 5s5p $^3P_1$ metastable state, this scheme strongly resembles previously studied double-resonance excitations with CO_2 laser photoionization (1). Similar to that work, we could easily saturate the photo-ionization step with 10W of CO_2 laser power. However, in contrast, only a weak onset of saturation could be observed in the third resonance step, even with 200 mW of power available from the dye laser. This is attributed to the fact that in the previous work the intermediate state decays only within the excitation ladder and thus the final resonance excitation need take place only on the time scale of the beam crossing, whereas in this case, the final excitation must be competitive with the nanosecond time scale decay of the intermediate 5s6s $^3S_1$ state into trapping metastable states.

Because of this need for higher power in the final resonance excitation step and the fact that the operation of diode lasers at 625 nm is marginal (requiring cooling of a nominally 635 nm diode to ~-10°C, and the available powers are only a few mW), we have considered a slightly different triple-resonance scheme. It is similar to that described above: the final bound state is still a 5snp $^3P_2$ Rydberg, but now with $n = 10$ instead of 14. This requires an excitation wavelength of 679 nm, which can be produced by single-mode diode lasers with powers up to 500 mW. Also the oscillator strength, proportional to n^{*3}, is expected to be about 4 times greater for $n = 10$ than $n = 14$. Although the diode laser currently used at 679 nm only provides 10 mW, we have been able to observe the onset of saturation using moderate one-dimensional cylindrical focussing.

One of the primary goals of using a triple-resonance scheme is to improve optical isotopic selectivity. Figure 3 shows isotope shift measurements for the stable even isotopes ^{84,86}Sr with respect to ^{88}Sr. In each of the scans shown, the mass spectrometer was set to the isotope of interest, the first two lasers were offset by the known isotope shift for the corresponding first two resonances, and the third diode laser was scanned over an identical frequency range in all cases. Thus, the observed position differences correspond to the 5s6s $^3S_1$ – 5s10p $^3P_2$ transition isotope shifts (TIS). Using well know isotope shift formulae (5) with known changes in nuclear charge radii (6), this is sufficient information to predict the ^{90}Sr TIS to be +180 MHz in the given transition. When combined with the observed lineshape (see ^{88}Sr spectra in Figure 3) the predicted optical selectivity is ~10^5 in the third transition alone, and, when combined with the >10^4 selectivity already demonstrated for the first two transitions, it is expected that overall optical isotopic selectivity for ^{90}Sr of ~10^9 can be achieved using this triple-resonance scheme.

While going to the $n = 10$ Rydberg level is favorable in terms of the diode lasers used, it is not possible to use a CO_2 laser for the ionization step. In the "brute force" approach, with fixed wavelength lasers, this could possibly be remedied by using a cw Nd:YAG laser. However, we are currently investigating a "finesse" approach that takes advantage of known even parity $J = 1,2$ autoionizing levels (7) with energies above 5s10p $^3P_2$

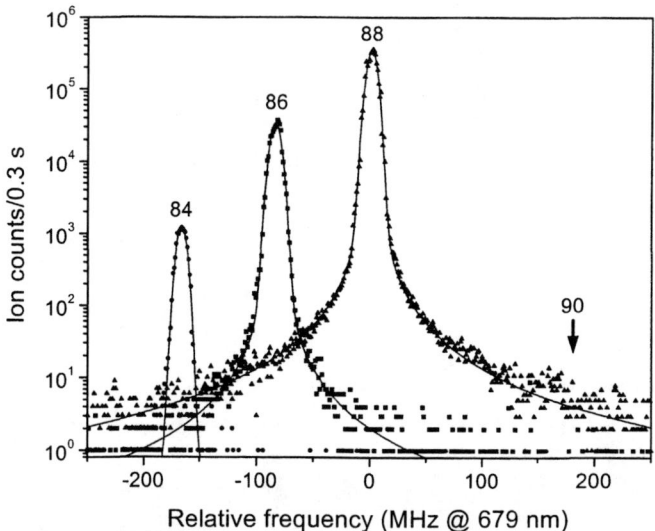

FIGURE 3. Transition isotope shifts of ^{84,86}Sr vs. ^{88}Sr in 5s5p $^3S_1$ – 5s10p $^3P_2$ observed in triple resonance ionization. Frequencies of the first two resonance lasers are offset by the know isotope shifts, mass spectrometer is set for isotope of interest, while the third resonance laser is scanned over same frequency range in all measurements. The predicted shift for ^{90}Sr is indicated.

corresponding to wavelengths between 600 and 850 nm. Specifically, we have investigated the region 615-640 nm with a cw dye laser and have thus far observed over thirty different narrow autoionizing states with widths ranging from 250 MHz to 20 GHz. Most of these can be assigned to the known $J = 1$ and $J = 2$ AI states, but several are previously unidentified and are tentatively assigned as $J = 3$ levels. Figure 4A shows a spectrum of the strongest of these, the $[4d_{3/2},14d_{5/2}]_1$ state, which is well described by a Lorentzian profile of 800 MHz FWHM and an ionization rate that is ~1500 times as large as the background continuum ionization rate. Figure 4B shows a similar spectrum of the $[4d_{5/2},11g_{9/2}]_2$ state, but because of interference with the underlying broader $[4d_{3/2},15d_{5/2}]_2$ state, it exhibits a distinctly asymmetric profile. Preliminary studies of the 820-850 nm region, where the density of AI states is much lower, has thus far only revealed two AI resonances, both previously unidentified. However, this work is still in preliminary stages and more comprehensive survey scans of the 820-850 nm, as well as the 660-680 nm region, which can also be reached with tapered amplifier diode lasers, are planned for the future.

CONCLUSIONS

RIMS using double-resonance excitation of Sr using two diode lasers and an Ar ion laser for excitation has already been shown capable of isotopic selectivity $>10^{10}$ and sub-femtogram detection limits. In this paper we have considered, and presented preliminary results on, a number of possible improvements to this scheme. Frequency modulation has been shown effective in tailoring laser lineshapes to residual Doppler profiles and

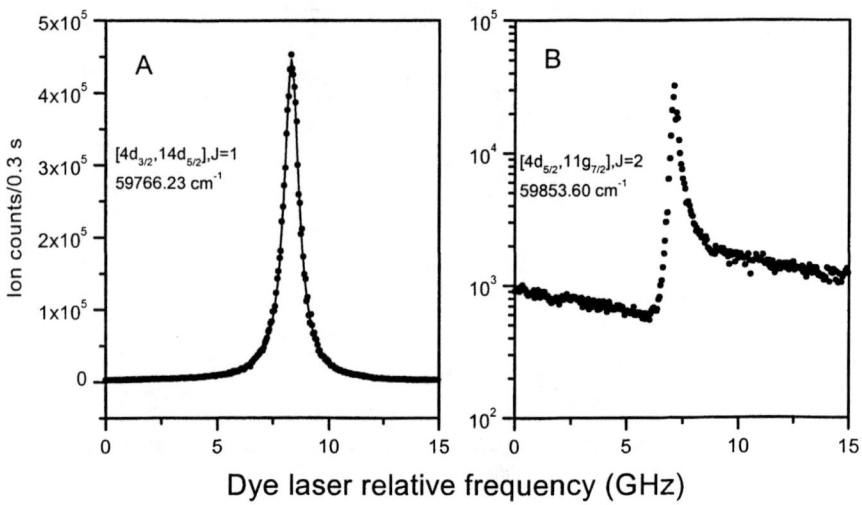

FIGURE 4. Autoionizing resonances observed with excitation out of the 5s10p $^3P_2$ Rydberg state using a cw dye laser. AI state designations and energies are inset. The profile in (A) is well fit by a Lorentzian with 808 MHz FWHM, while (B) exhibits an asymmetric Fano profile.

attaining modest increases in excitation efficiency. Triple-resonance schemes have been considered which significantly improve optical isotopic selectivity, may improve overall efficiency, and remove the need for a large Ar ion laser for the ionization step. Continuing work will further quantify the these initial results and address test measurements on ^{90}Sr, with an ultimate goal of producing a RIMS ionization scheme that can be performed exclusively with diode lasers (including the ionization step), has detection limits in the low attogram range, and yields overall isotopic selectivity of greater than 10^{12}.

ACKNOWLEDGEMENT

This work was supported by the Office of Basic Energy Sciences of the U.S. Department of Energy under contract DE-AC06-76RLO. Pacific Northwest National Laboratory is operated by Battelle Memorial Institute.

REFERENCES

1. Bushaw, B. A., *Inst. Phys. Conf. Ser.* **128**, 31-36 (1992).
2. Monz, L., et al., *Inst. Phys. Conf. Ser.* **128**, 225-228 (1992).
3. Wendt, K., et al., *AIP Conf. Proc.* **388**, 361-364 (1997).
4. Bushaw, B. A. and Cannon, B. D., *Spectrochim. Acta B* **52**, 1839-1854 (1997).
5. Heilig, K. and Steudel, A., *At. Data Nucl. Data Tables* **14**, 613-638 (1974).
6. Buchinger, F., et al., *Phys. Rev. C* **41**, 2883-2897 (1990).
7. Goutis, S., et al., *J. Phys. B* **25**, 3433-3461 (1992).

Determination of the First Ionization Potential of Nine Actinide Elements by Resonance Ionization Mass Spectroscopy

G. Passler, M. Nunnemann, G. Huber

Institut für Physik, Universität Mainz, D-55099 Mainz, Germany

R. Deißenberger, N. Erdmann, S. Köhler, J.V. Kratz, N. Trautmann, A. Waldek

Institut für Kernchemie, Universität Mainz, D-55099 Mainz, Germany

J.R. Peterson

Department of Chemistry, University of Tennessee, Knoxville, TN 37996-1600, USA

Abstract. Due to its high sensitivity, RIMS enables the precise determination of the first ionization potential of actinide elements with a sample size of $\leq 10^{12}$ atoms. The actinide atoms under investigation are ionized in the presence of an electric field by multiple resonant laser excitation, and the ions are mass-selectively detected in a time-of-flight spectrometer. The first ionization potential is obtained by scanning the wavelength of the laser used for the last excitation step across the ionization threshold W_{th} – indicated by a sudden increase of the ion count rate – at various electric field strengths. Extrapolation of W_{th} to zero electric field strength leads directly to the first ionization potential. The first ionization potentials (IP) of Am, Cm, Bk, Cf and Es were determined for the first time as $IP_{Am} = 5.9736(3)$ eV, $IP_{Cm} = 5.9914(2)$ eV, $IP_{Bk} = 6.1979(2)$ eV, $IP_{Cf} = 6.2817(2)$ eV, $IP_{Es} = 6.3676(5)$ eV with samples of 10^{12} atoms. Furthermore, the ionization potentials of Th, U, Np and Pu were remeasured.

INTRODUCTION

The first ionization potential (IP) of an element is a fundamental physical and chemical property. As it is directly connected to the atomic spectra precise measurements of this quantity are helpful in understanding the electronic structure of an element and identifying systematic trends in binding energies. In particular, for the heaviest elements reliable information about the electronic structure is necessary to predict deviations from the regularities of the periodic system of elements (1). As a result of the relativistic mass increase of the inner electrons in heavy elements relativistic effects are expected (2). Precise experimental values for the ionization potentials may help for a better understanding of these effects, and, particularly, for the actinide elements enable testing the predictions of multi-configuration Dirac-Fock calculations (3), one of the most successful theoretical approaches for heavy multi-electron atoms.

For the lanthanides, U and the lighter transuranium elements Np and Pu, high precision values of the first ionization potentials were determined by the study of the convergence of long Rydberg series (4-7). As Rydberg series may be perturbed by configuration interactions, however, a large amount of data is required for a correct interpretation of the spectra. For the determination of the ionization potential of Pu by Worden et al. (7) as much as 2 g of the radioactive isotope ^{239}Pu were used throughout the measurements which shows the limits of this method:.

From extrapolated spectral properties, the first ionization potential was derived for the actinide elements up to Es (8,9). The analogy between lanthanides and actinides was used in another approach based on the semiempirical Slater-Condon method and ab initio Hartree-Fock calculations (10).

Resonance Ionization Mass Spectroscopy allows the precise experimental determination of the first ionization potential of an element with a total amount of only $\leq 10^{12}$ atoms (≈ 400 pg for the actinides). Using such small samples, we were able to measure the first ionization potential of Am, Cm, Bk, Cf and, recently, of Es (11) for the first time. Furthermore, we did remeasure the first ionization potentials of Th, U, Np and Pu.

EXPERIMENTAL

Set-up

The element under investigation is electrolytically deposited onto a thin metal backing foil in form of its hydroxide and covered with a thin metal layer. By resistive heating of such a sandwich filament inside a vacuum chamber, the hydroxide is converted to the oxide which is reduced to the elemental state during diffusion through the covering layer. An atomic beam is created by evaporation of the actinide element from the surface of the sandwich filament. A combination of a tantalum backing with a titanium covering layer obtained by sputtering has proven to be most efficient for the production of an atomic beam for the transprotactinium elements (12).

The laser system for optical excitation and ionization of the atoms consists of three dye lasers, which are pumped by two copper vapor lasers of high pulse repetition rate (6.5 kHz, 30 W and 50 W average output power, 30 ns pulse duration) (13). The bandwidth of the dye lasers is typically several Gigahertz and can be reduced to 1 GHz by means of an intracavity etalon. A pulsed wavemeter (Burleigh, model WA 4500, or - more recently - ATOS, Lambdameter) serves for the determination of the wavelengths with a precision of $\Delta\lambda/\lambda = 10^{-6}$. The dye laser beams are focused into the apparatus either by quartz fibers or prisms and cross the atomic beam perpendicularly. The resonantly produced ions are accelerated by electric fields, then pass a field free drift tube and, finally, are detected with a multichannel plate detector. This time-of-flight mass spectrometer is described in (14).

Photoionization Threshold Method

The first ionization potential results from the determination of photoionization thresholds in the presence of an external static electric field. According to the classical saddle point model (15), the excitation energy $W(r)$ relative to the electronic ground state of an atom with one highly excited electron, located in an external electric field F, is – in a one-dimensional approximation – given by

$$W(r) = IP - \frac{Z_{eff} e^2}{4\pi\varepsilon_0 r} - eFr, \tag{1}$$

where e is the electric charge of the electron, Z_{eff} the effective charge number of the core, r the distance of the excited electron from the nucleus, ε_0 the permittivity of the vacuum, and IP the first ionization potential. The ionization threshold W_{th}, which is the maximum value of $W(r)$, depends on the electric field strength as follows:

$$W_{th}(F) = IP - 2 \cdot \sqrt{\frac{Z_{eff} e^3}{4\pi\varepsilon_0}} \cdot \sqrt{F} = IP - const \cdot \sqrt{F} \tag{2}$$

By a one- (in the case of Np) or two-step (all other cases) resonant laser excitation in the presence of a known electric field the atoms are promoted to a high-lying state. The wavelength of a further laser beam is scanned across the ionization threshold W_{th} which depends linearly on the square root of the electric field strength. If the total excitation energy exceeds the threshold W_{th}, the atoms are ionized and mass-selectively detected in the time-of-flight spectrometer with the multichannel plate unit. Thus the ionization threshold is indicated by a sudden onset of the ion signal. The first ionization potential is obtained by repeating this procedure at various electric field strengths and extrapolating W_{th} to zero electric field strength.

TABLE 1. Excitation schemes for the determination of first ionization potential of actinide elements by RIMS, starting from the ground state of the atom. All wavelengths are vacuum wavelengths.

Actinide element	λ_1 [nm]	1st step energy level [cm^{-1}]	λ_2 [nm]	2nd step energy level [cm^{-1}]	Ionizing laser [nm]
Th	580.58	17,224.3	623.07	33,273.8	≈ 568
U	639.72	15,631.9	{ 591.64 586.02	32,534.1 32,696.3	≈ 577 ≈ 582
Np	311.90	32,061.3	—	—	≈ 541
Pu	649.07	15,406.6	629.75	31,285.9	≈ 579
Am	640.68	15,608.5	654.60	30,885.1	≈ 578
Cm	655.64	15,252.2	640.74	30,859.1	≈ 573
Bk	566.06	17,666.0	{ 720.71 664.70	31,541.3 32,710.3	≈ 544 ≈ 581
Cf	572.76	17,459.2	625.21	33,453.7	≈ 583
Es	561.69	17,803.5	661.31	32,924.9	≈ 544

MEASUREMENTS AND RESULTS

The excitation schemes (16) used for the various actinides are summarized in Table 1. As an internal control of the method two different second excitation levels differing in the J quantum number were involved in the measurements of U and Bk. Figure 1 shows the shift of the photoionization threshold as a function of the electric field strength for ^{249}Cf ($T_{1/2}$ = 350.6 a), for example. Depending on the counting statistics, the threshold energy at the onset of the ion signal can be determined with an accuracy between 0.5 and 2 cm^{-1}.

For five transplutonium elements, the ionization thresholds obtained in that way are plotted versus the square root of the electric field strength F in Figure 2. Linear extrapolation to zero field strength by means of weighted least squares fits yields the first ionization potentials. The results for the actinide elements investigated so far are compiled in Table 2 in comparison with data from literature. The statistical errors (2σ) determined in the least squares fit including the weighted errors for each datapoint are given as uncertainties of the experimental values derived in this work. For Am, a second experimental value was obtained by Rydberg convergence which is also listed in Table 2. For more details see (18-22).

Figure 3 illustrates the development of the first ionization potentials, normalized to $5f^N 7s^2 \rightarrow 5f^N 7s$ plotted versus N. The experimental values are below the extrapolated data for the actinide elements (8,9) and always slightly above the Hartree-Fock calculations (10). According to (10), the actinide ionization potentials should follow the trend for binding energies of the s-electrons by forming two straight lines, changing the slope at the half-filled 5f-shell (Am). While

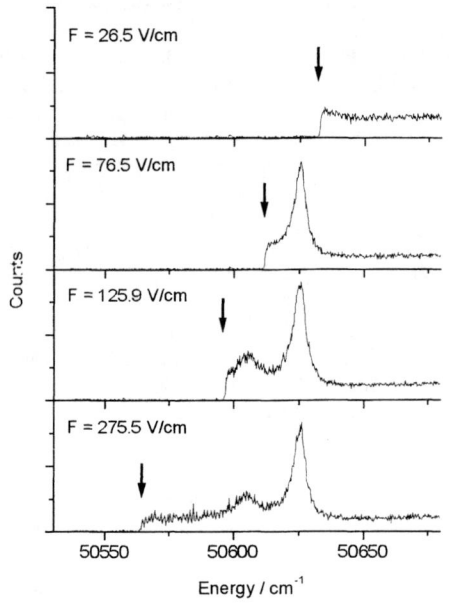

FIGURE 1. Ionization thresholds of ^{249}Cf for four different electric field strengths F.

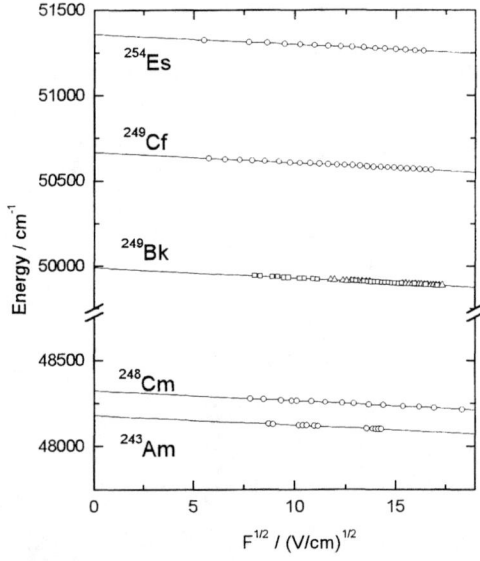

FIGURE 2. Plot of ionization thresholds versus square root of the electric field strength F for five actinide elements. The first ionization potential is obtained by extrapolation to zero field strength.

TABLE 2. First ionization potentials of nine actinide elements. Compiled are the experimental data IP_{exp} obtained in this work, experimental data IP_{exp}^{b} previously known (4,6,7,17), and the predictions IP_{th} by extrapolated spectral properties (8,9).

Actinide element	IP_{exp} / cm^{-1} this work	IP_{exp} / eV this work	IP_{exp}^{b} / cm^{-1}	IP_{th} / cm^{-1} (8,9)
Th	50867(2)	6.3067(2)	50890(20) *(Ref. 17)*	49000(1000)
U	49957(3)	6.1939(3)	49958.4(5) *(Ref. 4)*	48800(600)
Np	50535(2)	6.2655(2)	50536(4) *(Ref. 6)*	49900(1000)
Pu	48601(2)	6.0258(2)	48604(1) *(Ref. 7)*	48890(200)
Am	{ 48180(3) 48183(2)	5.9736(3) 5.9739(2)		48340(80)
Cm	48324(2)	5.9914(2)		48560(200)
Bk	49989(2)	6.1979(2)		50240(200)
Cf	50665(2)	6.2817(2)		50800(200)
Es	51358(5)	6.3676(5)		51800(200)

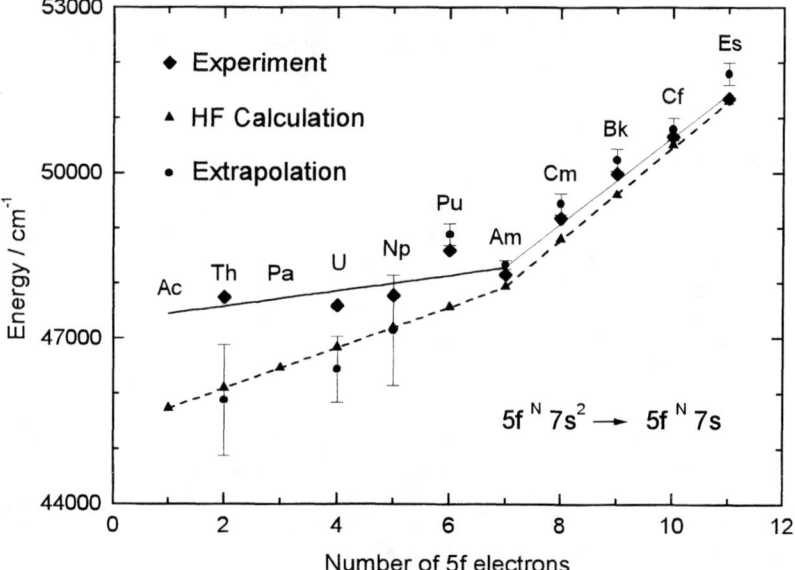

FIGURE 3. Normalized first ionization potentials of actinide elements for the ionization process $5f^{N}7s^{2} \rightarrow 5f^{N}7s$ as a function of N, the number of 5f electrons. Straight lines are drawn through our experimental (solid) and the calculated data (dashed). Plotted are the normalized experimental data ◆ by our group, the normalized values from extrapolated spectral properties ● (8,9), and the results from Hartree-Fock calculations ▲ (10).

the experimental data for the actinides from Am to Es, in fact, support the assumption of a linear dependence, the ionization potentials of the lighter actinides show quite strong deviations from a straight line.

OUTLOOK

The still unknown ionization potentials of Pa and Ac will be measured in the near future. Furthermore, the isotope dependence of the ionization potential will be investigated as a further test of theoretical calculations. The difference in the ionization potential for neighbouring isotopes is expected to be of the order of 0.1 cm^{-1} (23). For elements which offer a great range in neutron number, this effect might be observable, e. g. for isotopes such as ^{232}U ($T_{1/2}$ = 70.0 a) and ^{238}U ($T_{1/2}$ = 4.468×10^9 a) or ^{236}Pu ($T_{1/2}$ = 2.858 a) and ^{244}Pu ($T_{1/2}$ = 8.00×10^7 a).

ACKNOWLEDGEMENTS

This work is funded by the Deutsche Forschungsgemeinschaft. The authors are indebted for the use of ^{249}Bk and ^{254}Es to the Office of Basic Energy Sciences, U.S. Department of Energy, through the transplutonium element production facilities at the Oak Ridge National Laboratory, managed by Lockheed Martin Energy Research Corporation.

REFERENCES

1. P. Pyykkö and J.P. Desclaux, *Acc. Chem. Res.* **12**, 276 (1979).
2. P. Pyykkö, *Chem. Rev.* **88**, 563 (1988).
3. B. Fricke, E. Johnson and G.M. Rivera, *Radiochim. Acta* **62**, 17 (1993).
4. R.W. Solarz et al., *Phys. Rev. Part* **A 14**, 1129 (1976).
5. E.F. Worden, R.W. Solarz, J.A. Paisner and J.G. Conway, *J. Opt. Soc. Am.* **68**, 52 (1978).
6. E.F. Worden and J.G. Conway, *J. Opt. Soc. Am.* **69**, 733 (1979).
7. E. F. Worden et al., *J. Opt. Soc. Am.* Part **B 10**, 1998 (1993).
8. J. Sugar, *J. Chem. Phys.* **59**, 788 (1973).
9. J. Sugar, *J. Chem. Phys.* **60**, 4103 (1974).
10. K. Rajnak and B.W. Shore, *J. Opt. Soc. Am.* **68**, 360 (1978).
11. J.R. Peterson et al., *J. Alloys Comp.*, in press
12. B. Eichler et al., *Radiochim. Acta* **79**, 221 (1997).
13. G. Passler et al., *Kerntechnik* **62**, 85 (1997).
14. F.-J. Urban et al., in: Resonance Ionization Spectroscopy, *Inst. Phys. Conf. Ser.* **128**, 233 (1992).
15. B.H. Bransden and C.J. Joachain, *Physics of Atoms and Molecules*, Longman, London (1983).
16. J. Blaise, J.-F. Wyart, Energy Levels and Atomic Spectra of Actinides. *Tables Internationales de Constantes Sélectionnées*, Université P. et Marie Curie, Vol. **20**, Paris (1992).
17. S.G. Johnson et al., *Spectrochim. Acta* Part **B 47**, 633 (1992).
18. N. Trautmann, *J. Alloys Comp.* **213/214**, 28 (1994).
19. R. Deißenberger et al., *Angew. Chemie Int. Ed. Engl.* **34**, 814 (1995).
20. J. Riegel et al., *Appl. Phys.* Part **B 56**, 275 (1993).
21. S. Köhler et al., *Angew. Chemie Int. Ed. Engl.* **35**, 2856 (1996).
22. S. Köhler et al., *Spectrochim. Acta* Part **B 52**, 717 (1997).
23. B. Fricke, private communication.

SESSION IX
NEW LASER DEVELOPMENTS AND APPLICATIONS

Laser Ion Source for Nuclear Spectroscopic Studies

Yuri Kudryavtsev, Andrej Andreyev, Bart Bruyneel, Serge Franchoo,
Johnny Gentens, Mark Huyse, Kirill Kruglov, Wilhelm Mueller,
Riccardo Raabe, Ils Reusen, Paul Van den Bergh, Piet Van Duppen,
Jan Van Roosbroeck, Ludo Vermeeren and Leonid Weissman.

Instituut voor Kern- en Stralingsfysika, K.U.Leuven
Celestijnenlaan 200 D, B-3001 Leuven, Belgium.

Abstract. An element selective laser ion source has been used to produce beams of exotic radioactive nuclei and to study their decay properties. The operational principle of the ion source is based on selective resonant laser ionization of nuclear reaction products thermalized and neutralized in a noble gas at high pressure. The ion source has been installed at the mass separator (LISOL) which is coupled on line to the cyclotron accelerator at Louvain-la-Neuve. Exotic nickel and cobalt nuclei were produced in proton-induced fission of ^{238}U. The β decay of $^{68-74}$Ni and $^{67-70}$Co isotopes has been studied by means of β-γ and γ-γ spectroscopy.

INTRODUCTION

There is an increasing interest in studies of exotic radioactive nuclei away from the valley of stability. Neutron-rich isotopes close to the neutron drip-line are interesting from the nuclear physics as well as astrophysics point of view. Some of the most interesting nuclei are in the neighborhood of doubly magic nucleus ^{78}Ni.

Development of efficient and element-selective ion sources is a key point in modern studies of exotic nuclei far from stability. Such nuclei are produced in nuclear reactions in very small quantities and usually are overwhelmed by much more abundant isobars. Laser resonant ionization can provide a very efficient and highly selective way to ionize the exotic atoms only.

DESCRIPTION OF THE LASER ION SOURCE

We have developed an element selective ion source based on the laser resonance multistep ionization of nuclear reaction products stopped in a high pressure noble gas (1-7). It has been used at the front end of the LISOL mass separator, which is coupled on-line to the cyclotron CYCLONE at Louvain-la-Neuve.

Fig.1 shows a general view of the Ion Guide Laser Ion Source and the SextuPole Ion Guide (SPIG). In the present setup the skimmer electrode has been replaced by the SPIG, which transports the laser produced ions to the mass separator. This allowed us to improve essentially the ion beam quality and the stopping efficiency by increasing the buffer gas pressure inside the ion source. Recently, a gas purifier (MonoTorr Phase II 3000) has been installed on the gas handling system allowing reduction of the impurity level in the buffer gas down to ppb level.

Radioactive isotopes were produced in a fission reaction of ^{238}U, induced by a 30 MeV proton beam with average intensity up to 6μA. The nuclear reaction products recoiling out of the targets are thermalized and neutralized in buffer gas, He or Ar, and move together towards the exit hole (0.5 mm). A new design of the ion source has been used, specially adopted for fission reactions. Since the recoil nuclei have a high energy of 90 MeV, the inner diameter of the source was increased up to 5 cm. This allowed the stopping of the major part of fission products in 500 mbar of argon. To increase the target thickness for the projectile particles without increasing the possibility to stop the fission products in the targets, two thin ^{238}U targets (10mg/cm^2) are tilted at 20°. Stable nickel or cobalt atoms can be produced inside the source by resistive heating of the corresponding filament. The laser beams can enter the cell longitudinally (as it is shown on the figure) or transversely and ionize the neutralized reaction products in the laser ionization region. The optical system consists of two dye lasers pumped by two XeCl excimer lasers. The laser pulse repetition rate is 200 Hz. Two-color, two-step scheme is used for ionization of cobalt and nickel atoms through the autoionizing state (3). The ions leaving the source are captured by the SPIG and transported towards the extraction electrode (fig.1). The SPIG has been described in (6). It consists of 6 rods (124 mm long and a diameter of 1.5 mm) cylindrically mounted on a sextupole structure with an inner diameter of 3 mm. An oscillating voltage V_{rf} with

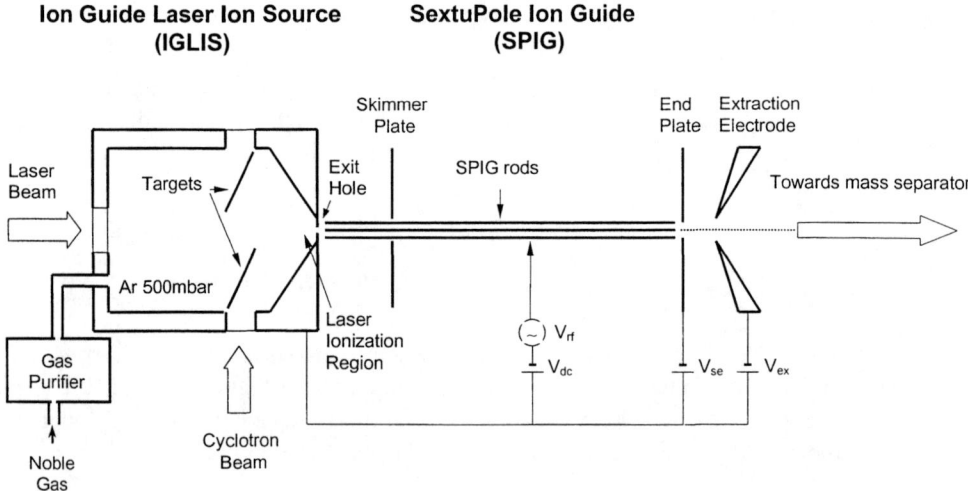

FIGURE 1. General layout of the Ion Guide Laser Ion Source (IGLIS) and the SextuPole Ion Guide (SPIG).

fixed frequency of 4.7 MHz and variable amplitude of 0-350 V is applied to the rods with every rod in antiphase to the neighboring rods. The buffer gas is pumped out efficiently trough the gaps between the rods, while the ions are confined and transported to the extraction electrode with the gas jet velocity. The skimmer plate separates the high vacuum chamber of the separator and low vacuum part around the gas jet. The main difference from the previous skimmer setup consists in the fact that ions can be guided towards the mass separator without applying a DC voltage in a high-pressure zone between the ion source and the SPIG rods. In this case the real ion composition inside the ion source can be studied since molecular ions formed in fast ion-molecular reactions are not decomposed (7). By ionization of stable cobalt (or nickel) atoms and subsequent detection of molecular ions, impurity molecules such as H_2O, O_2, N_2 can be detected in noble gases below the ppb level.

ON-LINE EXPERIMENTS

The IGLIS was used successfully for the study of nuclear decay properties of neutron-rich $^{68-74}$Ni (8) and $^{66-70}$Co isotopes. New nuclear spectroscopic information has been obtained. The mass separated radioactive isotopes were implanted in a movable tape. Two high efficiency germanium γ-ray detectors and thin plastic detectors for β

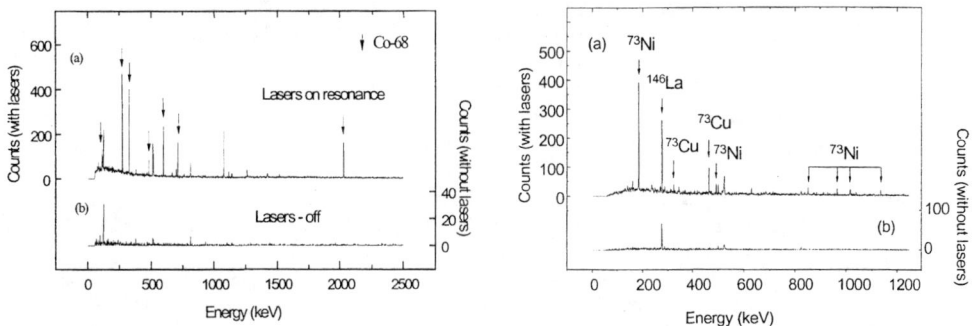

FIGURE 2. β-gated γ-spectra of cobalt-68 (left panel) and nickel-73 (right panel) with lasers tuned on resonance (a) and off resonance (b).

particles were used to measure γ-spectra.

The right panel in figure 2 shows the β-gated γ-spectrum on mass 73 with lasers tuned on- and off-resonance with nickel atoms; no information on the β-delayed γ decay of the nucleus existed till now. The γ-lines at 166 keV and 479 keV are present only in the on-resonance spectrum. The observed Cu-activity in the on-resonance spectrum is produced by the decay of ^{73}Ni. The selectivity of the source is > 35. The measured production cross section of ^{73}Ni is equal to 3 μbarn while the total fission cross section is equal to 2 barn. The line at 258 keV, observed as well in the on- as in the off-resonance spectra, is due to the double charged ^{146}La^{++} ions produced with a very high cross section. The half-lives as well as the experimental reaction cross sections have been measured for the chain of nickel isotopes (A=68-74).

The left panel in the figure 2 shows a β gated ϒ spectrum on mass 68 with lasers tuned on resonance with cobalt atoms (a) and off resonance (b). The arrays indicate peaks that can be associated to the β-decay of ^{68}Co. The comparison of these spectra clearly shows the laser enhancement of Co production.

ACKNOWLEDGEMENTS

We would like to thank the cyclotron group at Louvain-la-Neuve for running and maintaining the accelerator. This work is supported by Inter-University Attraction Poles (I.U.A.P.) Research Programs, financed by Belgian Government and by the Funding for Scientific Research - Flanders (FWO, Belgium). M.H. is Research Director, L.V. Postdoctoral Researcher and S.F. Research Assistant of the FWO.

REFERENCES

1. Van Duppen P., Dendooven P., Huyse M., Vermeeren L., Qamhieh Z.N., Silverans R.E., Vandeweert E., *Hyperfine Interactions* **74**, 193 (1992).
2. Vermeeren L., Bijnens N., Huyse M., Kudryavtsev Yu., Van Duppen P., Wauters J., Qamhieh Z.N., Thoen P., Vandeweert E., Silverans R.E., *Phys. Rev. Lett.* **73**, 1935-1938 (1994).
3. Kudryavtsev Yu., Andrzejewski J., Bijnens N., Franchoo S., Gentens J., Huyse M., Piechaczek A., Szerypo J., Reusen I., Van Duppen P., Van den Bergh P., Vermeeren L., Wauters J., Wöhr A., *Nuclear Instruments and Methods in Physics Research* **B114,** 350-365 (1996).
4. Kudryavtsev Yu., Andrzejewski J., Bijnens N., Franchoo S., Huyse M., Piechaczek A., Szerypo J., Reusen I., Van Duppen P., Vermeeren L., Wauters J., Wöhr A., *Review of Scientific Instruments* **67**, 938-940 (1998).
5. Vermeeren L., Andreyev A., Bijnens N., Breitenbach J., Franchoo S., Gentens J., Huyse M., Kudryavtsev Yu., Piechaczek A., Reusen I., Van Den Bergh P., Van Duppen P., Wöhr A., *Nuclear Instruments and Methods in Physics Research* **B126,** 81-84 (1997).
6. Van den Bergh P., Franchoo S., Gentens J., Huyse M., Kudryavtsev Yu., Piechaczek A., Raabe R., Reusen I., Van Duppen P., Vermeeren L., Wöhr A., *Nuclear Instruments and Methods in Physics Research* **B126,** 194-197 (1997).
7. Kudryavtsev Yu., Franchoo S., Gentens J., Huyse M., Raabe R., Reusen I., Van Duppen P., Van den Bergh P., Vermeeren L., Wöhr A., *Review of Scientific Instruments* **69,** 738-740 (1998).
8. Franchoo S., Huyse M., Kruglov K., Kudryavtsev Yu., Mueller W.F., Raabe R., Reusen I., Van Duppen P., Van Roosbroeck J., Vermeeren L., Wöhr A., Kratz K.-L., Pfeiffer B., Walters W.B. Submitted to *Phys. Rev. Lett.*

SESSION X
SURFACE APPLICATIONS AND SPUTTERING

Resonance Ionization Spectroscopy Investigations of Electronic Processes during Ion-Beam Sputtering of Metal Atoms

Roger E. Silverans and Peter Lievens

Laboratorium voor Vaste-Stoffysica en Magnetisme, K.U. Leuven
Celestijnenlaan 200 D, B-3001 Leuven, Belgium

Abstract. The application of resonance ionization spectroscopy to study the electronic mechanisms involved in the emission of atoms in different electronic states following ion-beam sputtering of metals is discussed. It is shown that resonance ionization of the sputtered atoms offers a very sensitive and generally applicable technique for the measurement of ground and metastable state population partitions and state specific kinetic energy distributions, including those of higly excited metastable states. Experimental results on the sputtering of clean polycrystalline Co and Ni are used to emphasize that information on the role of both the electronic configuration of the atomic states and the bulk electronic structure can be extracted from this kind of data. The observations are discussed in view of resonant electron transfer as a mechanism for the emission of metal atoms in excited states.

INTRODUCTION

The emission of particles in different electronic states during ion-beam sputtering of solid surfaces, although routinely employed for the quantitative analysis of the composition of solids, involves electronic mechanisms that are still poorly understood. In particular the relative importance of several possible processes leading to the ejection of atoms in various atomic excited states is very unclear. This motivated extensive experimental and theoretical investigations aiming at understanding the fundamental aspects of sputtering via characterization of quantum state dependent sputtering yields and emission velocity distributions. In the case of sputtering of metals, mechanisms such as collisional excitation at or above the surface, bond-breaking of quasimolecules, nonradiative deexcitation, and electron transfer between the metal and the sputtered particle have been involved to explain specific observations (1).

Early experiments focussed on the atoms sputtered in short-lived high-lying states detected via their radiative decay. The fluorescence experiments suffered however from the small yields of atoms sputtered in these states, the limited efficiencies of fluorescent light detection, and especially the cumbersome interpretation of the experimental data. These investigations were followed by experiments probing the atoms emitted in metastable states which are much more abundant in the cloud of sputtered particles and are expected to reflect the electronic mechanisms leading to their population with less disturbance. First, laser induced fluorescence (LIF) and Doppler shifted LIF were used. As low detection efficiencies are also inherent to these techniques, they were primarily used to detect atoms in low lying metastable states. An overview is given in ref. 1. Finally, resonance ionization mass spectrometry (RIMS) was introduced in this research domain (2).

In this article we show how RIMS is turned into a generally applicable and at the same time very sensitive technique for the quantitative determination of population partitions and state specific kinetic energy (KE) distributions of all metastable states involved. The method, based on double resonant two-step two-colour laser ionization, is demonstrated by a study of the emission of metastable Co and Ni atoms during ion-beam sputtering of clean polycrystalline surfaces. Both Co I and Ni I have many metastable states in a broad excitation energy range with two distinct types of outer shell electronic configuration, which makes them excellent probes to study electronic processes during emission. The results are interpreted in view of resonant electron transfer as a mechanism for the emission of atoms in excited states.

RIMS APPARATUS FOR SPUTTERING STUDIES

The RIMS apparatus for the sputtering studies reported here consists of a UHV chamber in which an ion gun directs 3 - 15 keV Ar^+ ions onto a centrally located target foil at 45° incidence, in continuous or pulsed mode for resp. population partition and KE distribution measurements. The plume of sputtered particles is intersected, parallel to the foil, by two overlapping laser beams from a pulsed optical parametric oscillator laser system and a pulsed dye laser system delivering laser pulses of ~6 ns in the wavelength range from 225 to 1600 nm with pulse energies up to 50 mJ. The ionized atoms emitted in a polar angle interval of ~10° around the surface normal are extracted into a time-of-flight mass spectrometer and counted. More information on the apparatus is given in ref. 3 and it is fully described in ref. 4.

DOUBLE RESONANT TWO-STEP TWO-COLOUR LASER IONIZATION

The experimental procedure is based on two-step two-colour resonance laser ionization processes. The first step sequentially saturates the excitation of atoms, in different metastable states, to *the same intermediate state*. A second independent laser ionization step is applied to ionize the atoms and count them. By this procedure, the relative photoion intensities directly reflect the population partition of the involved metastable states: only saturation of the excitation steps is required, which is already obtained with modest laser pulse energies. As only part of the metastable states can be coupled to a particular intermediate state due to angular momentum selection rules, population distributions of overlapping subsets of metastable states are combined.

The very high sensitivity of the measuring procedure is obtained by tuning the ionizing laser to resonant transitions into autoionizing states, situated just above the ionization limit. The high ionization cross sections of such transitions, comparable to those of bound - bound transitions, in most cases even allow to saturate the ionization step (5).

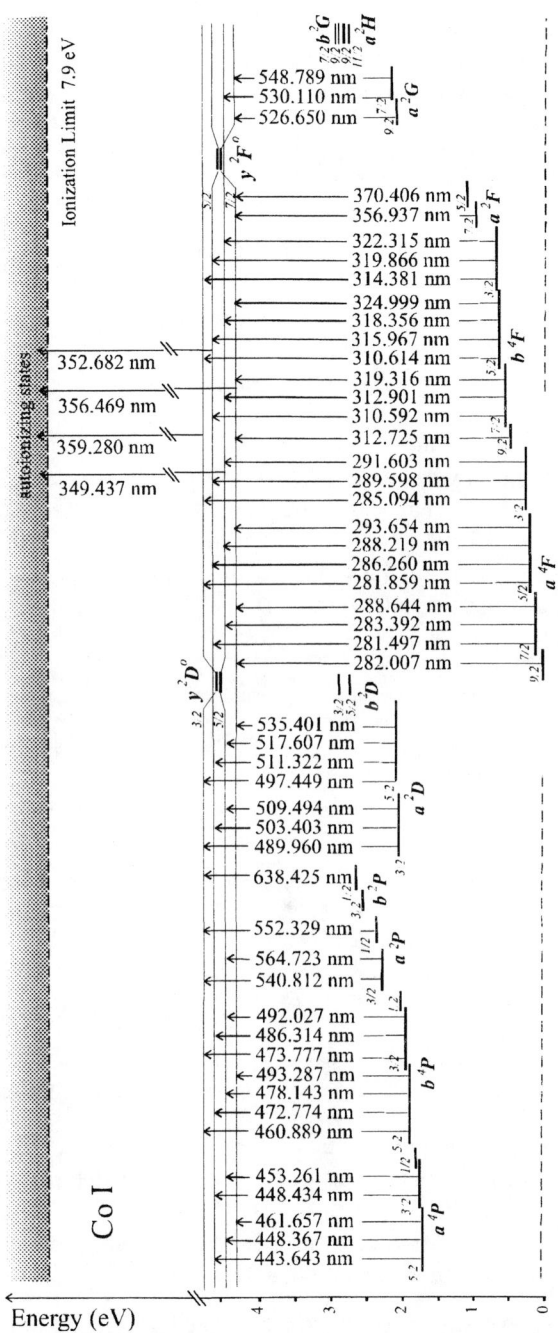

FIGURE 1. Partial level scheme of Co I. All metastable states and the employed double resonance ionization schemes including the wavelengths of the optical transitions are shown.

The procedure is exemplified in FIGURE 1 where a partial level scheme of Co I is presented, showing all metastable states and the triplets of intermediate states together with the excitation steps used in the experiments dealt with below.

RIMS STUDY OF ION BEAM SPUTTERING OF CO AND NI

Metastable state population partitions

The measured population distributions over the ground and metastable states of Co and Ni atoms emitted following 12 keV Ar ion-beam sputtering of clean polycrystalline Co and Ni target foils are presented in FIGURE 2, giving the relative populations of the states as a function of their energy above the ground state. These are populations integrated over all emission velocities. In FIGURE 2, Co states belonging to multiplets with a $[Ar]3d^74s^2$ electronic configuration are represented by filled dots (in order of increasing excitation energy: the ground state multiplet a^4F_J, and the a^4P_J and a^2G_J states), whereas open dots are used for states with a $[Ar]3d^84s^1$ electronic configuration (the b^4F_J, b^4P_J, and a^2D_J states). A similar representation is used for the Ni states: filled dots for states with a $3d^84s^2$ outer shell electronic configuration (the ground state multiplet a^3F_J, and the b^1D_J, $a^3P_{1,2}$ and a^1G_4 states), and open dots for $3d^94s^1$ (the a^3D_J and a^1D_2 states).

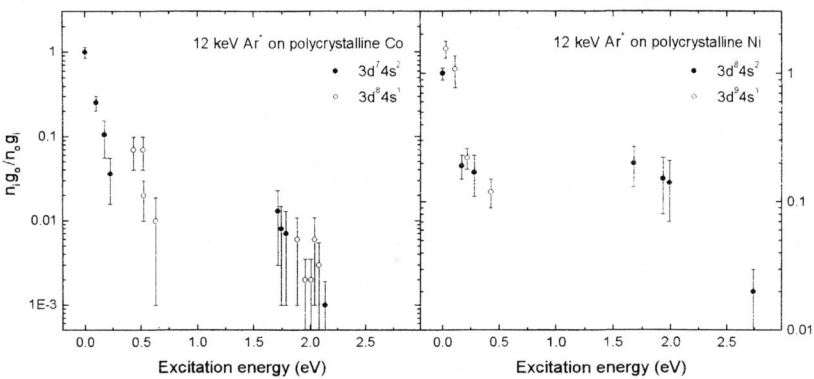

FIGURE 2. Population distribution of sputtered Co and Ni atoms versus excitation energy. The populations (n_i) divided by their statistical weight (g_i) are given relative to the ground state.

The Co and Ni population distributions show distinct common features. First, the high lying states with excitation from 1.5 to 2 eV exhibit a surprisingly high population. Second, the populations of the low lying states with $3d^x4s^1$ configuration (open dots) are systematically shifted to higher values with respect to the states with $3d^{x-1}4s^2$ configuration (filled dots).

State specific kinetic energy distributions

The extreme efficiency of the ionization schemes not only allows to measure the occupation of weakly populated states, but also to measure the state specific kinetic energy (KE) distribution of atoms in these states. They are derived from the photoion intensities measured as a function of delay time between the primary ion pulse impinging on the target and the ionizing laser pulses.

FIGURE 3 shows two examples of these KE distributions of Co states, examples for Ni states can be found in ref. 6, contribution to these proceedings. Again, as for the populations, the two types of low lying metastable states were found to exhibit a different behaviour: the states with the $3d^x4s^1$ configuration have kinetic energy distributions which peak at higher kinetic energy and fall off *less steeply* than those of the $3d^{x-1}4s^2$ states. The relative populations of the $3d^x4s^1$ states vs. the $3d^{x-1}4s^2$ states increase drastically as a function of emission energy. The KE distributions of the high-lying Co and Ni states (for Ni only $3d^84s^2$ states are present) are similar to those of the low lying $3d^{x-1}4s^2$ states.

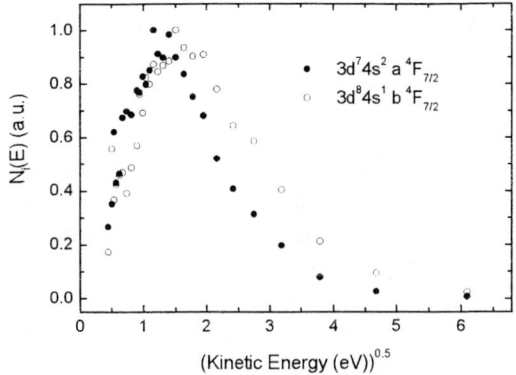

FIGURE 3. Examples of state specific KE distributions of sputtered Co atoms: the a $^4F_{7/2}$ ($3d^84s^2$) state (closed symbols) and the b $^4F_{7/2}$ ($3d^74s^1$) state (open symbols).

ELECTRONIC PROCESSES DURING EMISSION OF SPUTTERED PARTICLES

Several of our observations cannot be explained by the most popular models used earlier to interpret population partitions and KE distributions of atoms sputtered in ground and metastable states. The preferential sputtering into states with a specific electronic configuration and the high population of highly excited states do not fit to the model in which population of excited states is the result of collision induced excitation at or above the surface. The nonradiative deexcitation model including the shielded state concept is in conflict with, e.g., our observation that the states of the same multiplet with "open" $3d^x4s^1$ configuration all show the same KE distribution. Similar discrepancies were found in recent related RIMS investigations of sputtering of single crystalline Ni (7) and polycrystalline Ag (8). An overview of the success and failure of these models is given in ref. 9. The strongly enhanced population of the low-lying $3d^94s^1$ states of Ni and their broader KE distributions have been explained (7) by the argument that this electron configuration should be the predominant bonding state of the metal as the valence band electronic structure of bulk Ni has a similar character. We have interpreted our population partition data of Ni by involving resonant electron transfer in the emission process (3). In the following, we present a combined interpretation of the population partitions and KE distributions of both Co an Ni, based on multichannel resonant electron transfer using a qualitative and certainly oversimplified modelling.

Resonant electron transfer

The basic starting point of a resonant electron transfer (RET) model is that the metal consists of ions immersed in a sea of delocalized valence electrons of the individual atoms occupying, in atomic terminology, the metal valence band states. As a result of the collision cascade created by the incoming particles, a sputtered particle escapes from the metal surface as a positive ion, the same charge state as in the solid, which can become neutralized by picking up an electron from the valence band of the solid. Resonant electron transfer is by far the most probable electron pick-up process. So the sputtering of a neutral atom is regarded as the evolution in time of an initially strongly coupled system consisting of a metal and an ion, in which an electron evolves from an initial valence band state to a final atomic state (10).

For a crude qualitative description we consider a metal with a broad valence band and work function Φ. The free atoms have electron configurations a_i with excitation energies E_i and the ionization potential is I. Close to the surface the atomic levels are shifted upward due to the classical image potential (the effective ionization potential is lowered) according to $\delta E_i(z_s) \cong e^2 / 16 \pi \varepsilon_0 (z_s + k)$ with z_s the distance to the surface and k taking into account the finite screening length of the metal. The energy levels are broadened by the interaction with the electron states of the metal. The broadening is proportional to the interaction matrix element between the atomic state and the metal valence band states: a strongly coupled state will be strongly broadened. The broadening decreases with the distance to the surface due to the decreasing spatial overlap between atomic and metallic wave functions. Usually an exponential dependence is assumed: $\Delta E_i(z) = \Delta_i(0) \exp(-\alpha_i z)$. We take $z = 0$ as the distance above the surface at which the atomic electron wave function takes its identity, assuming that shorter to the surface the atomic states are too strongly disturbed and hybridized. The shift is typically a few eV close to the surface and the broadening $\Delta E_i(z_s = 0) \cong$ 1-2 eV for strong coupling (11). For an atom receding from the surface with a perpendicular velocity component v, they are time dependent according to $z = vt$.

The case for which only one atomic state (labeled state "1") is below the Fermi level can be treated quantitatively. If the shifted and broadened atomic level is completely below the Fermi level, even at very short distances, the resonant neutralization rate is correlated to the broadening of the level according to:

$$\Gamma_1 = \Delta E_1(vt) / \hbar = \frac{\Delta_1(0)}{\hbar} \exp(-\alpha_1 vt) \tag{1}$$

In this case, the occupation probability of the atomic state as a function of time is given by the rate equation:

$$\frac{dP_1(t)}{dt} = [1-P_1(t)]\Gamma_1 = [1-P_1(t)]\frac{\Delta_1(0)}{\hbar}exp(-\alpha_1 v t) \qquad (2)$$

Integration gives the occupation probability after complete decoupling at large distance:

$$P_1(\infty) = 1 - exp\left(-\frac{\Delta_1(0)}{\hbar \alpha_1 v}\right) \qquad (3)$$

Thus the electron transfer probability to this state depends on *i)* the coupling strength between the atomic state and the metal valence electrons, *ii)* the extent of spatial overlap between atomic and metallic electron wave functions, and *iii)* the perpendicular velocity of the sputtered particle.

If the atomic level is below the Fermi level at large distances but at short distances partially or totally above it, the neutralization rate close to the surface will be smaller than given by eq. 1 or zero, as the broadened level faces unoccupied electron states in the metal. The smaller the energy difference $\Delta E_{F,a_i}$ between the Fermi level and the atomic level, the smaller the final population of the atomic state.

The experimental determination of the *ionic fraction* in systems to which this case is applicable, and for which an exponential dependence on $\Delta E_{F,a_i} = I - \Phi - E_i$ has been observed, has led to the electron tunneling model for the emission of ions (12). Also, single channel resonant electron transfer is well documented, both theoretically and experimentally, to be the dominant charge exchange mechanism in *atom - metal scattering* (11). In the case of grazing incidence or hyperthermal scattering, the perpendicular velocities of the particles are comparable to sputtered particles, so both processes are very similar.

Modelling the case of several more or less equivalent electron transfer channels is problematic. Theoretical treatments, and their experimental verification, have been restricted to scattering of atoms with at most one atomic level far below the Fermi level (13-15). In our case of multichannel RET we have to deal with a set of coupled rate equations:

$$\frac{dP_i(t)}{dt} = P_{ion}(t)\Gamma_i = P_{ion}(t)\frac{\Delta_i(0)}{\hbar}exp(-\alpha_i v t) \qquad 1 \leq i \leq nr.\ of\ states$$

$$\sum_i P_i(t) + P_{ion}(t) = 1 \qquad (4)$$

The different transfer channels are competing with each other. Although solving this set of rate equations is in most cases impossible due to the lack of adequate knowledge about the differences in coupling strengths and spatial extension of the wave functions, some general trends can be formulated.

Concerning the population partitions, all states lying within the energy window of the metal valence band will have a substantial probability of being populated, and secondly, those atomic states with electronic wave functions similar to those of the valence electrons in the metal, will be preferentially populated due to the higher coupling strength.

The final occupation probabilities of the states also depend on the velocity of the escaping particle. For very low velocities, the electron transfer will be completed close to the surface, i.e. at a distance where all transfer channels are still open. The faster the particle recedes from the surface, the larger the mean distance at which the electron transfer occurs. As channels with smaller coupling strenghts (less broadening) and less extended overlap of the wavefunctions get blocked closer to the surface, higher velocity implies a higher population probability of the strongly coupled states.

It should be mentioned here that RET has been used in ref. 16 to explain the narrow KE distribution of a Ag state with 3.5 eV excitation energy. In that work it was assumed that the sputtered particles were ejected either as atoms in their ground state or as excited ions such that the result of a single channel electron transfer treatment (see eq. 3) could be used to interpret the data.

Comparison of experimental data with the RET model

The bulk work functions of Co and Ni are 5 eV and 5.2 eV resp., and the ionization potentials of the atoms are 7.9 eV and 7.7 eV, so their metastable states with excitation energy up to 2.9 eV and 2.5 eV resp. lie below the Fermi level at large distances. The overall picture of the population distributions, see FIGURE 2, therefore agrees with the RET model. All states within the energy window of the metal valence band are substantially populated, whereas the population of the Ni state with an excitation energy of 2.74 eV, the a^1G_4 state, is drastically lower. The observed decrease in population with excitation energy for states within each low-lying multiplet indicates that the competition between the different neutralization channels *within* one multiplet is governed by $\Delta E_{F,a_i}$, suggesting that the trend given by the single atomic level description survives a multichannel situation.

The observed dependence on the electronic configuration of the population and KE distributions can be linked to the resonant electron transfer picture considering the bulk metallic band structure of Ni and Co. The valence band electronic structure of Ni is calculated to be $3d^{9.4}4s^{0.6}$ and Co has a similar structure $\sim 3d^{8.4}4s^{0.6}$ (17,18). Thus there is a closer correspondance of the electron configuration of the $3d^x4s^1$ states with the valence band of the sputtered metals. The observed preferential population of these states is therefore in agreement with the RET model favouring population of states with high coupling strength. The observation that the KE distributions of the low-lying $3d^x4s^1$ states are shifted to higher energies compared to those of the low-lying $3d^{x-1}4s^2$ states is also compatible with the qualitative prediction of the RET model that the higher the velocity of the ejected particle, the higher the probability of electron transfer via channels with strong coupling and extended overlap of wave functions. For highly excited states, this effect might be blurred by the fact that at short distances, the broadened levels are partly above the

Fermi level. A more quantitative discussion of the KE distributions of Ni, involving a comparison with the KE distributions predicted by collision cascade theory (19), is given in a separate contribution to this conference (6).

In summary, we showed that RIMS is a generally applicable and at the same time very sensitive technique to measure the population partitions and the state specific KE distributions of sputtered metal atoms. The results of sputtered Co and Ni have been used to demonstrate that this type of experiments provides information on the electronic mechanism during the emission process. The observations were discussed in view of resonant electron transfer as a mechanism for the emission of metal atoms in excited states.

ACKNOWLEDGMENTS

This work is financially supported by the Fund for Scientific Research - Flanders (F.W.O.), the Flemish Concerted Action (G.O.A.) Research Programme and the Interuniversity Poles of Attraction Programme - Belgian State, Prime Minister's Office - Federal Office for Scientific, Technical and Cultural Affairs. P.L. is a Postdoctoral Research Fellow of the F.W.O.

REFERENCES

1. M.L. Yu, in *Sputtering by Particle Bombardement III*, eds. R. Behrisch and K. Wittmaack, (Springer-Verlag, Berlin, 1991) Chap. 3, *and refs. therein*.
2. B.J. Craig, J.P. Baxter, J. Singh, G.A. Schick, P.H. Kobrin, B.J. Garrison, and N. Winograd, Phys. Rev. Lett. **57**, 1351 (1986).
3. E. Vandeweert, V. Philipsen, W. Bouwen, P. Thoen, H. Weidele, R.E. Silverans, and P. Lievens, Phys. Rev. Lett. **78**, 138 (1997).
4. P. Lievens, E. Vandeweert, V. Philipsen, and R.E. Silverans, submitted to J. Chem Phys.
5. P. Lievens, E. Vandeweert, P. Thoen, and R.E. Silverans, Phys. Rev. A **54**, 2253 (1996).
6. V. Philipsen, J. Bastiaansen, E. Vandeweert, P. Lievens and R.E. Silverans, "Electron Configuration Dependence of Kinetic Energy Distributions of Ion-Beam Sputtered Ni Atoms studied by Double Resonant Laser Ionization", contribution to this conference, *Proceedings RIS-98*.
7. C. He, Z. Postawa, S.W. Rosencrance, R. Chatterjee, B.J. Garrison, and N. Winograd, Phys. Rev. Lett. **75**, 3950 (1995).
8. W. Berthold and A. Wucher, Phys. Rev. Lett. **76**, 2181 (1996).
9. B.J. Garrison, N. Winograd, R. Chatterjee, Z. Postawa, A. Wucher, E. Vandeweert, P.Lievens, V. Philipsen, and R.E. Silverans, Rap. Comm. Mass Spectrometry (in press).
10. E. Veje, Phys. Rev. B **28**, 5029 (1983), *and refs. therein*.
11. J. Los and J.J.C. Geerlings, Phys. Reports **190**, 133 (1990).
12. M.L. Yu and N.D. Lang, Phys. Rev. Lett. **50**, 127 (1983).
13. D.C. Langreth and P. Nordlander, Phys. Rev. B **43**, 2541 (1991).
14. J. B. Marston, D.R. Andersson, E.R. Behringer, B.H. Cooper, C.A. DiRubio, G.A. Kimmel, and C. Richardson, Phys. Rev. B **48**, 7809 (1993).
15. A.G. Borisov, D. Teillet-Billy, J.P. Gauyacq, H. Winter, and G. Dierkes, Phys. Rev. B **54**, 17166 (1996).
16. A. Wucher and Z. Sroubek, Phys. Rev. B **55**, 780 (1997).
17. J.W.D. Connolly, Phys. Rev. **159**, 145 (1967).
18. L.F. Mattheiss, Phys. Rev. **134**, A970 (1964).
19. P. Sigmund, Phys. Rev. **184**, 383 (1969).

Photoionization Studies of Small Biological Molecules Using Femtosecond Laser Pulses

V. Vorsa, K. F. Willey, T. Kono, N. Winograd

Department of Chemistry, The Pennsylvania State University
184 Materials Research Institute Building, University Park, PA 16802, USA

Abstract. We report on the fs photoionization and dissociation dynamics of small biological molecules sputtered from surfaces and in the gas phase. Femtosecond postionization studies of ion-beam desorbed dopamine at 800, 400, 267, and 200 nm are presented and compared to the fragmentation patterns observed for gas phase dopamine under similar laser conditions. Differences in the amount of fragment ions originating from the molecular ion suggest that dopamine undergoes extensive fragmentation during the sputtering process. We also present fs postionization data for the amino acid alanine at 267 and 200 nm along with its positive SIMS mass spectrum.

INTRODUCTION

Laser postionization of surface-desorbed neutral species has emerged as a highly sensitive surface analysis technique (1) that is particularly well suited for secondary ion mass spectrometry (SIMS) experiments. Most of the studies, thus far, have concentrated on resonance enhanced multiphoton ionization (REMPI) of the desorbed neutral species with nanosecond lasers. Although this REMPI has proven to be successful in the detection of atomic species (2), molecular species have proven to be more challenging. This is mainly due to a high degree of fragmentation particularly for larger organic and biological molecules.

Femtosecond postionization has shown great potential for producing large ionization efficiencies while at the same time keeping fragmentation to a minimum (3). In this paper we present the fs photoionization studies of the neurotransmitter dopamine at 800, 400, 267, and 200 nm and amino acid alanine at 267 and 200 nm.

EXPERIMENTAL METHOD

All experiments are carried out using a TOF-SIMS apparatus in conjunction with a Ti:sapphire-based laser system. The sample is desorbed from the surface with a pulsed 25 keV Ga^+ ion beam. The desorbed species are intersected with a laser beam in the extraction region of the mass spectrometer where they are pulse-extracted into the reflectron-based TOF mass spectrometer. The 1 kHz Ti:sapphire femtosecond laser system (3,4) (Clark-MXR, Inc.) produces 100 fs pulses centered on 800 nm containing 3.5 mJ of energy per pulse. This output is directed into a harmonic generator to produce 400, 267, and 200 nm wavelengths. The pulse widths for these wavelengths are ~200, 250, and 400 fs, respectively.

RESULTS AND DISCUSSION

Shown in Fig. 1(a-d) (top) are the gas phase mass spectra of dopamine following fs photoionization at the four wavelengths taken at near threshold laser conditions. Three major and one minor species are observed including the molecular ion M^+ at m/z 153, a fragment ion McL^+ at m/z 124 (see below), and a side chain fragment $CH_2NH_2^+$ at m/z 30. The aminomethyl side chain fragment has a particularly low ionization potential of 6.2 eV. A much weaker fragment ion channel $M^+-CH_2NH_2$ is also observed at m/z 123. The m/z 124 product ion appears to arise from a radical site rearrangement reaction. This type of rearrangement fragment (McL^+) is frequently observed in the electron impact mass spectra of phenylalkanes where the side chain is propyl or larger (5) and is commonly referred to as a McLafferty rearrangement Eq. (1). Note, this reaction takes place in the ion-state rather than in the neutral.

$$\text{(scheme)} \tag{1}$$

The minimum number of photons required to ionize dopamine (IP = 8.18 eV (6)) at 800, 400, 267, and 200 nm is 6, 3, 2, and 2, respectively. Energetically, this leads to an excess of 1.1 eV for the three longer wavelengths and 4.2 eV for 200 nm. From power dependence studies, we believe the 800 nm mass spectra results in more than 6 photons being deposited into the molecule. This extra energy, therefore, results in a mass spectrum that more closely resembles the 200 nm spectrum.

Figure 1(a-d) (bottom) displays the postionization spectra of dopamine taken at threshold laser intensities. A much higher degree of fragmentation and spectral congestion (particularly at lower masses) is observed for 800 and 400 nm postionization. The 267 nm data exhibits simpler a spectrum containing the characteristic dopamine fragments as seen in the gas phase data while 200 nm postionization produces mostly $CH_2NH_2^+$. The dominant peak at all excitation wavelengths and laser intensities is $CH_2NH_2^+$. The m/z 123 and 124 fragments are both present, however, the yield of the m/z 123 fragment is much greater than the McL^+ fragment. In the gas phase experiments the McL^+ fragment is one to two orders of magnitude greater. Gas phase power dependence studies show identical photon orders for the M^+ and McL^+ ions, while postionization power dependence studies show the M^+ ion exhibiting a substantially lower photon order than the McL^+ ion. This suggests that extensive fragmentation of dopamine is occurring prior to postionization. Moreover, the molecular ion yield is very low (<1% of the characteristic fragments). Power dependence studies of the fragments exhibit strikingly low photon orders (~3 for 800 nm and ~2 for 400 nm) suggesting that the

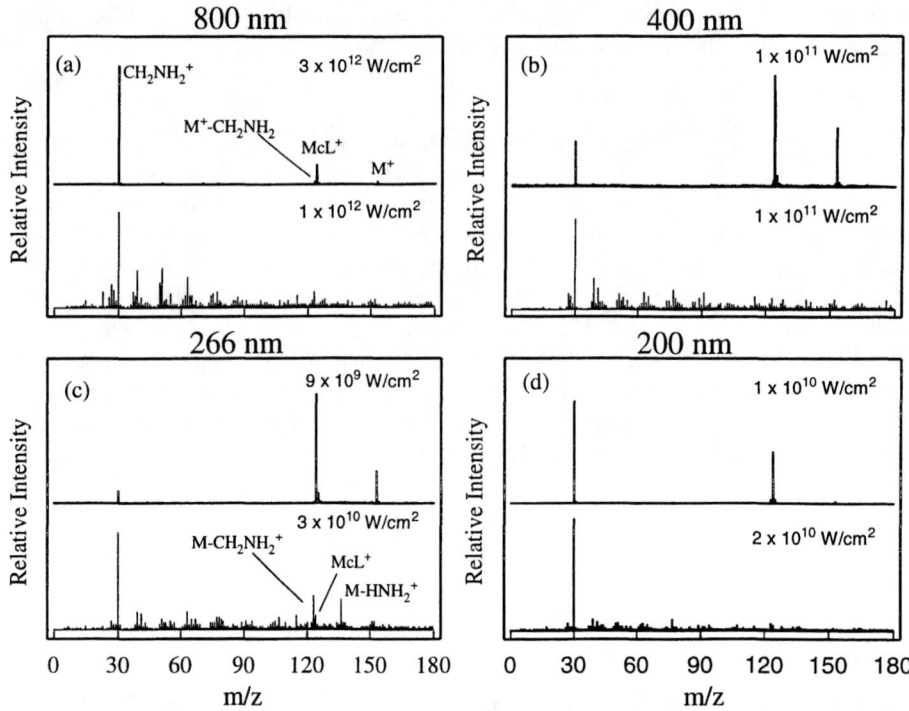

FIGURE 1. Femtosecond ionization-dissociation mass spectra of gas-phase dopamine (a) - (d) (top) and ion desorbed dopamine (a) - (d) (bottom) following 800, 400, 267, and 200 nm excitation.

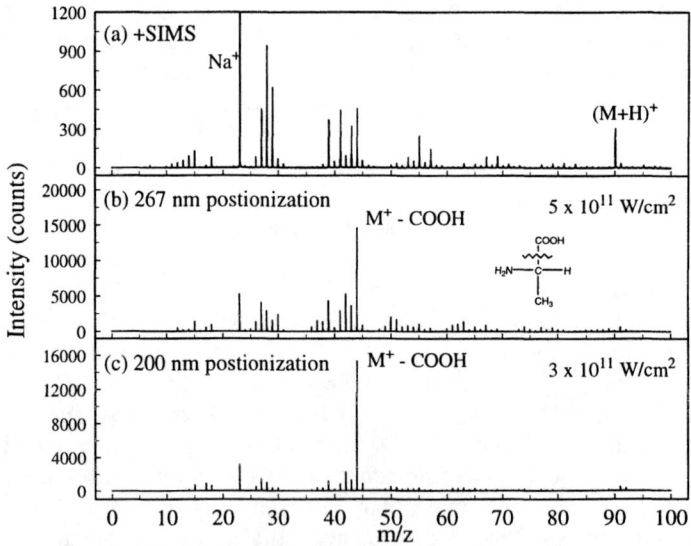

FIGURE 2. Mass spectrum of alanine (IP = 8.9 eV) under identical sputtering conditions. (a) +SIMS, (b) fs postionization at 267 nm, and (c) fs postionization at 200 nm.

desorbed species contain a high degree of internal excitation. A comparison of ion yields between postionization at 267 nm and SIMS was made. With a laser intensity of ~7 x 10^{11} W/cm^2 we measured up to a 20-fold increase in the amount of characteristic ions produced with postionization over SIMS under identical ion sputtering conditions.

Positive SIMS and postionization spectra (267 and 200 nm) of alanine recorded under identical sputtering conditions are shown Fig. 2. The protonated molecular ion $(M+H)^+$ can clearly be seen in the SIMS spectra at m/z 90. The postionization spectra, on the other hand, show a characteristic fragment at m/z 44 corresponding to the loss of the carboxylic acid group, but very little molecular ion. Comparing the molecular ion yield for SIMS to the M^+-COOH ion yield for postionization we find 1-2 orders of magnitude increase in signal. We observe a loss of carboxylic acid in other amino acids with alkyl side chains with 200 nm excitation, but only in certain cases for 267 nm excitation. This loss appears to be correlated with the removal of an electron from the nitrogen lone pair orbital of the amine group which corresponds to the first IP for amino acids with alkyl side chains.

CONCLUSIONS

We have carried out femtosecond photoionization studies on gas phase and ion-beam desorbed dopamine at 800, 400, 267, and 200 nm. Ion-desorbed dopamine is found to undergo extensive fragmentation and the fragments are found to contain a high degree of internal excitation. We find up to a 20-fold increase in ion yield with 267 nm postionization over SIMS. Postionization at 267 and 200 nm produces large yields of M^+-COOH in alkyl amino acids. We are currently investigating the fs postionization of other amino acids and biologically important molecules.

ACKNOWLEDGEMENTS

We gratefully acknowledge the financial support of the National Institutes of Health and the National Science Foundation.

REFERENCES

1. Winograd, N., *Anal. Chem.* **65**, 622 A (1993).
2. Parks, J. E., *Lasers and Mass Spectrometry*; Lubman, D. M., Ed., Oxford : Oxford University Press, 1990, ch. 2 pp. 37-64.
3. Brummel, C. L., Willey, K. F., Vickerman, J. C., Winograd, N., *Int. J. Mass Spectrom. Ion Processes* **143**, 257 (1995).
4. Willey, K. F., Vorsa, V., Braun, R. M., Winograd N., *Rap. Commun. Mass Spectrom.*, (in press).
5. McLafferty, F. W., Turecek, F., *Interpretation of Mass Spectra*, 4th ed., Sausalito, California: University Science Books, 1993, Ch. 9 pp. 238-240.
6. Choi, Y., Lubman, D. M., *Anal. Chem.* 64, 2726 (1992).

State-selective laser photoionization of neutral benzene molecules ejected from keV ion bombarded $C_6H_6/Ag\{111\}$

C. A. Meserole, E. Vandeweert, R. Chatterjee,[1] B.R. Chakraborty,[2] B.J. Garrison, and N. Winograd

Department of Chemistry, The Pennsylvanian State University, University Park, PA 16802

Z. Postawa

Institute of Physics, Jagellonian University, ul. Reymonta 4, PL 30-059 Krakow 16 Poland

Abstract. One-color two-photon ionization spectroscopy was used to probe state-selectively neutral benzene molecules desorbed from a benzene overlayer physisorbed on a $Ag\{111\}$ surface upon 8 keV Ar^+ bombardment. Time distributions were measured for benzene molecules ejected in the zero level of the molecular ground state and in the first state of the $v_6^{''}$ vibration. These distributions are found to show a strong dependence both on the internal energy of the ejected molecules and the degree of coverage of the Ag surface. Up to monolayer coverages, benzene molecules are ejected by direct collisions with Ag particles sputtered from the underlying substrate. Molecules with higher internal energy leave the surface with a distribution shifted towards lower flight times. At multilayer coverages, a second, thermal-like ejection mechanism gains significance. It is suggested that only molecules excited near the benzene-vacuum interface, survive the ejection process without being deexcited.

INTRODUCTION

Analytical techniques based on ion beam induced desorption of molecules are routinely used for the surface-analysis of complex chemical and biological systems (1). Up to now, it is still poorly understood how *intact* molecules can be desorbed from a surface upon bombardment with primary ions with kinetic energies which are many orders of magnitude higher than the energy contained in the chemical bonds. In this work, we study the desorption of *neutral* benzene molecules from a cold $Ag\{111\}$ surface during keV Ar^+ ion bombardment. Benzene/$Ag\{111\}$ provides a model system in which the organic molecules are physisorbed on a fairly unreactive metallic surface. The C_6H_6 exposure can be varied from submonolayer to multilayer coverages in order to probe the influence of the local molecular environment on the desorption process. Moreover, the energy level scheme for benzene is well known and its ultraviolet spectrum is experimentally easily accessible. We used resonant one-color two-step

---

[1] permanent address: Corporate Research Laboratories, 3M Center, Bldg. 201-2S-16, St. Paul, MN 55144

[2] on sabbatical leave from the National Physical Laboratory, New Delhi, India.

CP454, *Resonance Ionization Spectroscopy*
edited by J. C. Vickerman, I. Lyon, N. P. Lockyer, and J. E. Parks
© 1998 The American Institute of Physics   1-56396-810-X/98/$15.00

photoionization to detect selectively neutral benzene molecules in the zero level and in the first vibrationally excited state of the molecular electronic ground state. To our knowledge, these are the first experiments where the desorption process of neutral benzene molecules from molecular overlayers is state-selectively monitored.

EXPERIMENTAL SETUP

The experimental setup and procedure are described in detail elsewhere (2). In short, benzene was adsorbed on a clean Ag{111} surface cooled to 120 K. Several freeze-pump-thaw cycles were applied before dosing to remove dissolved contaminants from the benzene. The benzene exposure was controlled by monitoring the chamber pressure and the dosing time.

The desorption process was initiated by bombarding the sample with an 8 keV, 250 ns Ar^+ pulse, at 45° incidence, focused to a spot with a diameter of 3 mm. The primary ion dose was kept sufficiently low so as to eliminate effects of surface damage. Desorbed particles were detected by multiphoton ionization in combination with time-of-flight mass spectrometry using a gated microchannel plate detector. The mass spectrum showed prominent peaks at m/z 54 (C_4H_4 fragment), 78 (molecular C_6H_6), 108 (Ag) and 216 (Ag_2). The laser beam was focused to a ribbon shape approximately 1 cm above the crystal surface. The density of the particles in the laser ionization volume was recorded as function of time by systematically varying the time delay between the primary ion and laser pulses. Resonant two-photon one-color photoionization of the ejected benzene molecules was achieved by tuning the frequency-doubled output of a Nd:YAG pumped dye laser to drive the 6_0^1 transition at 259.01 nm originating from the zero level of the molecular ground state, and the 6_1^0 (266.82 nm) and $6_1^0 1_0^1$ (260.37 nm) transitions starting from the first vibrationally excited state of the v_6'' mode (3,4).

RESULTS AND DISCUSSION

In Fig. 1 the time density distributions are shown both for molecules ejected in the ground and the excited state upon 8 keV Ar^+ bombardment of C_6H_6/Ag{111} at a submonolayer (Fig. 1a) and a multilayer (Fig. 1b) coverage of the surface. The width and the position of the maximum of the time distributions are strongly dependent on the benzene coverage. We first discuss the distributions for molecules ejected in the *zero level of the molecular electronic ground state*. Since most molecules leave the surface in this state, it is not unexpected that their time distributions are very similar to those obtained when the desorbed molecules are non-resonantly photoionized (i.e. not state selective) (2). At submonolayer coverage the peak in the time distribution (labeled as A in Fig. 1) corresponds to benzene molecules that leave the surface with kinetic energies of about 1 eV. As the coverage increases to a complete monolayer, the most probable kinetic energy of the ejected molecules decreases to 0.25 eV. These results, in combination with experimentally obtained angular distributions and molecular dynamics simulations (5), indicate that the ejection process at these coverages is largely ballistic in nature. As the primary ion hits the sample, a collision cascade in the Ag crystal is initiated and the benzene molecules are ejected by direct collisions with the departing Ag particles. As the thickness of the organic overlayer increases, a second maximum (peak B in Fig. 1b) appears in the time distribution. This maximum corresponds to benzene molecules that desorb from the surface with an extremely low kinetic energy of about 0.04 eV. With increasing coverage, the importance of prompt ejection of benzene molecules by substrate originating particles decreases while other mechanisms become more likely. Plausible scenarios include desorption of the molecules due to exothermic reactions between molecular fragments created by the impact of the primary particle or the ejection upon collisions between molecules and fragments.

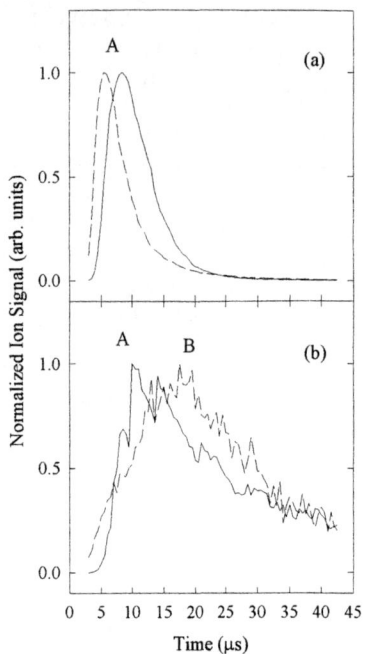

FIGURE 1. State-selective time density distributions of neutral benzene molecules desorbed from C_6H_6/Ag {111} upon 8 keV Ar^+ bombardment at (a) 1L exposure, resulting in a submonolayer coverage of the Ag crystal surface and (b) 200 L exposure yielding a multilayer coverage. The solid line represents the distributions obtained for molecules ejected in the zero level of the molecular ground state, while the dashed line shows the distributions for vibrationally excited molecules.

The time distributions of benzene molecules ejected in the *vibrationally excited state* show that these molecules are more sensitive to underlying ejection mechanisms. At low coverages, the time distributions of the excited molecules are narrower and peak earlier compared to the one for molecules ejected in the ground state. The kinetic energy distributions are thus shifted to higher energies for molecules with higher internal energies. These experiments confirm the trend predicted by molecular dynamics simulations that have delineated the ejection mechanisms of molecules from C_6H_6/Ag{111} from a microscopic point of view (5).

The time distributions obtained for vibrationally excited molecules desorbed from thick organic overlayers show almost solely a peak corresponding to molecules departing with low kinetic energies. The mechanisms that govern slow desorption might thus favor the emission of vibrationally excited molecules. Alternatively, since molecules with high kinetic energies are predominantly created near the Ag surface by collisions with substrate particles, the absence of these fast molecules from the time distributions might be indicative that only molecules that get excited close to the benzene-vacuum interface have a sufficiently large probability to survive the process in that vibrationally excited state. Collisional quenching or unimolecular decomposition of the molecules before detection are among the possible deexcitation mechanisms.

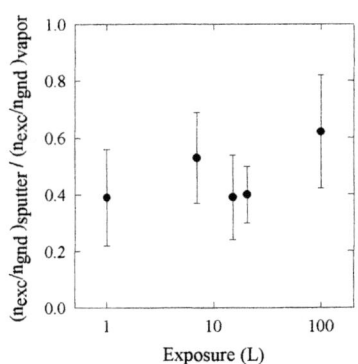

FIGURE 2. Ratio of the relative benzene populations on the vibrationally excited state (n_{exc}) the ground state (n_{gnd}) for the ion induced desorption from C_6H_6/Ag {111} and the vapor phase, as function of the benzene exposure.

The state-selective time distributions can be used to determine the ratio of the vibrationally excited state relative to the ground state population. Since one-color two-step ionization schemes are used, the final continuum state will be different for the different ionization channels. A pronounced continuum structure can result in dramatic variations in the overall ionization efficiency. Also photofragmentation of the benzene molecules was found to be wavelength dependent. To correct for such effects, vapor phase experiments were performed under identical experimental laser conditions. The resulting ratio between the population ratio obtained from the sputter experiment and the vapor phase population ratio is shown in Fig. 2 as function of the benzene exposure. Within the accuracy of the measurement, no real trend can be observed. With increasing coverage, the reduction of the population of the excited state induced by substrate originating particles seems to be compensated by molecules excited by phenomena occurring primarily in the organic overlayer.

In conclusion, we have proven that state-selective time distributions of neutral benzene molecules ejected from keV bombarded C_6H_6/Ag{111} can be determined. In combination with molecular dynamics simulations, these experiments provide insight to the underlying ejection mechanisms. Future experiments will be performed to determine state-selective angular distributions of desorbed benzene at coverages low enough to retain the {111} character of the crystal surface.

ACKNOWLEDGMENTS

The financial support of the National Science Foundation, the National Institutes of Health, the Office of Naval Research, and the Polish Committee for Scientific Research Fund No. PB 1128/T08/96/11 and Maria Sklodowska-Curie Fund MEN/NSF-97-304 is gratefully acknowledged. E.V. is partially supported by the Fulbright-Hayes Association and a NATO Research Fellowship.

REFERENCES

1. A.W. Czanderna and D.M. Hercules (*ed.*), *Ion spectroscopies for surface analysis* (Plenum, New York, 1991).
2. R. Chatterjee, D.E. Riederer, Z. Postawa, and N. Winograd, accepted for publication in J. Phys. Chem.
3. J.H. Callomon, T.M. Dunn, and I.M. Mills, Phil. Trans. Roy. Soc. London **259A**, 499 (1966).
4. G.H. Atkinson and C.S. Parmenter, J. Mol. Spectrosc. **73**, 20 (1978).
5. R. Chatterjee, Z. Postawa, N. Winograd, and B.J. Garrison, in preparation.

SESSION XI
SURFACE APPLICATIONS AND SPUTTERING II

Detection of Large Neutral Clusters in Sputtering

C. Staudt and A. Wucher

Fachbereich Physik, Universität Kaiserslautern, 67653 Kaiserslautern, FRG

Abstract. Neutral silver clusters Ag_n which are released from a polycrystalline silver surface under bombardment with rare gas ions were detected by 157 nm-VUV non resonant single photon ionization and time-of-flight mass spectrometry. Using Xe^+ primary ions with a bombarding energy of 15 keV, we were able to detect sputtered clusters consisting of up to $n = 60$ atoms. For $n \leq 47$, the laser intensity dependence of the mass spectral signals could be followed to demonstrate saturation of the cluster photoionization process. As a result, the cross section σ for single photon absorption is determined as a function of the cluster size. It is found that clusters containing ten or more atoms σ does not depend on the cluster size. For clusters with $n \geq 17$, doubly ionized species are detected and in some cases shown to represent a significant fraction of the total yield. The saturated signals are used to determine the relative cluster yields $Y(Ag_n)$, for which a power law decay according to $Y \propto n^{-\delta}$ with an exponent $\delta = 3.7$ is observed. The resulting contribution of clusters to the total sputtered flux is significantly larger than in previous experiments where primary ions of lower mass and kinetic energy were used.

INTRODUCTION

When a solid is bombarded by energetic ions, a variety of particles emerge from the surface. It is well known that this sputtered flux contains agglomerates of several or many atoms. The formation of these clusters in the course of the collision processes leading to the ejection of sputtered particles represents one of the most interesting phenomena in sputtering physics. Since their detection about four decades ago /1/, sputtered clusters have therefore drawn much attention in the literature /2,3/. While it is relatively straightforward to detect *charged* clusters leaving the surface, it is clear that the charge fraction, i. e. the fraction of sputtered clusters of a given size which leave the surface in a charged state, generally depends in a non predictable way on the cluster size as well as on the chemical state of the bombarded surface. Hence, in order to gain insight into the cluster formation mechanisms during sputtering, it is mandatory to detect the sputtered *neutral* clusters. Due to the apparent difficulty to efficiently ionize a cluster without fragmentation, corresponding experimental data was for a long time limited to very small clusters containing less than five atoms. Only fairly recently, the detection of larger aggregates became possible by the use of UV /4,5,6,7/ or VUV /8/ lasers to post-ionize the sputtered neutral species subsequent to the sputtering event and thus render them accessible to mass analysis and detection. In these experiments, it is of great importance that photoionization is achieved by absorption of a *single* photon, since this technique combines a non-resonant, i.e. non state selective ionization scheme (which is needed due to the large internal excitation of sputtered clusters /9/) with a high ionization efficiency and relatively low fragmentation rates. For the specific case of silver clusters investigated here, a VUV laser operated at 157 nm provides an ideal postionization tool, since the photon energy of 7.9 eV /8/ is sufficient to ionize all investigated species. Using this technique, we have in the past been able to detect sputtered neutral Ag_n clusters up to a size of n = 19. The present work documents an attempt to extend the range of observable cluster sizes by increasing the both the mass and energy of the primary ions.

EXPERIMENTAL

The experimental setup employed in the present work has been described in great detail elsewhere /5,8,10/. In short, neutral particles which are sputtered from a polycrystalline metallic silver sample by a pulsed rare gas ion beam impinging under 45° with respect to the surface normal are ionized by a pulsed laser beam directed closely above and parallel to the sample surface. During the primary ion pulse, the sample was usually kept at a negative potential in order to prevent positively charged ionic species from leaving the surface and, hence, to minimize the secondary ion background. The photoions created during the laser pulse are extracted toward a reflectron-type time-of-flight mass spectrometer (ToF-MS) operated at a mass resolution of about $m/\Delta m = 1000$ (at m = 108) by an electric field which is switched on shortly (~ 20 ns) after the laser pulse.

The ionizing laser employed in the present experiments is a conventional excimer laser (Lambda Physik model LPX 120i) operated with an F_2/He gas fill. Under optimized conditions, this laser produces pulses of up to 6 mJ energy and about 20 ns duration at a wavelength of 157 nm. In connection with a beam cross section of about 1 mm², this corresponds to peak power densities of about 10^7 W/cm² in the ionizing region above the sample surface. The laser beam was transported into the ultrahigh vacuum (UHV) chamber housing the experiment by means of a flow box flushed with nitrogen. This beam line also contained the CaF_2 lens used to focus the laser beam into the interaction region above the surface as well as other optical elements. In order to study the dependence of the photoion signals on the intensity of the ionizing laser, the laser pulse energy could be reduced in a controlled fashion by a stack of two variable dielectric attenuators located in front of the lens, which were tilted in opposite directions to compensate for the beam walkoff with increasing tilting angle. The laser intensity was monitored in situ by a home made photoelectric detector described elsewhere /11/ which was mounted directly in the UHV chamber. The measured pulse energies were converted to peak power densities by assuming a rectangularly shaped spatial and temporal laser profile.

RESULTS AND DISCUSSION

Figure 1 shows a typical mass spectrum of neutral silver atoms and clusters which are sputtered from a polycrystalline silver foil. The spectrum was recorded under bombardment with 15-keV Xe^+ ions using 157-nm VUV radiation (photon energy 7.9 eV) with a laser power density around 10^6 W/cm^2 for post-ionization. It is evident that Ag_n clusters consisting of up to n = 60 atoms can be identified. To

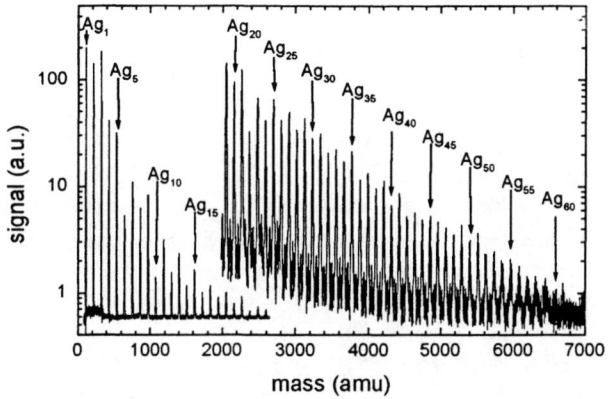

Figure 1. Mass spectrum of neutral silver atoms and clusters sputtered from a silver surface by 15-keV Xe^+ ions.

the best of our knowledge, these comprise the largest neutral clusters generated directly by sputtering which have been detected so far. In addition to the peak groups representing singly ionized clusters, a number of small intermediate peaks are seen which correspond to doubly charged Ag_n^{2+} clusters with odd values of *n*. Assuming equal detection efficiency for singly and doubly charged cluster ions, one can use these signals to estimate the fraction of Ag_n^{2+}. It is found that only at the highest value of P_L this contribution may in some cases reach non negligible values. However, one should be aware that the detection efficiency for doubly charged ions is higher due to their higher kinetic energy when hitting the photoion detector. In the following, we therefore chose to disregard the influence of multiply charged cluster ions.

For a quantitative interpretation of the measured signals it is important to study their dependence on the intensity of the ionizing laser. Figure 2 shows the integrated ToF signals of a number of arbitrarily selected clusters as a function of the laser power density P_L. The observed behavior is typical for all investigated clusters: In the regime of low P_L the signals increase linearly with increasing laser intensity, as is expected for a single photon absorption process. It is of interest to note that the signals of doubly ionized clusters increase more strongly with increasing P_L in this regime than those of singly charged ions, as would be expected from the multiphoton character of the double ionization process. At power density values of several 10^6 W/cm^2, the signals level off due to saturation of the photoionization process, and at higher P_L a slight decrease is observed which is due to multiphoton induced fragmentation. Since the magnitude of fragmentation loss is apparently small and, on the other hand, increases with increasing P_L, we assume that in the region where the signal maxima are observed this influence can be neglected. For P_L values up to this region, we therefore fit the measured signals to the theoretically expected saturation behavior of a one-photon ionization process

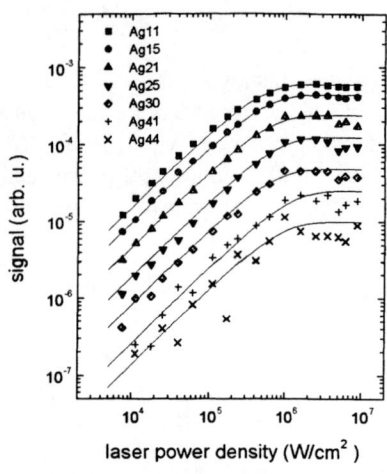

Figure 2. Integrated signal of selected Ag$_n$ clusters vs. power density of the ionizing laser

$$S(P_L) = S_{sat} \cdot \left[1 - exp\left(-\sigma \cdot \frac{P_L}{h\nu} \cdot \Delta t\right)\right], \qquad (1)$$

(Δt: effective laser pulse duration) by treating the photoabsorption cross section σ and the saturated signals S_{sat} as fitting parameters. The corresponding fitting curves are included in Figure 2 as solid lines. The cluster size dependence of the resulting cross sections is depicted in Figure 3. The error bars included in the figure

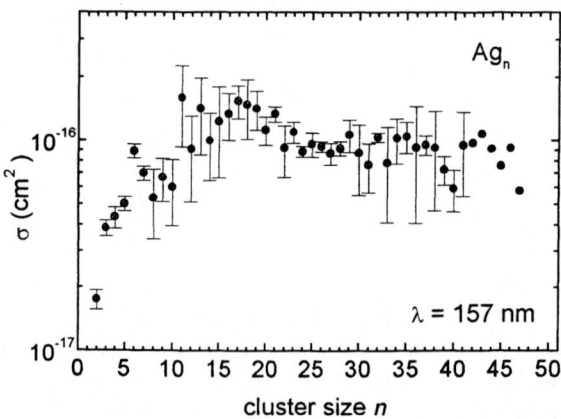

Figure 3. Single photon absorption cross section at $\lambda = 157$ nm of neutral silver clusters vs. cluster size

represent the reproducibility of the experiment. In addition, the scaling of the absolute values may include a systematic error of up to a factor of two which is due to the uncertainty of the laser power density scaling. It is seen that for small clusters σ increases with increasing n, until at $n \approx 10$ a value around 10^{-16} cm^2 is reached which for larger clusters becomes essentially independent of the cluster size. This latter finding is surprising, since particularly for larger clusters one would from simple geometrical considerations expect an increase with increasing cluster size according to $n^{2/3}$ /12/.

The saturated signals S_{sat} can be used to quantitatively estimate the contribution of clusters to the total flux of sputtered particles. In this respect, it is important to note that the laser postionization experiment detects the number density of the neutral particles within the ionization volume rather than their flux /5/. This is particularly important if the velocity distribution of sputtered neutral clusters depends on the cluster size. In this case, a correction with respect to the average inverse emission velocity $<v^{-1}>$ of the clusters must be applied to the measured signals in order to convert from density to sputtered flux. Previous measurements of the velocity distribution of sputtered neutral silver clusters have revealed that $<v^{-1}>$ roughly scales with $n^{0.8}$ /8/. Figure 4 shows the relative sputtering yields, i. e.

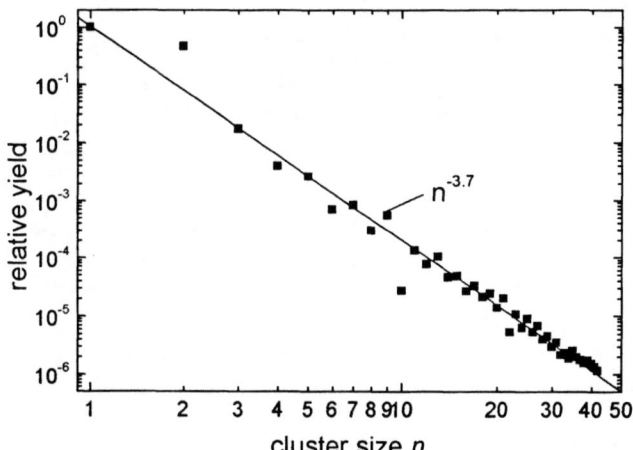

Figure 4. Relative yields of neutral silver clusters sputtered from a polycrystalline silver surface by 15-keV Xe$^+$ ions.

the saturated signals divided by $n^{0.8}$ and normalized to the value of monomers ($n = 1$), as a function of the cluster size. In agreement with previous experiments employing primary ions of lower mass and energy, we find a power law dependence according to $Y(n) \propto n^{-\delta}$. The value of the exponent $\delta = 3.7$, however, is significantly lower than those observed in our previous work /8/. This finding is presumably connected with the total sputtering yield Y_{tot} of the silver surface. The tabulated value of $Y_{tot} \approx 17$ /13/ for bombardment with 15-keV Xe$^+$ is by at least a factor of two larger than those corresponding to the bombarding conditions employed in ref. 8. Hence, the data presented here represents a clear indication that the inverse correlation between δ and Y_{tot} which had been observed earlier /6,8,14/ is continued to even higher values of Y_{tot}. In fact, we feel that this is the reason which enables us here to detect sputtered neutral clusters of much larger size than previously possible.

References

1. R.E. Honig, *J. Appl. Phys.* **29**, 549 1958)
2. H.M. Urbassek, W.O. Hofer, *Det kongelige Danske Vid. Selsk. Mat. Fys. Medd.* **43**, 97 (1993)
3. W.O. Hofer in *Sputtering by Particle Bombardment III*, eds. R. Behrisch and K. Wittmaack (Springer 1991), pp. 97
4. S.R. Coon, W.F. Calaway, J.W. Burnett, M.J. Pellin, D.M. Gruen, D.R. Spiegel, J.M. White, *Surf. Sci.* **259**, 275 (1991)
5. A. Wucher, M. Wahl, H. Oechsner, *Nucl. Instr. Meth.* **B 82**, 337 (1993)
6. S. R. Coon, W.F. Calaway, M.J. Pellin, J.M. White, *Surf. Sci.* **298**, 161 (1993)
7. Z. Ma, S.R. Coon, W.F. Calaway, M.J. Pellin, E.I. von Nagy-Felsobuki, *J. Vac. Sci. Technol.* **A 12**, 2425 (1994)
8. M. Wahl, A. Wucher, *Nucl. Instr. Meth.* **B 94**, 36 (1994)
9. A. Wucher, *Phys. Rev.* **B 49**, 2012 (1994)
10. A. Wucher, M. Wahl, H. Oechsner, *Nucl. Instr. Meth.* **B 83**, 73 (1993)
11. D. Koch, M. Wahl, A. Wucher, *Z. Phys.* **D 32**, 137 (1994)
12. M.M. Kappes, *Chem. Rev.* **88**, 369 (1988)
13. H. H. Andersen, H. L. Bay in *Sputtering by Particle Bombardment I*, ed. R. Behrisch (Springer 1981), p. 182
14. A. Wucher, M . Wahl, *Nucl. Instr. Meth.* **B 115**, 581 (1996)

SESSION XII
RIS APPLIED TO NUCLEAR AND PARTICLE PHYSICS

Hyperfine Structure Studies with the COMPLIS Facility

J.E. Crawford and J.K.P. Lee
Physics Department, McGill University, Montréal, Canada H3A 2T8

F. Le Blanc, D. Lunney, J. Obert, J. Oms, J.C. Putaux, B. Roussière,
J. Sauvage, S. Zemlyanoi, and D. Verney
Institut de Physique Nucléaire, IN2P3-CNRS, 91406 Orsay Cedex, France

J. Pinard, L. Cabaret, and H.T. Duong
Laboratoire Aimé Cotton, 91405 Orsay Cedex, France

G. Huber, M. Krieg, and V. Sebastian
Institut für Physik der Universität Mainz, 55099 Mainz, Germany

M. Girod and S. Péru
C.E.A., Service de Physique Nucléaire, B.P. 12, 91680 Bruyères-le Châtel, France

J. Genevey and F. Ibrahim
Institut des Sciences Nucléaires, IN2P3-CNRS, 38026 Grenoble Cedex, France

J. Lettry and the ISOLDE Collaboration
CERN, 1211 Genève 23, Switzerland

Abstract. COMPLIS is an experimental facility designed to carry out spectroscopic studies on radioisotopes produced by disintegration of elements available at CERN's Booster-ISOLDE on-line isotope separator. During recent series of experimental runs, hyperfine structure measurements have yielded information on nuclear moments and deformations of platinum and iridium isotopes. For the first time, population by α-decay from Hg was exploited to investigate $^{178-181}$Pt - the most neutron-deficient Pt isotopes yet studied. Successful measurements have recently been carried out on $^{182-189}$Ir.

INTRODUCTION

The hyperfine interaction is one of the most useful tools for the determination of nuclear properties: the Isotope Shift (IS) along an isotopic series reveals the change in nuclear charge radius; the detailed Hyperfine Structure (HFS) can yield nuclear magnetic moments μ_I and electric spectroscopic quadrupole moments Q_S. The Q_S value and sign in turn provide information on the character of the nuclear deformation. The discovery — 25 years ago (1) — of the huge change in charge radius and large odd-even staggering beginning at the isotope ^{185}Hg sparked great interest in the progression of nuclear shapes in this region, close to the neutron mid-shell number N=104. Calculations of potential energy surfaces have been carried out for even-A (2) and odd-A Pt isotopes (3,4); these suggest shapes ranging from oblate for isotopes near A=196, becoming triaxial above the mid-neutron-shell

(A=182), and prolate for lighter isotopes. Recently Esser et al (5), using systematics based on a rigid rotor model find all even-A Pt isotopes in this region to be triaxial, with increasing β deformation for lighter isotopes close to A=180. Calculations using a particle + triaxial rotor model have been carried out by Hilberath et al (3); these suggest a similar trend for odd-A Pt, with ground-state shapes varying from prolate through triaxial to oblate as A increases. However, the previous laser spectroscopic studies extend only down to A=183. There is therefore great interest in the variation for still lower masses; this is the main motivation for the present study.

EXPERIMENTAL METHOD AND RESULTS

Isotopes of $_{80}$Hg are produced at very high yields at CERN's Booster ISOLDE facility, and through radioactive decay many daughter isotopes of the refractory metals become accessible to laser studies. Recent experiments by the COMPLIS group (COllaboration for Measurements with a Pulsed Laser Ion Source) have been done on the most neutron-deficient isotopes of Pt and Ir yet studied. ISOLDE's COMPLIS facility, shown in Fig. 1 is an RIMS (Resonance Ionization Mass Spectrometer) system. A selected Hg isotope beam is implanted at low energy (~ 1 keV) in a slowly rotating graphite target wheel, and allowed to decay. When the population of the daughter isotope (or a subsequent generation) has reached a maximum, the implanted isotope is desorbed by pulses from a Nd:YAG laser, producing an

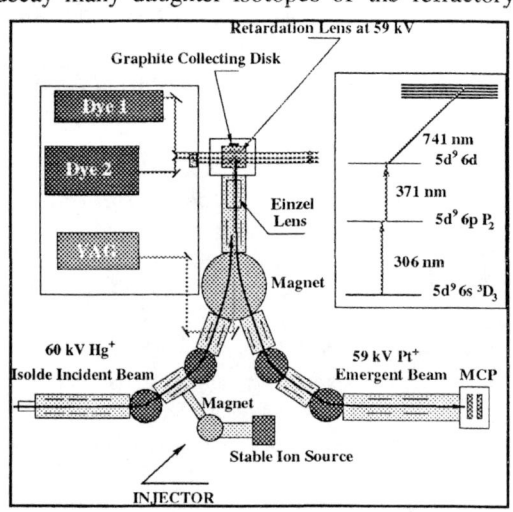

FIGURE 1. The COMPLIS injection and desorption lines. The inset shows the RIS transitions used for platinum isotopes.

atomic cloud near the graphite. This is probed by RIS beams that ionize the isotope through a three step excitation. For Pt, the first step is at 306 nm ($5d^96s\ ^3D_3 \rightarrow 5d^96p\ ^3P_2$); the second is 371 nm, derived by frequency doubling a 741 nm beam, which itself is used at high intensity for the final ionizing step. The target rotation rate, implantation and desorption sequences, and digital oscilloscope averaging are all computer controlled. For long-lived isotopes (e.g., ^{186}Pt, with $T_{1/2}$ = 2h), the entire implantation can precede a later desorption scan; for isotopes with short lives, (e.g., ^{178}Pt, with $T_{1/2}$ = 21s) step-by-step implantation-desorption sequences are carried out. For Pt isotopes, β decay was used to populate isotopes with A ≥ 182, but below this mass, the ISOLDE yield becomes too low. For the first time, we have used α-decay from Hg to study the $^{178-181}$Pt isotopes. For example, ^{178}Pt can be produced by two β-decays from ^{178}Hg, or by α–decay directly from ^{182}Hg. However, ISOLDE's yield is 8 x 10^6 atoms/s of ^{182}Hg, but only 38 atoms/s of ^{178}Pt. Although the α branch from ^{182}Hg is only ~15%, we gain a factor of ~10^4 in the production of ^{178}Pt by the α-decay route compared to β-decay, for comparable implantation times.

RESULTS AND DISCUSSION

The shapes of nuclear ellipsoids are described by two parameters: β (the major axis 'stretch') and γ, the degree of triaxial deformation. In this representation a sphere is represented by β = 0, γ = 0°, a prolate spheroid has β > 0, γ = 0°, an oblate spheroid has β > 0, γ = 60°, and a triaxial nucleus has β > 0, 0°< γ <60°. The β deformation can be obtained in several ways. The IS data provide the field shift from the reference ^{194}Pt isotope δν ^{194,A}; this is related to the δ<r^2> change in mean square nuclear charge radius, and from this the mean square deformation parameter change δ < β^2 > ^{194,A} can be deduced. For odd A isotopes with spin I>1/2, the Q$_s$ value can be calculated from the HFS ; if the coupling scheme for the nuclear spin is known, the intrinsic Q

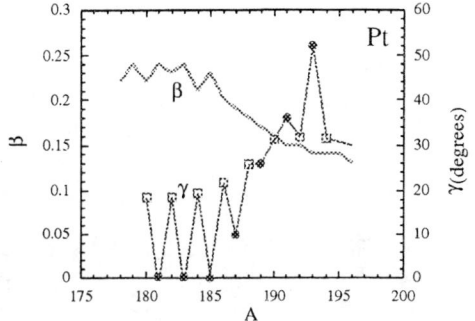

FIGURE 2. β and predicted γ values.

value and deformation β can be determined. (Details of these calculations are given in Ref. 6 , with tabulated values of measured moments μ and Q$_s$.) Nuclear B(E2) transition probabilities, if available, can also be used to calculate rms < β^2 >$^{1/2}$ values. Estimates of the γ parameter can be obtained from nuclear level systematics, and nuclear model calculations. The β measurements from the present and previous experiments (3,7), and selected predicted γ values are shown in Fig. 2. The even-A γ values (open squares) are from Ref. 5; odd-A γ values (circles) are from the mean-field calculation of Ref. 3. The β-deformation increases at lower A; the nuclei become less triaxial and more prolate, but with shape oscillations below A=185.

The δ < r^2 > values for this experiment, combined with previous results for higher A now extend from A=178 to A=198. These are plotted in Fig. 3a, and compared with a microscopic Hartree-Fock-Bogolyubov (HFB) calculation using the Gogny force (8,9), and assuming axial symmetry (γ = 0°). This calculation overestimates < r^2 > for A<188; it also predicts normal odd-even staggering (OES), with higher <r^2> values for even-even isotopes — interpreted as a result of the nuclear pairing force (10). In contrast, the experimental OES is inverted for A<185. An HFB calculation for even-A isotopes assuming triaxiality is shown in Fig. 3b. The general trend is well reproduced with calculated <r^2> values closer to the experimental ones, compared to the calculation of Fig. 3a. The triaxial HFB calculations also predict smaller β (by 0.01) than the axial ones. The inverted OES for the A<185 region therefore appears to reflect the transition between prolate shape for the odd A isotopes and prolate triaxial for the even A isotopes. This is similar to the interpretation for the spectacular variation in the Hg isotopes — in that case indicating a transition between prolate and triaxial-oblate.

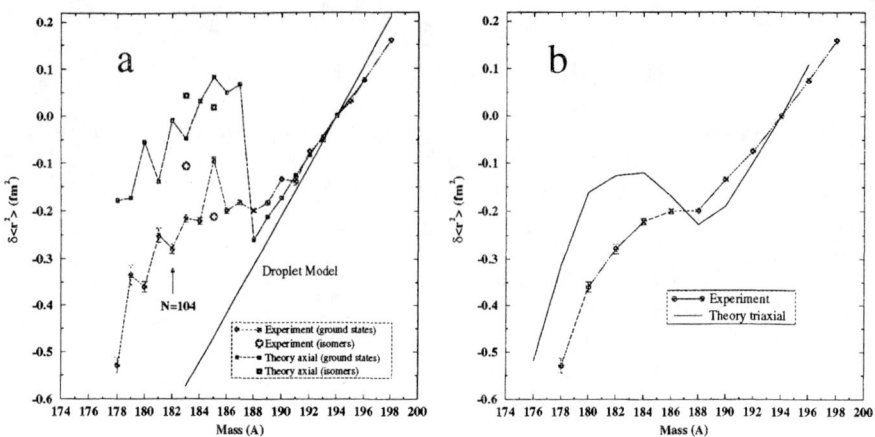

FIGURE 3. (a) $\delta\langle r^2\rangle$ measurements, compared with an HFB calculation assuming axial symmetry. **(b)** Even A $\delta\langle r^2\rangle$ values, compared with a triaxial HFB calculation.

Recently, we have carried out successful measurements on $_{77}\text{Ir}^{182-189}$, populating these isotopes by three β decays from implanted Hg. In this case the first RIS excitation step was the 351.5 nm transition from the $5d^7 6s^2$ (J=9/2) ground state to the $5d^7 6s 6p$ (J=11/2) level. These IS shift measurements are currently being analyzed.

REFERENCES

1. Bonn, J., et al., E.W. *Phys. Lett.* **38B**, 308 (1972).
2. Bengtsson, R., et al.,*Phys. Lett.* **183B**, 1-6 (1987).
3. Hilberath, T. et al., *Zeit. Phys.* A **342**, 1-15 (1992).
4. Cederwall, B. et al., *Zeit. Phys.* A **337**, 283-292 (1990).
5. Esser, L. et al., *Phys. Rev.* **C55**, 206-210 (1997).
6. Le Blanc, F., et al., to be published (1998)
7. Duong, H.T., et al., *Phys. Lett.* **B217**, 401 (1989).
8. Dechargé, J. and Gogny, D. *Phys. Rev.* **C21**, 1568 (1980).
9. Girod, M., et al., *Phys. Rev.* **C37**, 2600 (1988).
10. Zawischa, D., *Phys. Lett.* **B155**, 309 (1985).

Laser Induced Nuclear Reactions

Ken Ledingham, Tom McCanny, Paul Graham, Xiao Fang and Ravi Singhal
Dept of Physics and Astronomy, University of Glasgow, Glasgow G12 8QQ, Scotland.
Joe Magill
European Commission, Institute for Transuranic Elements, Karlsruhe, Postfach 2340, D-76125, Germany
Alan Creswell and David Sanderson
Scottish Universities Research and Reactor Centre, East Kilbride, Glasgow G75 0QU.
Ric Allott, David Neely and Peter Norreys
Rutherford Appleton Laboratory, Chilton, Didcot, U.K.
Marko Santala, Matthew Zepf, Ian Watts, Eugene Clark, Karl Krushelnick Michael Tatarakis and Bucker Dangor
Blackett Laboratory, Imperial College, London SW7 2BZ, UK
Antonin Machecek and Justin Wark
Clarendon Laboratory, Dept of Physics, University of Oxford, OX1 3PU, UK.

Abstract. Dramatic improvements in laser technology since 1984 have revolutionised high power laser technology. Application of chirped-pulse amplification techniques has resulted in laser intensities in excess of 10^{19}W/cm^2. In the mid to late eighties, C.K.Rhodes and K.Boyer discussed the possibility of shining laser light of this intensity onto solid surfaces and to cause nuclear transitions. In particular, irradiation of a uranium target could induce electro- and photofission in the focal region of the laser. In this paper it is shown that μCi of ^{62}Cu can be generated via the (γ,n) reaction by a laser with an intensity of about 10^{19}Wcm^{-2}.

Introduction

The mechanism of the interaction of charged particles in intense electromagnetic fields has been considered for more than fifty years. This was one of the first explanations put forward by the early workers to explain the origin and energies of cosmic rays (1-3). Simply the idea is as follows: a charged particle in an intense electromagnetic field is accelerated initially along the direction of the electric field. The vxB force causes the particle's path to be bent into the direction of travel of the wave. In large fields the particle's velocity rapidly approaches the velocity of light and tends to travel with the EM wave gaining energy from it. In astrophysical situations the solar corona was thought to be one of the sources of the electromagnetic waves. These ideas were the counterparts of the machines built on the earth to accelerate particles to high energies.

In 1971 the possibility of accelerating electrons in focussed laser fields was first proposed by Feldman and Chiao (4). They showed that an electron could gain energies as high as 30 MeV after a single pass through the focus of a diffraction limited laser beam of power 10^{12} W and wavelength 1μ. Chan (5) similarly calculated that an intense laser beam could be used as an energy booster for relativistic charged particles showing that a 10 MeV electron can absorb 40 MeV from a laser beam of 1μ wavelength and an electric field of 3×10^{10} V/cm in a distance of 1.3 mm.

Recently Tajima and Dawson (6) discussed a laser electron accelerator which could be created when an intense laser pulse produced a wake of plasma oscillations through the action of the non-linear ponderomotive force. They demonstrated through computer simulations that existing glass lasers of 10^{18} Wcm^{-2} shining on plasmas of densities 10^{-18}cm^{-3} could yield electrons of GeV energy per cm of acceleration.

More recently the importance of Raman forward scattering in short pulse high intensity lasers has been realised by a number of authors e.g. (7-10) in generating electrons of very energies in very short distances. The probability of using this technology for the construction of compact accelerators which might find applications where 2-200 MeV electrons or photons are needed was considered.

Many probes have been used for producing nuclear reactions e.g. neutrons, protons, alpha particles and heavier nuclei as well as non nuclear beams such as electrons and gamma rays. It is therefore not surprising that one can induce nuclear reactions using intense light beams. In the mid to late eighties the possibility of producing laser induced fission using laser intensities as high as 10^{21}W/cm^2 was suggested (11-13). Application of chirped-pulse amplification techniques has revolutionised high power laser technology and have resulted in laser intensities in excess of 10^{19}W/cm^2 being generated. A number of high intensity lasers have been built around the world e.g. Vulcan at the Rutherford Appleton Laboratory, UK and Nova at the Lawrence Livermore National Laboratory, USA which are capable of reaching such intensities. Indeed recently (14) it was announced that intense light beams are capable of transmuting elements and specifically gold was transmuted to platinum and mercury.

This paper is presented as a proof of concept that an intense laser beam can be used to induce nuclear reactions.

Experimental

The laser used in these experiments was the Vulcan Nd glass laser at the Rutherford Appleton Laboratory with a wavelength of 1.053μ. The pulse width was typically 1 ps and with some 40J in the pulse. The laser beam of p-polarised radiation had dimensions 150 x 88mm and was focussed on axis using a f /1.7 parabolic mirror. The focal spot had a diameter of about 12-15 μ resulting in laser intensities in excess of 10^{19}Wcm^{-2}. The crater caused by the laser in a lead target of dimensions 10x10x3 mm is shown in Fig 1 and has a diameter of about 2mm. This photograph was taken with a scanning electron multiplier and it is felt the dimensions of the crater and the height and width of the rim can reveal a great deal about the plasma processes involved. When the lead target was replaced with a copper sample of similar dimensions the crater size was reduced by a factor of two. It was noticed that if the laser interacted with the copper target without a lead irradiator, no radioactivity was generated in the Cu target which suggests that the high energy electrons were generating a bremsstrahlung beam in the lead target which then caused subsequent γ induced nuclear reactions. (γ,n) reactions are typically nearly two orders of magnitude larger than (e⁻,n) raections.

The targets were housed in the ultra-short pulse interaction chamber on Vulcan which was evacuated to about 10^{-5} torr and the laser could deliver a pulse every 30 minutes. It took several minutes to take the chamber up to atmospheric pressures and to extract the samples for analysis and thus at this stage we can only analyse radioactive samples with half lives greater than a few minutes. This is a temporary limitation which will be rectified by rapid transport systems in the future.

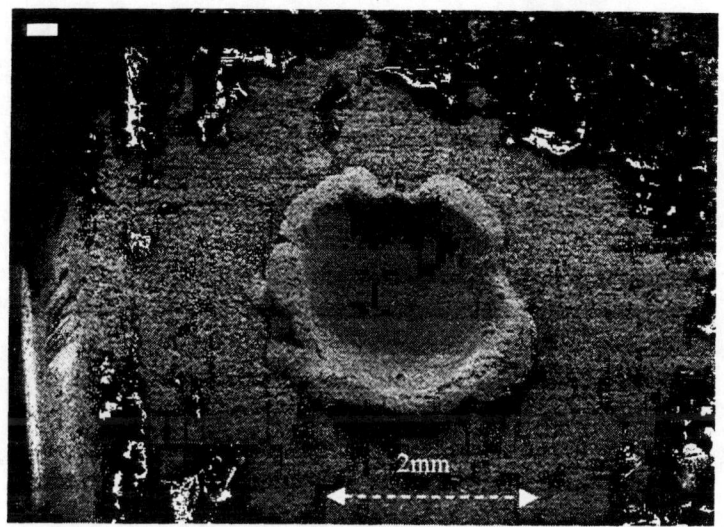

Fig 1 The crater caused by a laser pulse of 40J interacting with a lead target. The γ beam generated at these intensities is in a direction 180° to the crater and has a continuous distribution energies of many MeV which can induce nuclear reactions.

Results and Discussion

One of the most fundamental photo-nuclear reactions is the (γ,n) reaction which has a Q value of typically > 8 MeV. Such reactions produce proton rich nuclei which decay normally by positron emission or EC. A positron emitting nucleus is very easy to identify unambiguously since the positron annihilates with electrons to produce two 511 keV γ-rays back to back.

In our experiment the laser beam was incident on a lead target (3 mm thick) at an angle of 45° as shown in Fig 2. Behind the lead target were placed a number of Cu samples of thickness 3 mm from which the angular distribution of the laser induced γ radiation could be determined. The laser beam incident on the lead target generates an MeV γ-ray beam heavily forward collimated as shown in Fig 3. [this is the subject of an extended paper to be published elsewhere (15)]. The γ rays irradiated the copper target to produce ^{62}Cu via a $^{63}C(γ,n)Cu^{62}$ which has a half life of about 10 minutes.

Target arrangement for activation studies

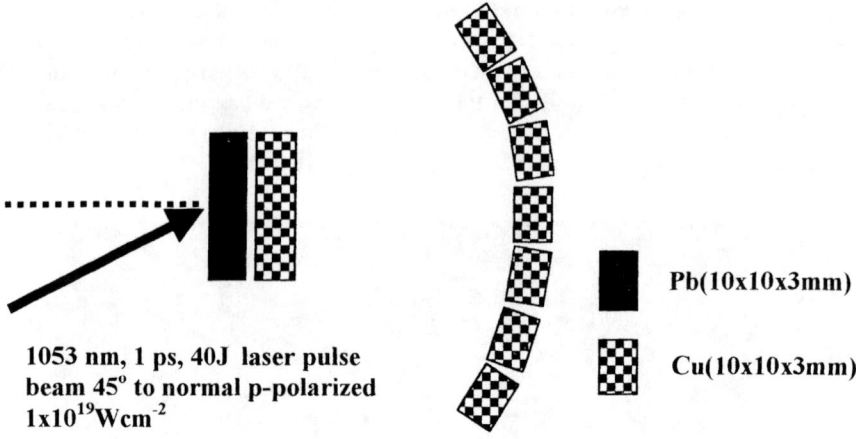

Fig 2 The target arrangement for activation studies. The p-polarized laser beam was incident on the lead target at 45°. The γ beam emerged from the radiator to activate the copper samples.

γ-Ray angular distribution above 10MeV by Cu activation

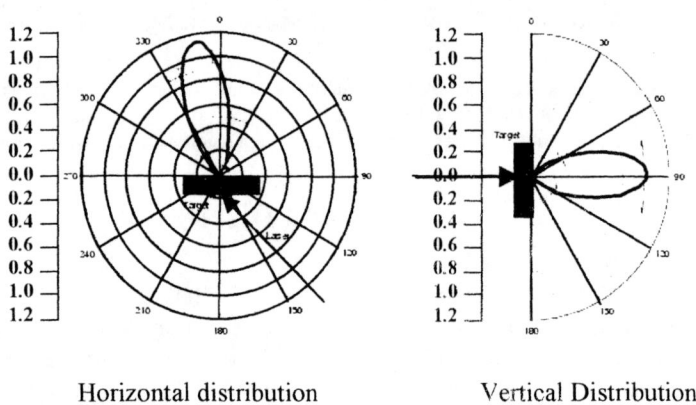

Horizontal distribution   Vertical Distribution

Fig 3 The γ-ray angular distribution above 10 MeV determined by ^{62}Cu activation. The beam is heavily collimated in the forward direction i.e. at right angle to the target plane.

Two large volume NaI scintillators (based on an existing mobile gamma spectrometer) were used in conjunction with a coincidence system. The system was calibrated with ^{58}Co positron activities ranging from 50 to 3200 Bq with the limit of detection of pure β^+ emitters estimated to be a few Bq. The copper samples were placed between the detectors and the summed coincidence spectra were recorded using 120s live times over periods of several tens of minutes and the net counts in the 1024 keV peak were determined. These were plotted as a function of time and fitted to exponential decays and the half lives determined (Fig 4). The mean half life for four data sets was 9.63±0.24 minutes which agrees well with the generally accepted value of 9.7 minutes for ^{62}Cu. For the data set with the maximum activity, a long lived component was also visible which was estimated to be several hours presumably from ^{64}Cu.

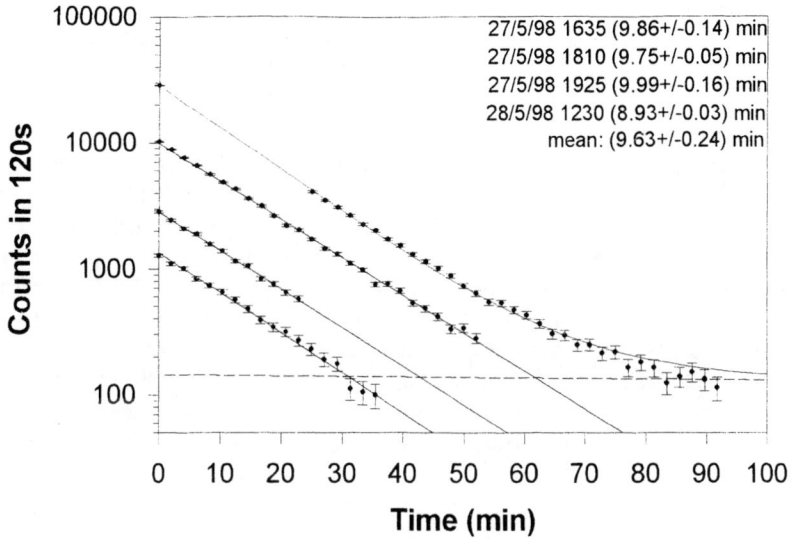

Fig4 The half lives of Cu^{62} as measured from four different laser shots of nominally the same pulse energy. The mean half life agreed well the accepted value of 9.7 mins.

The lead target was also investigated for activity but the (γ,n) reactions in this element produced isotopes with either very short or very long half lives and hence could not be detected. In reference (14) the author discusses that the Nova laser transmuted a gold target to both platinum and mercury. This is simply the laser producing a γ ray beam which caused a ^{197}Au (γ,n)^{196}Au which decays by β^+ to ^{196}Pt and by β^- to ^{196}Hg. A possible application of a laser accelerator is the production of the important short lived positron emitters ^{11}C, ^{13}N, ^{15}O and ^{18}F as biological tracers.

Conclusions

It has been shown that a high power laser with pulse widths of 1 ps and intensities of about 10^{19}W/cm^2 at a wavelength of 1μ when allowed to interact with a lead target produces a well collimated beam of γ rays of energies in excess of 10 MeV. This high energy radiation has been shown to produce a (γ,n) reaction in ^{63}Cu to produce μCi activity of ^{62}Cu. Clearly however many problems have yet to be solved. For example as shown in Fig 4 the radioactivity from four different laser pulses, with nominally similar energies, appears to differ by more than an order of magnitude. It was reported in (9) in similar experiments for X-ray production using a much lower energy beam of 120 fs pulses, that a low intensity prepulse which arises from amplified spontaneous emission precedes the main pulse in time and according to these authors this prepulse appears to be necessary for efficient high energy X-ray production. The question of the prepulse must be investigated further.

It has thus been demonstrated that high power lasers can be used to induce nuclear reactions. It has been shown that μCi of ^{62}Cu can be generated at 10^{19}Wcm^{-2}. It is hoped that Vulcan will be upgraded to produce laser intensities in excess of 10^{21}Wcm^{-2} and if the activity scales as the 3/2 power (9) then mCi of activity can be generated with many possible areas of application.

The possibility also exists to produce (e$^-$,fission) or (γ,fission) reactions and to verify the calculations of ref.13. It is therefore very exciting to contemplate the possibility of laser light beams producing radioactive isotopes and point sources of neutrons and fission fragments from the fission reaction. Although promising, much more research is necessary to determine whether plasma particle accelerators can compete with existing accelerators in energy, intensity and quality and particular whether table top high energy accelerators can be produced using lasers. Even now rather small lasers of pulse energies less than a joule and with pulse lengths of about 50fs are capable of being focused to laser intensities of about 10^{19}Wcm^{-2}. These lasers have the advantage of being able to be pulsed to repetition rates of about 10Hz and by integrating many pulses intense radioactive sources may be generated. Finally as in many scientific endeavours however, the most important results on plasma acceleration may be totally unexpected. The data presented in the present paper is the subject of an extended publication dealing specifically with the applications of a laser accelerator and its application (16).

References

1) Menzel D H and Salisbury W W *Nucleonics* **2**, 67, (1948)
2) Fermi E *Phys.Rev.* **75**, 1169, (1949)
3) McMillan E M *Phys. Rev.* **79**, 498, (1950)
4) Feldman M J and Chiao R Y *Phys.Rev.A*, **4**, 352, (1971)
5) Chan Y W *Phys Lett* **35A**, 305, (1971)
6) Tajima T and Dawson J M *Phys.Rev.Lett* **43**, 267, (1979)
7) Joshi C, Tajima T, Dawson J M, Baldis H A, and Ebrahim N A, *Phys.Rev.Lett.* **47**, 1281, (1981)
8) Dawson J M *Scientific American* **260**, 34, (1989)
9) Kmetic J D, Gordon III CL, Mackin J J, Lemoff B E Brown G S and Harris S E *Phys.Rev.Lett* **68**, 1527, 1992
10) Modean A, Najmudin Z, Dangor, A E, Clayton C E, Marsh K A, Joshi C, Malka V, Darrow C B, Danson C, Neely D, Walsh F N *Nature* **377**, 606, (1995).
11) Rhodes C K *Science* **229**, 1345 (1985)
12) Lynn J E *Nature* **333**, 116 (1988)
13) Boyer K, Luk T S and Rhodes C.K. *Phys,Rev.Lett.* **60**, 557 (1988)
14) Irion, R, *New Scientist*, **18th April**, 30, (1998)
15) Santala M et al to be published
16) Ledingham K W D et al to be published

Spectral Studies Related to the 3.5 eV Isomeric State of Th-229

J. P. Young*, R. W. Shaw*, and Oren F. Webb†

*Chemical & Analytical Sciences Division and †Chemical Technology Division
Oak Ridge National Laboratory
Oak Ridge, TN 37831

Abstract. There have been reports (1,2) of photon emission (possibly gamma rays) from the relaxation of the 3.5 eV excited nuclear state of Th-229 generated in the alpha decay of U-233. We have carried out studies using U-233 of very high isotopic purity, less than 1 ppm U-232, and did not observe such emission. Experimental details and discussion of our negative results are given.

INTRODUCTION

Th-229 has a very low energy nuclear isomeric state, 3.5 ± 1 eV.(3) It has been reported (1,2) that photon emission exhibiting a broad (2.9 to 4.1 eV) energy range was observed for U-233 samples as they alpha decay. At face value, this reported emission is of great interest in that it represents a unique situation: a gamma ray at an optical frequency. It potentially provides a link between nuclear and electronic excitations. For a solid $UO_4 \cdot 2H_2O$ sample, the reported emission from 295 to 427 nm is structured and believed to be the result of the de-excitation of the 3/2[631] Nilsson state, 3.5 ± 1 eV, to the 5/2[633] ground state of the Th-229 nucleus.(1) For this sample a broad green visible emission at 522 nm is also observed that is attributed to a red-shifted γ-ray emission resulting from an electronic bridging mechanism of the γ-ray. This mechanism excites the thorium atom from the $6d_{3/2}$ electronic state to the $7p_{1/2}$ state. Emission spectral studies of a sample of $^{233}UO_2(NO_3)_2$ dissolved in 3M HNO_3 yielded similar results.(2) The authors of this report were less certain of the origin of the emissions, however.

Our intention is to study the possibility of coupling a laser to the highest energy structure of the reported UV emission for a sample of separated Th-229 in an effort to generate the isomeric state optically; the output of a XeCl excimer laser, for example, is at this energy.

Workers at the Oak Ridge National Laboratory possess a relatively large quantity of U-233, the Th-229 parent, in varying degrees of isotopic purity. Also at ORNL, Th-229 is extracted from this supply for use as a "cow" to ultimately generate an interesting medical isotope, Bi-213.(4,5) The isotopes and equipment are at hand to carry out studies of this unique system. Our first task was to confirm the previously reported optical emission results using our apparatus.

EXPERIMENTAL

We wished to repeat the first reported experiment (1) where emission was observed; so we prepared a sample of $UO_4 \cdot 2H_2O$ using a procedure similar to one previously reported.(6) Based on a mass spectral analysis, our U-233 had less than 1 ppm mass-232 impurity; 1 ppm is the minimum detectable amount by the thermal ionization mass analysis scheme used and represents any U or Th impurity at that mass. We received the sample 2.5 weeks after the U was separated from daughter products that had been generated in storage. The peroxide product was light yellow in color.

The solid sample was loaded into a closed-end quartz tube, 2 mm I.D., 4 mm O.D., 2.7 cm long, and sealed with Torr Seal epoxy (Varian Vacuum Products, Lexington, MA). This tube was then positioned in a quartz-windowed holder that was modified to hold the sample in a required double-confinement arrangement. A photograph of the sample container is shown in Figure 1. As shown at the right, a sample could be loaded into the optical cell. It was then sealed and placed in the secondary container; the SiO_2 windows of that container are held in place and also sealed with Torr Seal. A screw cap was then placed over the container. After loading the cell with 40 mg of sample, the outside of the cell read 20000 dpm/cm² at contact.

Figure 1. Photograph of the fused quartz sample containment assembly and sample tube loading operation.

The holder was designed to fit on a commercial optical rail system (part # 1700F) such that the image of the sample was focused on the entrance slit of a Spex polychromator (Model HR 460), both of these items available from Instruments SA, Inc., Edison, NJ. A 150 groove/mm diffraction grating, blazed at 500 nm, was the dispersing element in the spectrometer. The dispersed image was focused onto a 1024 by 256 pixel, liquid-nitrogen-cooled, UV coated, CCD detector (Model EEV CCD15-11) also available from Instruments SA. By this optical arrangement, the spectral range from 250 to 650 nm could be continuously observed in an experiment. CCD exposures up to 24 hours have been collected using this arrangement. The sample compartment was enclosed with standard lead bricks to shield the CCD.

Data were collected via Spectramax (version1.1D), a DOS program available from Instruments SA. Limitations of the program currently prevent a single 24 hour observation, but the program does allow multiple one hour exposures that are averaged after each individual exposure. Our 24 hour exposure is then the averaged spectral output of 24 exposures of one-hour duration each.

RESULTS

A typical spectral integration, of 24 hours duration, is shown in Figure 2. We see no evidence of any emission in the near ultraviolet region, where the quantum

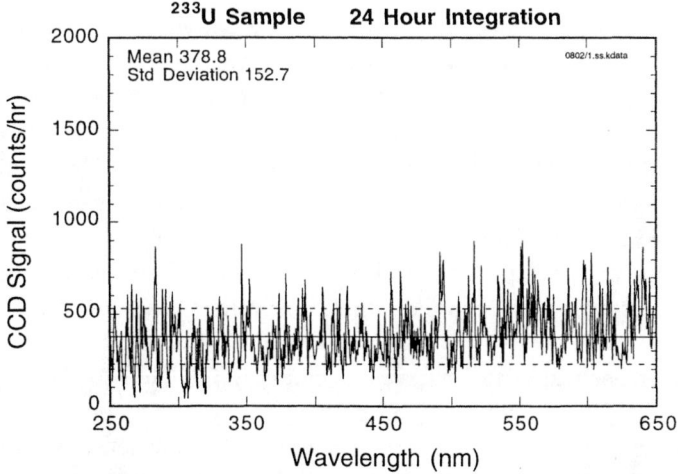

Figure 2. CCD detector response for a 24-hour exposure imaging isotopically-pure $233UO_4 \cdot 2H_2O$ powder.

efficiency of the CCD detector is 20% or in the visible region where the QE approaches 100%. A number of extended exposures such as that shown in Figure 2 have been acquired with similar negative results. Occasionally in the data set, there is a prominent spike that is several pixels wide; these spikes appear at random wavelengh locations, however.

The reason for our different results when compared with reported data (1,2) is not clear. Possible differences in the various experimental procedures are summarized in Table I.

There are a variety of experimental parameters represented in these studies. Time of data collection is quite different in comparing study 1 vs. 2 or 3; study I used a scanning spectrometer, 2 and 3 did not. The most obvious differences in our experiment, study 3, when compared to either study 1 or 2 is the U-232 impurity level and the age of the uranium since purification. Our sample was essentially devoid of the relatively short-lived U-232 impurity (70 year half life). By mass analysis the U-232 content was below detectable limits (<1 ppm), and it was estimated to be 0.1 ppm by radiation field measurements. The U-232 impurity level of the samples used in the other experiments was not specifically mentioned but could be inferred to be present in ppm concentration levels from their experimental description. A nominal U-232 concentration in U-233 is 2 to 10 ppm.

Our U-233 was freshly separated (2.5 weeks old), while the separation of uranium for the other studies occurred months to years before those experiments. U-232 comes into transient equilibrium with its progeny in less than 10 years; the uranium used in study 2 was 6 years old. Th-229 and its progeny come into secular equilibrium in a matter of several months; Th-228, the daughter of U-232, and its progeny come into equilibrium in a much shorter time period. The different thorium isotopes yield different progeny elements, and these elements have their own unique properties.

Another parameter to consider is sample composition. The $UO_4 \cdot 2H_2O$ sample in the first report (1) was not well described; the $UO_2(NO_3)_2$ solution sample and the experimental arrangement in the second publication (2) was described; the authors pointed out, however, that the spectral results were obtained by subtraction of two large numbers. Interestingly, the reported emission spectra of both experiments are similar.

We have not yet been able to observe any emission related to the expected presence of the Th-229 isomeric state [populated to 2 percent by the uranium decay(1)]. The emission seen in the other studies might be due to some experimental parameter that exists in only those studies. This common parameter could be U-232 impurity and its radioactive progeny; it is mentioned (2) that the emission seen might even be related to excitation of the Th-229 progeny by U-232 activity leading to subsequent fluorescence. The $UO_4 \cdot 2H_2O$ sample was one that remained from earlier experiments. The $UO_2(NO_3)_2$ solution was prepared from U-233 produced in 1992.

Table I

Comparison of Three U-233 Emission Experiments

#	Sample Type	U-232 Content	Time Since Separation of U-233	Container	Experimental Set-Up	Observation Duration	Emission Observed
1(1)	$UO_4 \cdot 2H_2O$ Solid	?	Months?	Kaptan, covered with transparent tape	Scanning monochromator, PMT	320 sec*	yes
2(2)	$UO_2(NO_3)_2$ solution	?	6 years	SiO_2	Monochromator, position-sensitive PMT	7 days	yes
3(This work)	$UO_4 \cdot 2H_2O$ Solid	<1 ppm‡	2.5 weeks	SiO_2	CCD Spectrograph	1 day	no

*Time on a resolution element is 0.4 sec x 50 x 16 scans = 320 sec.

‡0.1 ppm estimated from radiation field.

Our U-233 was freshly separated and should be relatively free of decay products.

We currently feel that the emission previously reported (1,2) is due to the presence of U-232 or some other artifact common to the two published works. A 10 year old U-233 sample with 2 ppm U-232 impurity will be available to us in the next several months. We will then be able to study any effect that U-232 or an aged sample has on sample emission.

ACKNOWLEDGEMENT

Research sponsored by Laboratory Directed Research and Development Program. U-233 and Th-229 provided by U.S. Department of Energy Isotope Production and Distribution Program. Oak Ridge National Laboratory is managed by Lockheed Martin Energy Research Corporation under Contract DE-AC05-96OR22464 with the U.S. Department of Energy.

REFERENCES

1. Irwin, G. M. and Kim, K. H., *Phys. Rev. Lett.*, **79**, 990-993 (1997).

2. Richardson, D. S.; Benton, D. M.; Evans, D. E.; Griffith, J.A.R.; and Tungate, G., *Phys. Rev. Lett.*, **80**, 3206-3208 (1998).

3. Helmer, R. G. and Reich, C. W., *Phys. Rev. C*, **49**, 1845-1857 (1994).

4. Kennell, S. J. and Mirzadeh, Saed, *Nuclear Medicine & Biology*, **25**, 241-246 (1998).

5. Boll, R. A.; Mirzadeh, Saed; Kennell, S. J.; DePaoli, D. W.; and Webb, O. F., *Journal of Labelled Compounds & Radiopharmaceuticals*, **40**, 341-4343 (1997).

6. Tipton, C. R. (Ed.), *Reactor Handbook*, 2nd Edition, New York, Interscience Publishers, 1961, p. 407.

POSTER SESSION I
RIS AND CLUSTERS

Threshold Photoionization Spectroscopy of Li_nO and Li_nC Clusters

P. Lievens, P. Thoen, S. Bouckaert, W. Bouwen, F. Vanhoutte, H. Weidele, and R.E. Silverans

*Laboratorium voor Vaste-Stoffysica en Magnetisme, K.U. Leuven
Celestijnenlaan 200 D, B-3001 Leuven, Belgium*

Abstract. We report on the measurement of ionization potentials of lithium monoxide and lithium monocarbide clusters by threshold laser ionization spectroscopy. The clusters are produced by a laser vaporization source, mass selected, and detected by a reflectron time-of-flight mass spectrometer. The values obtained for the ionization potentials of the small clusters are consistent with hypervalent molecular bonding mechanisms. For the larger clusters the evolution of the ionization potentials with size is in agreement with metallic behaviour as evidenced by several distinct steps and a pronounced odd-even staggering superposed on a smoothly decreasing trend. The steps in the ionization potentials correspond to the magic numbers predicted by cluster shell models (2, 8, 20, 40, ...), provided that n-2, respectively n-4, electrons are delocalized in the clusters Li_nO and Li_nC.

INTRODUCTION

Since the discovery of shell structure in the abundance spectrum of Na clusters (1), there has been an ever increasing interest in the study of fundamental properties of simple metal clusters (2). While there is abundant information on the physical properties like stabilities, ionization potentials, electron affinities, photoabsorption cross sections, etc. for single element clusters, experiments on binary clusters are scarce. Nevertheless, also for binary systems the evolution of chemical and physical properties from the molecule towards a macroscopic scale can only be addressed by studying clusters. A particular subset of binary clusters that got much attention lately are substoichiometric combinations of alkali metals with electronegative elements (e.g., O, H, F). In several cases such systems, e.g., Cs_nO_m (3), Li_nO_m (4), Li_nH_m (5), Na_nF_m (6), were described as segregated structures consisting of an ionically bound bulk-like crystallite and a metallic part with properties very similar to bare metal clusters. In the case of O-doped clusters, recent studies suggest that whether there is segregation or not can be strongly dependent on the dopant concentration (7,8). For example, for Ba_nO_m a structural transition as a function of the O content was related to a possible absence of segregation for suboxides (8).

Furthermore, both theoretical and experimental evidence exists for high stability of non-stoichiometric doped alkali clusters with high geometrical symmetry. The increased stability is attributed to so-called hypervalent binding mechanisms, i.e., an at least formal violation of the octet rule for second period elements. Well studied examples are Li_6C and Li_4O, with the Li atoms forming a regular octahedron, resp. tetrahedron, surrounding a central C or O atom (9-12). The enhanced stability with respect to Li_4C and Li_2O is due to the surplus of valence electrons occupying outer orbitals which are Li-C antibonding, but Li-Li bonding. Up to now many more hypervalent molecules are examined (13).

EXPERIMENT AND RESULTS

In this contribution we report on the measurement of the ionization potentials of lithium monoxide and lithium monocarbide clusters by threshold photoionization spectroscopy. The experimental setup is shown in FIGURE 1. Lithium monoxide and monocarbide clusters were produced by a laser vaporization source (14), using isotopically enriched ^7Li. Reflectron time-of-flight (RTOF) mass spectrometry was used for the mass-selected detection of the photoionized clusters. When ablating Li metal, the most intense clusters that appear in the mass spectrum are Li_nO (14), with also a significant amount of Li_nC present. These clusters were selected for the present investigation.

The clusters were photoionized by laser light in the wavelength range from 225 to 400 nm, provided by an optical parametric oscillator laser system and a dye laser. In order to take possible cluster production fluctuations into account, photoionization efficiencies were calibrated using reference measurements employing excimer laser light with $\lambda=193$ nm.

Photoionization efficiency curves for each cluster mass are then obtained by plotting the integrated mass peak intensities as a function of the photon energy. The ionization potentials are deduced by linear extrapolation of the ionization thresholds (15). The values for the ionization potentials of Li_nC and Li_nO clusters up to n=70 are plotted in FIGURE 2.

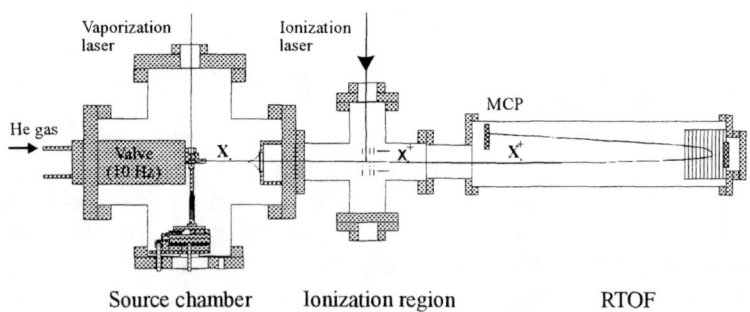

FIGURE 1. Experimental setup

DISCUSSION AND CONCLUSIONS

The evolution of the ionization potential with size evidences metallic behaviour for the larger clusters. Superposed on a gradual decrease of the ionization potentials, pronounced discontinuities at n=10, 22 and 42 for Li_nO and at n=24 and 44 for Li_nC are present. This is consistent with the magic numbers of cluster shell models, if it is assumed that the oxygen and carbon atoms localize 2, resp. 4 Li valence electrons. This leaves the remaining Li valence electrons delocalized in the metallic cluster, resulting in closed shells for 2, 8, 20 and 40 electrons. This interpretation is also consistent with the clear odd-even staggering (i.e. higher ionization potentials for clusters with an even number of delocalized electrons) observed for clusters containing up to approximately 40 Li atoms.

The interpretation of the ionization potentials of the small clusters (Li_nO (n<6) and Li_nC (n<16)) is less obvious. Deviations from the metal-like behaviour are attributed to the existence of very stable, geometrically-rigid multi-coordinated structures, at least partially preventing the Li valence electrons from delocalization throughout the cluster. The stability of these small clusters is attributed to so-called

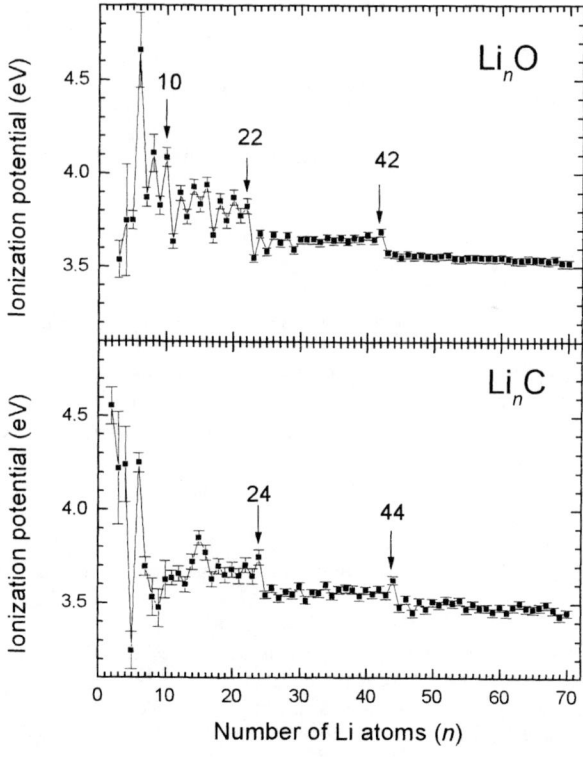

FIGURE 2. Ionization potentials of Li_nO and Li_nC clusters as a function of cluster size.

hypervalent (in our case hyperlithiated) bonding mechanisms. For such systems the metal-metal bonding between the Li atoms arranged in a regular polyhedron around the dopant, ensures the hypervalent cluster to be more tightly bound than the molecules with bindings satisfying the octet rule. Quantum chemical predictions indeed show that these hypervalent structures have quite low ionization potentials (13,15). In the case of C-doped clusters the deviation from metal-like behaviour continues up to larger sizes. This is consistent with quantum chemical calculations predicting high stability of multi-coordinated Li_nC clusters at least up to n=12 (16).

ACKNOWLEDGEMENTS

This work is financially supported by the Fund for Scientific Research - Flanders (F.W.O.), the Flemish Concerted Action (G.O.A.) Research Programme and the Interuniversity Poles of Attraction Programme - Belgian State, Prime Minister's Office - Federal Office for Scientific, Technical and Cultural Affairs. P.L. would like to thank the F.W.O., W.B. the Flemish Institute for Scientific and Technological Research (I.W.T.), and H.W. the European Community Training and Mobility of Researchers Programme (T.M.R.), for financial support.

REFERENCES

1. W.D. Knight, K. Clemenger, W.A. de Heer, W.A. Saunders, M.Y. Chou, and M.L. Cohen, Phys. Rev. Lett. **52**, 2141 (1984).
2. W.A. de Heer, Rev. Mod. Phys. **65**, 677 (1993).
3. H.G. Limberger and T.P. Martin, J. Chem. Phys. **90**, 2979 (1989).
4. C. Bréchignac, Ph. Cahuzac, F. Carlier, M. de Frutos, J. Leygnier, and J.Ph. Roux, J. Chem. Phys. **99**, 6848 (1993).
5. R. Antoine, Ph. Dugourd, D. Rayane, E. Benichou, and M. Broyer, J. Chem. Phys. **107**, 2664 (1997).
6. P. Weis, C. Ochsenfeld, R. Ahlrichs, M.M. Kappes, J. Chem. Phys. **97**, 2553 (1992).
7. C. Bréchignac, Ph. Cahuzac, M. de Frutos, and P. Garnier, Z. Phys. D **42**, 303 (1997).
8. V. Boutou, M.A. Lebeault, A.R. Allouche, C. Bordas, F. Paulig, J. Viallon, and J. Chevaleyre, Phys. Rev. Lett. **80**, 2817 (1998).
9. P.v.R. Schleyer, E.-U. Würthwein, and J.A. Pople, J. Am. Chem. Soc. **104**, 5839 (1982).
10. H. Kudo, Nature **355**, 432 (1992).
11. C.H. Wu, H. Kudo, and H.R. Ihle, J. Chem. Phys. **70**, 1815 (1979).
12. P.v.R. Schleyer, E.-U. Würthwein, E. Kaufmann, T. Clark, and J.A. Pople, J. Am. Chem. Soc. **105**, 5930 (1983).
13. E. Rehm, A.I. Boldyrev, and P.v.R. Schleyer, Inorg. Chem. **31**, 4834 (1992).
14. P. Lievens, P. Thoen, S. Bouckaert, W. Bouwen, E. Vandeweert, F. Vanhoutte, H. Weidele, and R.E. Silverans, Z. Phys. D **42**, 231 (1997).
15. P. Thoen, S. Bouckaert, W. Bouwen, F. Vanhoutte, H. Weidele, P. Lievens, R.E. Silverans, A. Navarro-Vázquez, and P.v.R. Schleyer, to be published.
16. J. Ivanic and C.J. Marsden, J. Am. Chem. Soc. **115**, 7503 (1993).

POSTER SESSION II
ATOMIC RIS

Control of the final state photoionization products by Laser-Induced Continuum Structure

S. Cavalieri*, R. Eramo*, L. Fini*, M. Materazzi*
O. Faucher†, and D. Charalambidis‡

*Istituto Nazionale di Fisica della Materia; Dipartimento di Fisica and
European Laboratory for Non Linear Spectroscopy (LENS)
Università di Firenze, Largo E. Fermi 2, I-50125 Firenze
†Laboratoire de Physique, UPRESA CNRS 5027, Faculté des Sciences Mirande,
Université de Bourgogne, BP 400, 21011 Dijon Cedex, France
‡ Foundation for Research and Technology, Hellas, Institute of Electronic Structure and Laser,
P. O. Box 1527, Heraklion, 71110, Crete, Greece, and
Department of Physics, University of Crete

Abstract. We report on the observation of the modification of the final status products in a photoionization of the Xe atom through the coherent process of laser-induced continuum structure (LICS). The photoionization decay of the Xe ground state into the two electronic continua, corresponding to the two fine structure levels of the ground state of the Xe ion, is varied by embedding a bound state into the two continua through a laser field (dressing laser). The dressed continua, probed from the atomic ground state through three-photon absorption, exhibit different induced structures. The experiment is performed on an atomic beam apparatus which uses a time-of-flight spectrometer for energy analyzing the emitted photoelectrons. The modified final state amplitudes are controlled by the wavelength of the dressing laser. No significant modification of the $^2P_{3/2}$ continuum has been observed, in the limit of the detection efficiency. The effect is characterized by a window resonance in the slow electron continuum as the dressing laser is tuned over the LICS resonance.

The possibility of controlling bound-free transitions has been predicted theoretically and investigated experimentally in atomic and molecular gases and semiconductors [1–10]. A way to achieve the control of photoionization is to use two competing laser ionization pathways that interferes each other: the interference is controlled by the relative phase of the laser fields. The first experimental evidences of this effect in the total photoionization yield have been obtained in the H_2S molecule [11] and Na atom [12].

A different way to get coherent control is to induce a resonance into the ionization continuum. It is now well established that the coupling of a discrete state of an atom to one of its continua by means of a coherent electromagnetic field induces a structure in the continuum which may be probed through the absorption or scattering of a second field by the atom. The effect established as laser induced continuum structure (LICS) [13] has been in recent years demonstrated in different experiments utilizing a variety of coupling schemes including bound [14–16], as well as autoionizing states [17]. The existence of continuum structure and its characteristics are conventionally considered to be determined by the atomic parameters only. Hence the ability of LICS to provide control in position and shape of the induced or modified continuum structure makes it of particular interest. In this paper we report on the observation of the modification through LICS of a quantity which is commonly meant to depend only on the atomic structure parameters, namely the ratio of the decay rate of the xenon atom in two different continua, when the atom is excited above the second ionization threshold, i. e. the first excited state of the ion.

The LICS structure is induced in the two xenon continua by dressing them with strong laser field, the latter establishing a coupling between the continua and an atomic bound state. This structure can then be probed through a second field that couples the ground state, which initially carries all the population, to the same continua and consequently ionizes the atom. The LICS can be observed separately for each continuum by employing energy resolved photoelectrons spectroscopy.

The coupling scheme is depicted in Fig. 1. The ground state of xenon is three

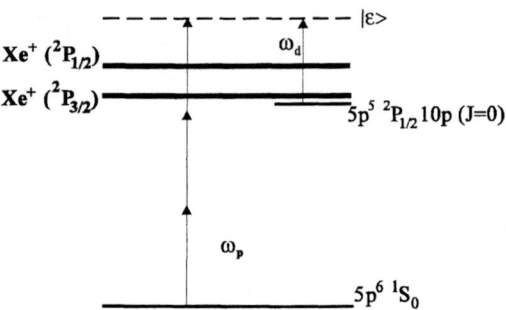

FIGURE 1. Schematic energy level diagram of the LICS process in Xenon. The ground state population is multiphoton ionized ($3\hbar\omega_p$) above the second ionization threshold. A second laser field ($\hbar\omega_d$) is used to dress the two continua with the $10p[\frac{1}{2}]_{J=0}$ excited state.

photon coupled (photon energy $\hbar\omega_p$) to both the $^2P_{\frac{3}{2}}$ and $^2P_{\frac{1}{2}}$ continua, corresponding to the two fine structure levels of the xenon ion, while a second dressing laser field (photon energy $\hbar\omega_d$), is induced to embed the $5p^5\ ^2P_{1/2}10p(J=0)$ state in both continua. Since the transition moments involved in the excitation of the two continua are not the same, the characteristics of the two induced structures are expected to be in general different and hence to result in a detuning $\Delta = E_{5p^6} + 3\hbar\omega_p - E_{5p^510p} - \hbar\omega_d$ dependent branching ratio, where E_{5p^6} and E_{5p^510p}

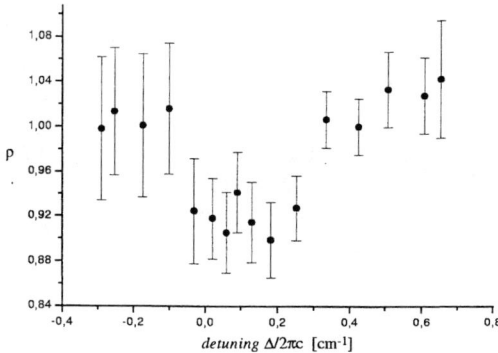

FIGURE 2. Ratio $\rho = r_{1/2}/r_{3/2}$ between the normalized ionization yields (see the text) as function of the LICS detuning Δ.

are the energies of the ground and the $5p^5\ ^2P_{1/2}10p(J=0)$ state. The electrons of the two continua of 1.85 and 0.54 eV are discriminated by means of the time of flight spectrometer which also discriminates against background processes.

In our experiment the dressing laser beam ω_d is provided by a tunable dye laser producing 5 ns duration pulses with a wavelength around 573.2 nm. As probing field ω_p, with duration of 6 ns, (λ_p=266.1 nm) is used the fourth harmonic of a Nd:YAG laser. The xenon gas is introduced in the interaction region in the form of a supersonic jet. We have measured the frequency dependence of the photoelectron yield ratio $r_{1/2}$ ($r_{3/2}$) for the decay in the $^2P_{\frac{1}{2}}$ ($^2P_{\frac{3}{2}}$) continuum as a function of the detuning Δ. These ratios are defined as the ionization yield along the laser polarization axis in the presence of the dressing field divided by the yield in the absence of the dressing field. In the two spectra we have clearly seen that whereas the photoelectron yield of the decay into the $^2P_{\frac{3}{2}}$ continuum remains unchanged within the experimental errors, there is a reduction in the decay rate into the $^2P_{\frac{1}{2}}$ around $\Delta = 0$. As there is no absolute yield calibration for different photoelectron energies the branching can not be deduced from the experimental data. Nevertheless the ratio $\rho = r_{1/2}/r_{3/2}$ is indicative of the effect of LICS on the branching of ionization in the two continua. Fig. 2 shows the ration ρ as a function of the detuning Δ, which exhibits a LICS minimum as expected from the constant value of the fast electron yield. These results show an application of the recently experimentally demonstrated quantum interference effect of LICS in the control of the photoabsorption products in an atomic continuum. Control of the branching of ionization into the $^2P_{\frac{1}{2}}$ and $^2P_{\frac{3}{2}}$ continua of xenon atom, along the polarization axis, has been achieved. The relative amplitudes of the two final outgoing waves are altered by dressing the two continua with $5p^5(^2P_{1/2})10p(J=0)$ bound state. Although the present experiment is restricted to a narrow energy region in the continuum, product control through LICS can be achieved in a wide continuous energy range by using two tunable laser fields, in contrast to interference schemes that involve more than two bound states.

Due to the general applicability of the method in different types of continua, other than the electronic continuum of an atom, photoabsorption product control through these schemes can be extended to molecular electronic or dissociative continua or as well as to laser condensed matter interactions.

The research has been carried out at LENS (TMR Program, contract number ERBFMGGECT950017). R. E. acknowledges support from INFM.

REFERENCES

1. M. Shapiro, J. Hepburn, and P. Brumer *Chem. Phis. Lett.* **149**, p. 2416, 1988.
2. Y.-Y. Yin, C. Chen, and D. S. Elliott *Phis. Rev. Lett.* **69**, p. 2353, 1992.
3. N. B. Baranova, I. M. Beterov, B. Y. Zel'dovich, I. I. Ryabtsev, A. N. Chudinov, and A. A. Shul'ginov *JEPT Lett* **55**, p. 439, 1992.
4. B. Sheehy, B. Walker, and L. F. D. Mauro *Phys. Rev. Lett.* **74**, p. 4799, 1995.
5. D. W. Schumacher, F. Weihe, H. G. Muller, and P. H. Bucksbaum *Phys. Rev. Lett.* **73**, p. 1344, 1994.
6. T. Nakajima, P. Lambropoulos, S. Cavalieri, and M. Matera *Phys. Rev. A* **46**, p. 7315, 1992.
7. E. Dipont, P. B. Corkum, H. C. Liu, M. Buchanan, and Z. R. Wasilewski *Phys. Rev. Lett.* **74**, p. 3596, 1995.
8. A. Shnitman, I. Sofer, I. Golub, A. Yogev, M. Shapiro, Z. Chen, and P. Brumer *Phys. Rev. Lett.* **76**, p. 2886, 1996.
9. S. Cavalieri, R. Eramo, L. Fini, M. Materazzi, D. Charalambidis, and O. Faucher *Phys. Rev. A* **57**, p. 2914, 1998.
10. R. Eramo and S. Cavalieri *Opt. Comm.* **149**, p. 296, 1998.
11. V. D. Kleinman, L. Zhu, X. Li, and R. J. Gordon *J. Chem. Phys.* **102**, p. 5863, 1955.
12. S. Cavalieri, R. Eramo, and L. Fini *Phys. Rev. A* **55**, p. 2941, 1997.
13. P. L. Knight, M. A. Lauder, and B. J. Dalton *Phys. Rep.* **190**, p. 1, 1990 and reference therein.
14. Y. L. Shao, D. Charalambidis, C. Fotakis, J. Zhang, and P. Lambropoulos *Phys. Rev. Lett.* **67**, p. 3669, 1991.
15. S. Cavalieri, F. S. Pavone, and M. Matera *Phys. Rev. Lett.* **67**, p. 3673, 1991.
16. R. Eramo, S. Cavalieri, L. Fini, M. Matera, and L. DiMauro *J. Phys. B* **30**, p. 3789, 1997.
17. O. Faucher, D. Charalambidis, C. Fotakis, J. Zhang, and Lambropoulos *Phys. Rev. Lett.* **70**, p. 3004, 1993.

Superelastic Collisions [e + Mg*] Following Resonant, 2-Photon Ionization of Mg Atoms

S. A. Darveau[†] and R. S. Berry[‡]

[†]*Department of Chemistry, University of Nebraska at Kearney, Kearney, Nebraska 68849, U.S.A.*
[‡]*Department of Chemistry, The University of Chicago, Chicago, Illinois 60637, U.S.A.*

Abstract. Photoelectron energy spectra of electrons extracted from Mg vapor illuminated with resonance radiation (Mg ^{1}S → Mg ^{1}P → Mg$^+$ + e) show not only the expected 1.1 eV electrons from the (1+1) process. They also show two bands of faster electrons that have been produced by inelastic collisions of the second kind -- up-conversion -- of the 1.1 eV electrons with the excited ^{1}P Mg atoms. One band, at 2.7 eV, appears to be due to transfer of energy to a free electron as a magnesium atom de-excites from the ^{1}P to the ^{3}P state. Faster electrons, at 5.5 eV, appear to be produced by a 3-photon ATI process, or by a superelastic collision process de-exciting the ^{1}P magnesium atoms to their ground state or by both; which process or processes are involved is under investigation as this is written. Approximate cross sections will be reported.

INTRODUCTION

Photoionization studies of magnesium have ranged from single photon ionization (1) to multiphoton ionization through autoionizing resonances (2-5). Ionization studies involving the ^{1}P (3s3p) excited level of magnesium as an intermediate or as an initial state have been reported. (4-6) A recent investigation of magnesium ionization in intense non-resonant laser field has revealed above-threshold ionization behavior.(7)

This paper presents ionization behavior of magnesium in a low intensity (<100 W/cm^2) resonant laser field. In addition to the normal threshold (1+1) ionization, two other peaks in the electron energy spectrum have been observed. Characterization of and cross sections for each peak have been obtained.

EXPERIMENTAL

The experiment was initiated by the two-photon resonant ionization of magnesium. The magnesium atomic beam was produced by an effusive oven at temperatures from 300°C to 750°C. Laser radiation from a 20-Hz Nd:YAG pumped dye-laser was frequency doubled by a KDP crystal. A 500mm plano-convex lens slightly focused the beam to a spot size of 0.051 cm^2 at the intersection with the magnesium effusive atom beam. Optical density filters were used to reduce the pulse energy to a maximum of 50nJ/pulse in a 10ns beam, in other words, to maximum intensities of 100 W/cm^2. The laser wavelength was set to the resonance line of magnesium, 285.21nm, and fine-tuned at moderate intensities to yield maximum two-photon ionization.

The intersection of the laser and effusive beam occurs on the axis of a magnetic bottle spectrometer (MBS) which is a 4π-steridian collecting version (8) of the original Kruit and Read design (9) using a permanent magnet (10) as the high field source. Electrons originating in the interaction region at 600 gauss follow parallel trajectories down a 0.57m flight tube at 3 gauss to a chevron dual-multichannel plate (MCP) detector. The MCP is used at saturation voltage for pulse detection. The MCP signals were amplified, then passed through an amplifier/discriminator to produce clean pulses coinciding with electron impact. The pulses were sent to both a gated counter and to a time-to-amplitude converter (TAC). The digital signal from the counter and the analog signal of the TAC were collected by a data acquisition board in a personal computer. The counter signal monitored total electron counts while the TAC signal allowed for energy analysis of the electrons. The TAC can only measure one count per laser pulse; for this reason the count rate was held below 0.5 counts/laser pulse to avoid skewing the time-of-flight spectrum accumulated from the TAC data. The counter gate may be set to allow specific monitoring of a specific section of the time-of-flight spectrum.

Laser pulse energy was monitored in two locations. A Scientech laser calorimeter measured a fraction of the laser power split from the main beam before entering the vacuum chamber. A Molectron J4-05 laser meter measured the full beam emerging from the magnesium beam. Output signals from both devices were measured by the personal computer. The use of two laser meters, one before and one after the magnesium beam, allowed for a Beers' Law like analysis to determine the magnesium density.

The laser intensity dependence of various electron peaks was measured by varying laser intensity with fixed oven temperature. Temperature variation of the effusive oven with fixed laser intensity allows for magnesium density dependence determinations of the electron peaks. The slope of a log-log plot of the electron counts versus either the laser intensity or the magnesium density yields the value for the exponential dependence of the ionization on each. The resonance behavior of the ionization was confirmed by scanning the laser wavelength with fixed laser intensity and oven temperature.

DATA ANALYSIS

Magnesium density, ρ_{Mg}, was calculated from the relationship

$$\ln\left(\frac{I_0}{I}\right) = \sigma_A \rho_{Mg} l \tag{1}$$

where I is laser intensity after the beam intersection, I_0 is the initial laser intensity, l is the absorbing path length, and σ_A is the absorption cross-section. The geometry of the experimental apparatus gives l the value of 1.1 cm. The absorption cross-section of the magnesium resonance line calculated for linear polarization, and a laser width of 1.1 x 10^{11} Hz (3.7 cm^{-1}) is 1.44 x 10^{-13} cm^2. The excited state magnesium density, ρ_{Mg^*}, can be calculated from

$$\frac{\rho_{Mg^*}}{\rho_{Mg}} = 1 - \exp(-\sigma_A I \tau) \qquad (2)$$

Cross sections for superelastic processes, σ_{SES}, may be estimated from

$$P_{SES} = \frac{N_{e^*}}{N_e^0} = 1 - \exp(\rho_{M^*}\sigma_{SES} l) \approx \rho_{M^*}\sigma_{SES} l \qquad (3)$$

where N_{e^*} is the number of superelastically scattered electrons and N_e^0 is the number of electrons before any scattering occurs. The approximation holds if the probability for superelastic scattering, P_{SES}, is much less than unity. The number of electrons available for scattering is estimated both from the calculated ionization cross section (6) of excited magnesium at 285nm and from the monitored threshold ionization counts. The calculated ionization rate is used because at the magnesium densities used in the experiment the MCP detector saturates during the threshold ionization peak and does not reflect the full ionization rate.

RESULTS AND DISCUSSION

The electron time-of-flight spectrum of magnesium in the resonant radiation field is shown in Figure 1. The peak locations and energies are: A, 410ns (5.50eV); and B, 585ns (2.70eV). The Peak C location as determined at lower magnesium density is 875ns (1.21 eV). The ionization potential for ground state magnesium is 7.65eV. (11) Each 285nm photon carries 4.35eV. Peak C has the correct energy for a two-photon ionization process from the ground state. Peak A appears to have resulted from the absorption of three photons from the ground state, a possible above threshold ionization process. The peak B electrons have an additional energy above the Peak C electrons that closely matches the energy gap between the singlet and triplet $3s3p$ levels of magnesium. A plausible explanation is the superelastic scattering of the threshold ionization electrons as they exit the interaction region by Mg $^1P(3s3p)$, leaving the magnesium in the ^3P $(3s3p)$ state. Peak A could also be the result of superelastic scattering as above, but leaving the magnesium in the ground state.

Figure 2 shows the log-log plots of ionization counts versus intensity. The bars indicate the deviation of mean for each point. The slopes have values of 2.64 and 1.46 for peaks A and B, respectively. The 2.64 value indicates a three-photon process, while the 1.46 value is ambiguous. Least squares fits of the intensity dependence data for Peak B show much better correlation with squared laser intensity than with linear laser intensity, therefore, Peak B appears to be the result of a two-photon process. Both Peak A and B show linear dependence on the magnesium density.

Above-threshold ionization cross sections for Peak A are $(1.8_5 \pm 0.1_2) \times 10^{-64}$ cm^6-s^2 when calculated from the ground state or $(1.42 \pm 0.09_6) \times 10^{-43}$ cm^4-s when calculated from the ^1P state. The superelastic scattering cross section for 1.1eV electrons by

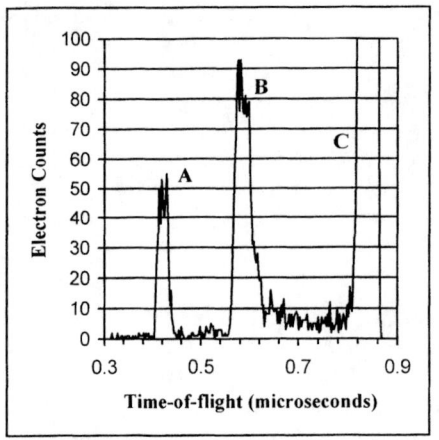

Figure 1: Photoionization Electron Time-of-Flight Spectrum

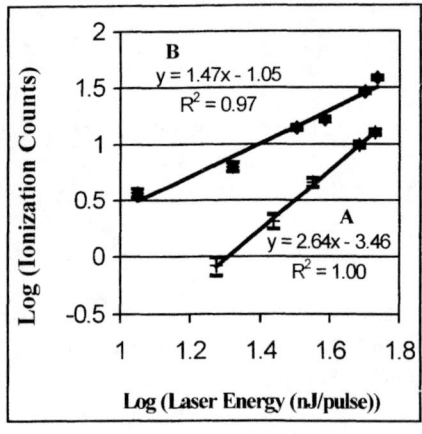

Figure 2: Laser Intensity Dependence of Photoionization Peaks

excited magnesium atoms (^{1}P $3s3p$) is $(2.8 \pm 0.5) \times 10^{-14}$ cm^2. The errors indicated are the estimated deviations in the mean for the random variation in the data sets.

More work must be done to confirm the assignments given to the photoionization peaks. One crucial experiment will be the use of a second color in addition to the resonance radiation. If the energy spacing between Peaks B and C remains constant, the assignment to a superelastic process will be more certain. Laser polarization studies, including circular polarization, will be important in characterizing the above-threshold ionization peak.

ACKNOWLEDGEMENTS

This work was supported by the National Science Foundation and by an NSF Graduate Fellowship (Darveau).

REFERENCES

1. Mehl-Balloffet, G. & Esteva, J. M. (1969) *Astrophys. J.* **157,** 945-56.
2. Chang, T. N. & Tang, X. (1992) *Phys. Rev. A* **46,** R2209-R2202.
3. Druten, N. J. v., Trainham, R. & Muller, H. G. (1994) *Phys. Rev. A* **50,** 1593-1606.
4. Shao, Y. L., Fotakis, C. & Charalambidis, D. (1993) *Phys. Rev. A* **48,** 3636-3643.
5. Bradley, D. J., Ewart, P., Nicholas, J. V., Shaw, J. R. D. & Thompson, D. G. (1973) *Phys. Rev. Lett.* **31,** 263-266.
6. Thompson, D. G., Hibbert, A. & Chandra, N. (1974) *J. Phys. B: Atom. Molec. Phys.* **7,** 1298-1305.
7. Kim, D., Fournier, S., Saeed, M. & DiMauro, L. F. (1990) *Phys. Rev. A* **41,** 4966-4973.
8. Chesnovsky, O., Yang, S. H., Pettiette, C. L., Craycraft, M. J. & Smalley, R. E. (1987) *Rev. Sci. Inst.* **58,** 2131-2137.
9. Kruit, P. & Read, F. H. (1983) *J. Phys. E* **16,** 313-324.
10. Tsuboi, T., Xu, E. K., Bae, Y. K. & Gillen, K. T. (1988) *Rev. Sci. Instrum.* **59,** 1357-1362.
11. Kaufman, V. & Martin, W. C. (1991) *J. Phys. Chem. Ref. Data* **20,** 83-148.

Studies on autoionization states of Sm by 3-step resonance photoionization

Hyunmin Park, Jong-hoon Yi, Jae-Min Han, Yong-joo Rhee, and Jongmin Lee

Laboratory for Quantum Optics, Korea Atomic Energy Research Institute P.O. Box, 105, Yusong, Taejon, 305-600, Korea

Abstract. Using 3-step resonance photoionization, we investigated the autoionization states of atomic samarium in the region of 48800 cm^{-1} - 52700 cm^{-1}. As an experimental result, we observed more than sixty autoionization states in the investigated range. In addition, We could obtain the resonance energies, the linewidths, and q-parameters of the observed autoionization states by fitting the spectra with Fano's formula.

INTRODUCTION

The investigation of the autoionization states by laser spectroscopic method has been of interest and studied over the years, because the states play an important role in the atomic physics. Especially, the efficiency of the ionization is determined mainly by the existence of an autoionization state to which the transition cross-section is large, because it induces a significant ionization. Therefore, searching for autoionization states has been one of the most important tasks in the photoionization research. Many papers have been devoted to the Sm spectroscopy since the work of King on the classification of approximately 4500 lines of Sm [1]. However, there are significantly few reports devoted to the autoionization states of Sm, as compared with the number of papers related to the other rare earth elements.

In this paper, we described the experimental results of autoionization states of Sm in the energy range of 48800 cm^{-1} - 52700 cm^{-1}, using three-step excitation. We observed new autoionization states and analyzed them by Fano's formula.

EXPERIMENT

Three-step excitation was used to investigate the autoionization state of Sm. The first excitation step was fixed at a transition from the ground state, $[4f^66s^2]$ $^7F_1$, to the first excited state $[4f^66s6p]$ $^7F_2^0$, of which the energy is 17190 cm^{-1}. Then the even states with the energies 34522 cm^{-1}, 34420 cm^{-1}, and 34399 cm^{-1}

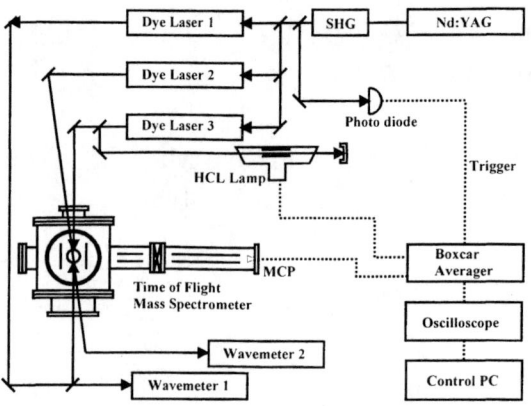

FIGURE 1. Schematic diagram of the experimental setup

were chosen as the intermediate levels. Finally, the energies of the autoionization state were obtained by scanning the wavelengths of the ionization laser in 550nm-700nm. The experimental setup for three-step excitation of Sm is schematically shown in Fig. 1 and is similar to the one described previously [2]. For the first and the second excitation step, we used two single-mode dye lasers pumped by a frequency-doubled Nd:YAG laser. The transition to an autoionization state from the second excited state was made by using another broadband dye laser pumped by the same Nd:YAG laser. Two wavemeters having resolutions of 200 MHz were used to measure the wavelengths of the two single-mode dye lasers. Additionally, to calibrate the ionization laser, a small portion of the dye laser pulses were used to observe the optogalvanic spectrum of a Sm-Ar hollow-cathode discharge lamp. The atomic beam of samarium was produced in a resistively heated tantalum boat containing the natural solid Sm. The photoions of Sm were detected at the end of the flight tube by a microchannel plate(MCP). Finally, the ion signals were averaged by a boxcar averager and the mass spectrum was recorded by a digitizing oscilloscope.

RESULTS AND DISCUSSIONS

Figure 2. shows a part of typical photoionization spectra by three-step excitation. The figure shows signals originating from the intermediate state with the energy value of 34399 cm^{-1} ($J=1$). Including the results shown in Fig. 2, we could observe more than sixty lines in the investigated region as new autoionization states. Neglecting the small transition lines, all experimental results in the investigated range are summarized in Table 1. The linewidths and Fano q-parameters in Table

TABLE 1. List of the investigated autoionization states

Intermediate. state(cm^{-1})	Ionizing waveln.(nm)	Autoionization state(cm^{-1})	Fano q	Width (cm^{-1})	signal[a] strength
	565.415	52080.0	59.0	36.0	s
	573.082	51843.4			w
	573.534	51829.7	21.0	8.6	m
	573.850	51820.1	-24.0	9.7	s
	575.226	51778.4			w
	575.313	51775.4			w
	575.720	51763.5			w
	579.252	51657.6			m
	582.662	51556.6			w
	583.151	51542.2			w
	590.591	51326.7			m
34399	593.256	51250.2	54.0	8.9	s
(J=1)	594.432	51216.9			w
	597.343	51134.9			w
	601.367	51022.9			m
	604.574	50934.8	32.7	23.7	s
	607.195	50863.4			m
	629.152	50288.9			m
	647.099	49848.1	52.9	5.1	s
	663.415	49468.1			m
	666.181	49405.6			m
	669.106	49340.0			m
	670.169	49338.5			m
	683.314	49029.3			m
	685.258	48987.8	48.6	5.0	s
	685.437	48984.0	56.2	6.4	s
	698.932	48702.4			w
	575.492	51792.2	80.0	13.4	s
	576.403	51764.8			m
	579.125	51683.2			m
34420	588.371	51412.0			w
(J=2)	592.128	51304.2			m
	594.540	51235.7	48.6	11.8	s
	602.733	51007.0			m
	615.171	50671.6	41.1	6.0	s
	615.967	50650.9	64.6	7.8	s
	642.951	49947.8			w
	578.851	51792.9			m
	578.950	51790.0			m
34522	624.082	50541.3			w
(J=3)	628.314	50433.4			m
	630.376	50381.4	109.5	8.1	s
	630.774	50371.3	22.4	6.0	s
	639.332	50159.2	-8.0	22.4	s
	652.319	49728.5	-87.8	9.1	s

[a] s, m, and w represent strong, medium and weak transition strength

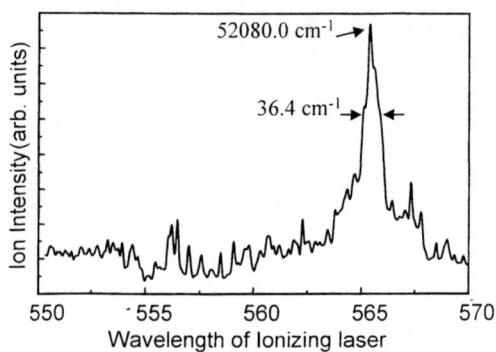

FIGURE 2. The autoionization spectrum in the scan range 550nm-570nm of the ionizing laser (intermediate level : 34399 cm^{-1}-1 (J=1))

1 were decided by fitting the experimental data against Fano's formula [3] which is generally used to analyze the interference characteristics between the autoionization states and the continuum states. Most q parameters recorded in Table 1 are much larger than 1. This means that the measured autoionization spectra exhibit nearly symmetric line profiles. From this fact, we can argue that the interference between the autoionization states and continuum states above the first ionization potential does not affect significantly the photoionization spectrum. The autoionization states which are located in 48800 cm^{-1} - 52700 cm^{-1} are placed between the $4f^66s$ and $4f^65d$ ionization limit. In Ref.[4], Dobryshin, $et\ al.$ calculated the energy values of the $4f^65dnp$ and $4f^65dnp$ configurations by using Hartree-Fork-Dirac method. In that paper, they said that the energy values of $4f^65d7p$ and $4f^65d8p$ configurations are located in 49951.6 cm^{-1} - 51948.5 cm^{-1}. Therefore, we can infer that our measured autoionization states also have configurations of $4f^65d7p$ and $4f^65d8p$. In the future we will identify the configuration of the observed states including total angular momenta by putting more efforts on the theoretical and experimental investigation.

REFERENCES

1. King, A.S., Astrophys. J., **82**, 644(1935).
2. Hyunmin Park, Hyun-chae Kim, Jong-hoon Yi, Jae-Min Han, and Jongmin Lee, Journal of the Korean Physical Society, **30**, 453(1997).
3. U. Fano, Phys. Rev. **124**, 1866(1961).
4. V. E. Dobryshin, N. A. Karpov, S. A. Kotochigova, B. B. Krynetskii, V. A. Mishin, O. M. Stelmakh, and V. M. Shustryakov, Opt. Spektro. **54**, 415 (1983).

Monitoring of Gd Photoionization Process by Detection of Fluorescence Characteristics

Yongjoo Rhee, Jonghoon Yi, Jin Tae Kim, Hyunmin Park, Jaemin Han, Jongmin Lee, and Moon-Gu Baik*

Laboratory for Quantum Optics, Korea Atomic Energy Research Institute
P. O. Box 105, Yusong, Taejon 305-600, Korea
*Department of Physics, Kyungwon University, Songnam 461-701, Korea

Abstract: Three-photon ionization of gadolinium (Gd) atom was investigated. The ions were detected by time-of-flight mass spectrometer. The amount of residual atoms in the excited state after the laser illumination in the interaction region was estimated by measuring the variation of the fluorescence signal. This method was used to estimate the ionization efficiency.

INTRODUCTION

For a sensitive detection of trace elements using RIS techniques, achieving high photoionization efficiency is essential and developing effective tools to monitor the ionization process is necessary. Resonant photons incident on atoms lead ground state atoms to ionic state through a series of excitation processes. Then the atoms left in the excited state after the laser irradiation eventually decay to lower states and the photons emitted in this decay process from the excited state after multi-photon ionization process indicates incomplete ionization of atoms. Hence the decrease of fluorescence light emitted from the excited states can be a good probe of the ionization efficiency since the decrease of the fluorescence signal can be regarded as the increase of the ionization efficiency and *vice versa*. Fluorescence measurement is usually used as a monitoring method in a plasma diagnosis as it provides a non-contact measurement tool with very high spatial resolution.(1) In this work, monitoring of three-photon ionization process of gadolinium (Gd) atom by this method was investigated. The effect of magnetic field on photoionization dynamics was also studied

EXPERIMENT AND RESULTS

The atoms in the ground state were ionized by three resonant laser beams. The

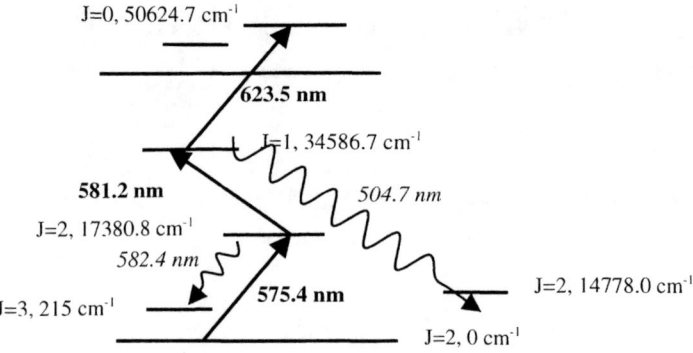

FIGURE 1. Schematic energy level diagram of photoionization scheme and fluorescence from excited states.

ionization pathway is shown in Fig. 1. This scheme was selected as the atomic parameters are well known from the previous work.(2) Ground state atoms effusing from a hole of high temperature oven were excited to the 17380.8 cm^{-1} level (J=2) by a resonant photon emitted from a tunable dye laser source (Lumonics HD-300) pumped by Nd:YAG laser (Lumonics, HY-750).

The fluorescence light from this excited state to lower state was collected and focused by a lens and transported through an optical fiber in a vacuum chamber to a monochromator (CVI, DK240T) as shown in Fig. 2. Vacuum-fiber interface flange was used to transmit the photon into a monochromator where the wavelength and intensity of fluorescence were measured. Since the number of photons arriving at the detector was very small, highly sensitive photon counter (Stanford, SRS430) was used for the lifetime measurement.

The measured lifetime of first excited state is 518 ns which is close to the previous results.(2-3) Branching ratios, wavelengths and oscillator strengths from the 17380.8 cm^{-1} level is listed in Ref. (2). Since most of the fluorescence lines has IR wavelength,

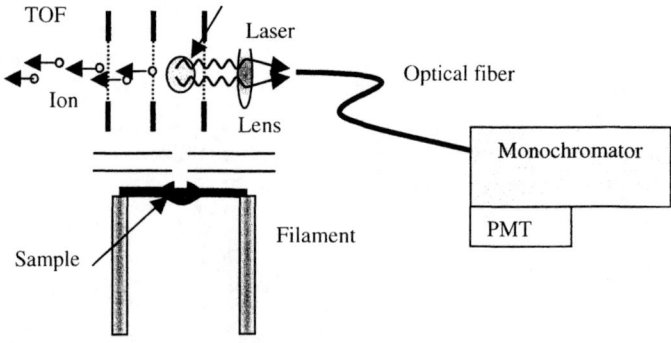

FIGURE 2. Schematic view of an experimental apparatus for simultaneous detection of ions and fluorescence light.

TABLE 1. Air Wavelength, Lower State Energy Level and Strength of Fluorescence Emitted from 34586.7 cm^{-1} Level

Air wavelength (nm)	Energy of lower level (cm^{-1})	Fluorescence strength
761.079	21450.164	Weak
757.121	21381.514	Medium
655.320	19330.624	Strong
653.386	19285.485	Weak
593.784	17749.978	Very Weak
518.042	15289.035	Strong
504.674	14777.975	Very Strong

photons with wavelength of 582.4 nm was detected with our visible sensitive PMT (Hamamatsu, R928) in addition to the resonance fluorescence photon of 575.2 nm.

When the second excitation laser with 581.2 nm wavelength entered into the reaction region at the same time as the first excitation laser, many fluorescence of visible frequency decaying from 34586.7 cm^{-1} level appeared as listed in table 1. Among the lines listed in table 1, 504.7 nm photon showed the most intense signal and Fig. 3 shows the measured photon decay from this state. A trial to measure the decrease of 582.4 nm fluorescence signal emitted from the first excited level was failed since the wavelength of the second excitation laser was too close to the fluorescence line to be distinguished by the monochromator.

The fluorescence signal from the second excited level to 14778 cm^{-1} level decreases, as shown in Fig. 4, when ionizing laser of 623.5 nm wavelength is irradiated. The variation of fluorescence signal indicates the change of excited state population. In the experiment saturation energies were used for the first and the second transitions, which were determined from the saturation curves of the ion signal, to be 0.3 mJ/cm^2

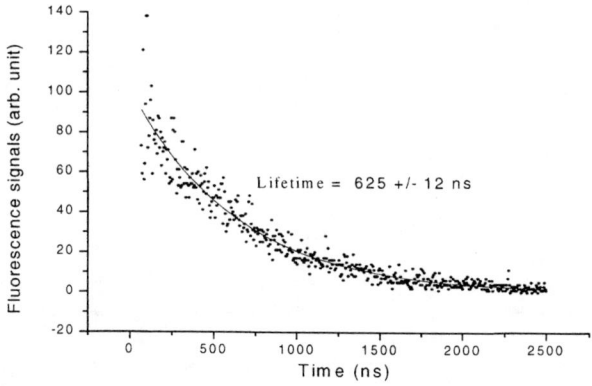

FIGURE 3. Lifetime of 34586.7 cm^{-1} level measured from counting of photon decay.

and 0.5 mJ/cm^2 respectively. Comparing the ion signal with the fluorescence data, it can be seen that the ion signal curve saturates as fluorescence curve saturates.

For the comparison of experimental data with theoretical calculation, level populations were calculated by density matrix equations obtained from Liouville equation. The equation was proved to describe the effect of magnetic field on population dynamics (4) successfully. In this ionization scheme, the angular momenta of the levels follow J=2-2-1-0 sequence. From the angular momentum selection rule, only the odd isotopes are ionized when linear and parallel polarization is used. However the selection rule cannot prevent the even isotopes from being ionized in the presence of external magnetic field that enables population transfer between magnetic sublevels.

In this experiment, filament heating current of 150 amperes produce magnetic field of ~20 gauss in the ionization region and the calculation results in Fig. 5 accounts for the magnetic field effect on photoionization of even and odd isotopes of Gd. In Figure 5, the ionization probability increases rapidly as the intermediate state population decreases. Although evolution of population in magnetic sublevels increases ionization probabilities of even isotopes, it dose not allow the ionization of all the even isotopes completely even at high intensities because Rabi oscillation interferes with population evolution between magnetic sublevels. As the fluorescence signal is proportional to the level population, it does not drop to zero even at high intensities as shown in Fig. 4. The theoretical calculation with the parameters of this experiment shows 60% of the ground state atoms are ionized. Once the system is calibrated with fluorescence signal measurement, fluorescence detection could be used for the ionization efficiency estimation. The more reliable estimation of ionization efficiency can be obtained by simultaneous monitoring of photons emitted from the first and the second excited levels.

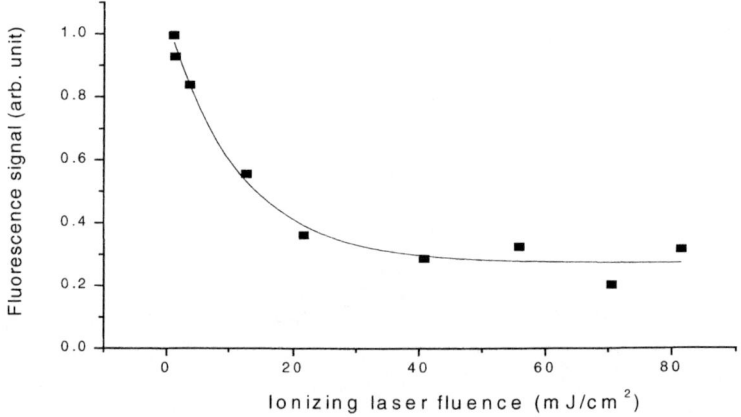

FIGURE 4. Fluorescence signal from 34586.7 cm^{-1} level *vs* energies of the ionizing laser pulse.

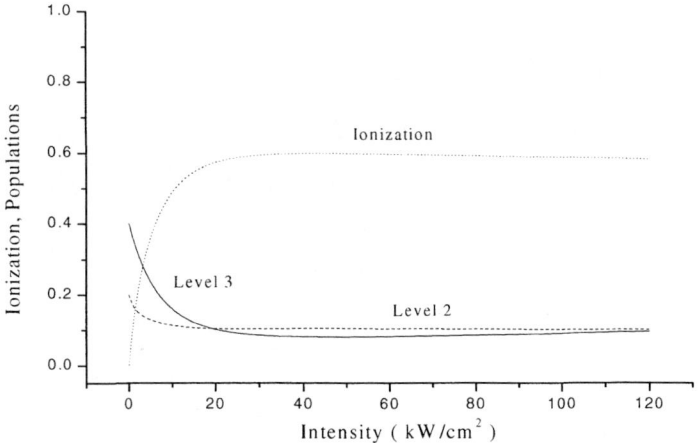

FIGURE 5. Calculation of fluorescence from excited state and ionization signal vs ionizing laser intensity.

CONCLUSION

Gd atoms in the ground state were ionized by three resonant lasers. Some useful lines for the monitoring of excitation process from the second excited state and the ionization process were proposed. Among these, fluorescence light of 504.7 nm was thought to be the most suitable. For the monitoring of the first excited state, IR detection was needed to avoid interference with resonance fluorescence from the second excited state. The result indicates that the degree of ionization of atoms in the second excited level can be estimated through the comparison of density matrix calculation and the fluorescence signal strength. Lifetimes of 17380.8 cm^{-1} level and 34586.7 cm^{-1} levels were measured and compared with the other results.

REFERENCES

1. V. M. Donelly, D. L. Flamm, and G. Collins, J. Vac. Sci. Technol. **21**, 817 (1982).
2. C. Haynam, B. Comaskey, J. Conway, J. Eggert, J. Glaser, E. Ng, J. Pasner, R.Solarz and E. Worden, SPIE **1859**, 24 (1993).
3. M. Miyabe, I. Wakaida, and T. Arisawa, Z. Phys. **D 39**, 181 (1997).
4. Jongmin Lee, Jonghoon Yi, and Moon-Gu Baik, J. Kor. Phys. Soc. **30**, 469 (1997).

POSTER SESSION III
RIS AND ULTRA-TRACE ANALYSIS

A RIS-TOF Instrument for the Measurement of Ultratrace Quantities of Uranium and Plutonium.

A.W. McMahon*, J.D. Gilmour[†], M.B. Hernandez* and M. Rateitzak*.

*Department of Chemistry and Materials, Manchester Metropolitan University, Chester St., Manchester M1 5GD

[†]Department of Earth Sciences, University of Manchester, Manchester M13 9PL

Abstract. A RIS-TOF MS instrument with a modified Wiley-McLaren ion source is described. Samples are loaded onto a rhenium filament from which they are vaporised, followed by single colour, three photon ionization. The instrument incorporates a Nd:YAG pumped dye laser. Two major limitations of such a design lie firstly in the chemistry of the ion source which yield both atomic and molecular species, the latter represents a signal loss or even an interference. The second limitation arises from the combination of a low duty cycle ionization laser (8 ns pulse width at 10 Hz) with a continuous atom source. Uranium has been ionized using the single-colour (591.54 nm) scheme described by Donohue et al. (1) The relative merits of different filament preparation schemes are discussed and a technique is proposed that addresses sample losses due to the low duty cycle of the 10 Hz laser.

INTRODUCTION

An instrument is being developed for the measurement of ultratrace quantities of uranium and plutonium. The device will be applied to the determination of U and Pu concentrations and isotope-ratios in environmental and biological samples. Initial investigative work has focused on uranium because of its relative ease of handling and the ready availability of aqueous standards and hollow cathode lamps. The instrument employs two Nd:YAG lasers (JK System 2000 and Lumonics HY400) and a dye laser (Lumonics HyperDYE 300). Uranium has been ionized using the single-colour (591.54 nm), three photon scheme of Donohue et al. (1). It is our intention to use the same authors' single-colour (588.04 nm), three photon scheme for ionization of plutonium. Laser wavelength calibration was performed using neon and uranium resonances by observing the optogalvanic effect in hollow cathode lamps. The samples were deposited onto Cathodeon 511 Re filaments (Cambridge, UK) or boat-shaped filaments (Cathodeon 527) in the resin bead work. Ions were generated in a modified Wiley-McLaren type source and detected by a Photek channel plate detector, in a linear, 2m time-of flight spectrometer. The combination of a filament, which is a continuous atom source, with a low duty-cycle, pulsed laser (8 ns, 10 Hz) time-of flight system, is extremely wasteful of very small samples. Attempts to electrically pulse the filament temperature have been reported (2) but the relatively high thermal mass of the filament prevents neutral beam modulation at the desired frequency. Our intention is to gently pulse the filament surface temperature using the fundamental emission of the second Nd:YAG laser, without heating the entire filament. This rapid surface heating procedure should allow more efficient sample use. Neutral species' kinetic energy distribution will be examined to determine the nature of the laser-hot surface interaction, in an attempt to distinguish pure thermal and laser desorption processes.

The accurate measurement of isotope ratios requires a technique that eliminates or can reproducibly compensate for isotopic biases. Direct resonance ionization of atomic uranium potentially suffers from isotopic biases due to fractionation at the filament and spectroscopic biases relating to the fine structure, isotope shifts and laser wavelength stability, power and bandwidth. These problems can be addressed by careful experimental procedure and by the use of standards and reference materials. It is our goal to work with this type of signal. However, non-resonant U^+ signals generated by photodissociation of molecular species will each be subject to a different isotope bias based on filament fractionation processes and isotopically sensitive photo-fragmentation cross sections. Since it is unlikely that the relative contribution of the neutral-atomic and molecular U^+ signals can be maintained at a constant level it is crucial to eliminate the molecular component of the U^+ signal. Accordingly, we are studying the response of the molecular and resonant signals to various methods of filament preparation.

RESULTS AND DISCUSSION

Samples can be loaded onto a filament in a variety of ways, including direct deposition of aqueous samples, co-deposition with colloidal graphite (AquaDag), electrodeposition or in the form of a sample-loaded ion-exchange resin bead. Two further sample modifications have been investigated, these involve the vacuum deposition of carbon and aluminium over uranium electrodeposits. Sample coating, in this manner, is expected to provide a degree of protection from air-oxidation and a reducing environment once the bead is heated in the vacuum system of the spectrometer.

Direct Deposition

Direct deposition onto the filament from uranyl solution, which are dried in air, gives only the molecular uranium ions UO^+ and UO_2^+ and molecular neutral-derived U^+. The wavelength dependence of ion signal intensity was recorded in the wavelength range 588 -598 nm (Figure 1) and showed similar profiles for U^+ and UO^+ suggesting a common source, presumably ionization of UO. The spectral profile for UO_2^+ is clearly different. This hypothesis is supported by the fact that co-deposition of the sample with colloidal graphite suppresses the UO_2^+ signal but has no significant influence on the $UO^+:U^+$ signal ratio off resonance, at 591.0 nm. The presence of a reducing carbon environment does, however, give rise to a resonant component (atom-derived) to the U^+ signal on resonance at 591.54 nm (Figure 2).

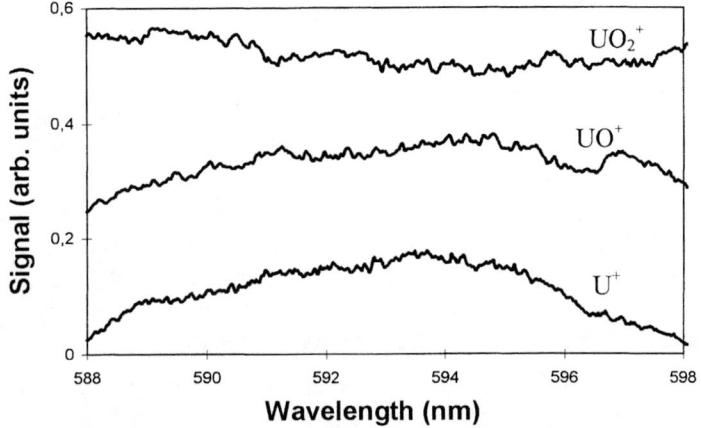

FIGURE 1. Molecular neutral-derived ion signals. The sample was a dried uranyl nitrate solution.

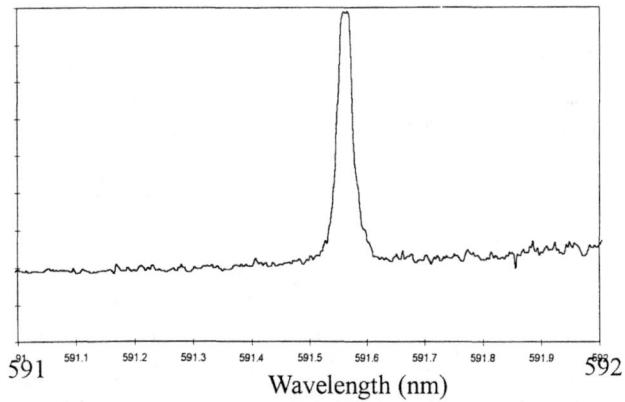

FIGURE 2. ^{238}U signal as a function of wavelength. The sample was a dried uranyl nitrate solution together with colloidal graphite.

Electrodeposition

Electrodeposition from bulk solution was achieved by masking all but the target area of the filament with clear nail varnish. This can subsequently be removed by washing with acetone and forms an adequately robust barrier during the electrodeposition process. Electrodeposition was from a 10 ml volume of solution (0.7 M NH_4Cl, 0.3 M HCl electrolyte), passing a current of 100 mA at ~ 5V for 1 hour.

Experiments in which large (μg) quantities of uranium have been electrodeposited from uranyl solution onto filaments lead to a dark, almost black deposit on the filament surface. Within an hour at atmospheric pressure, the deposit turns to a green colour before developing a yellow hue. The process presumably corresponds to the deposition of uranium as UO_2 (described as a dark brown material (3)), slow oxidation by air to U_3O_8 (very dark green (3)) and ultimately UO_3 (orange-yellow (3)). The oxidation process is clearly observed in the mass spectrum. If an electrodeposited sample is stored at reduced pressure (10^{-3} mbar in a vacuum dessicator) prior to measurement, thereby limiting its exposure to atmosphere to a few minutes, the UO_2^+ peak in the mass spectrum is almost absent, compared with the strong UO_2^+ peak observed in a sample stored at atmospheric pressure for several hours, see Figure 1.

Electrodeposited samples were vacuum coated with carbon, causing almost complete removal of the UO_2^+ signal (see Figure 3). However it is clear that there is still a molecular neutral derived U^+ signal, again presumably due to photodissociation of UO^+.

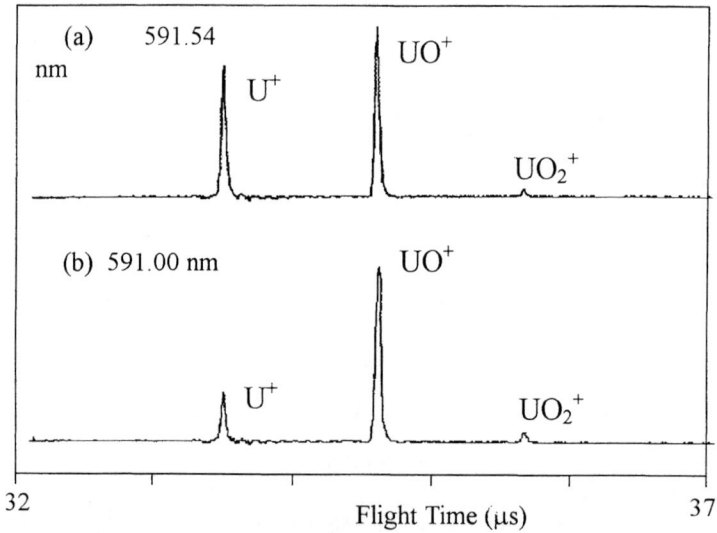

FIGURE 3. U^+, UO^+ and UO_2^+ ion signals derived from a carbon-coated electrodeposit, (a) on and (b) off resonance.

The Resin Bead Method.

Samples loaded onto anion exchange resin beads (Dowex 1X2-100), from 8M HNO_3 gave a pure U^+ signal, with only fleeting appearances of molecular ion signals, see Figure 4. However, tuning the dye laser off resonance, to 591.00 nm reduces the size of the signal but does not eliminate it, indicating that again a non-resonant, presumably molecular route is involved in ion formation. The parent neutral will not be observed unless it is efficiently ionized under the conditions chosen for uranium RIMS. In this instance, it is presumably not UO.

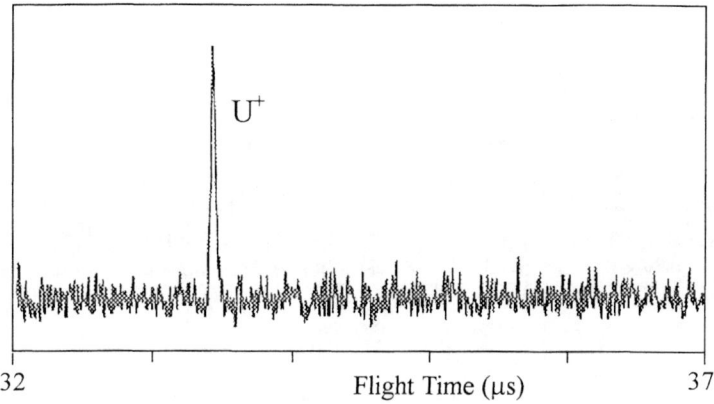

FIGURE 4. ^{238}U signal recorded on resonance (591.54 nm) from a resin bead.

Laser Power

It might be expected that the ionization of molecular species would require higher power densities than atomic resonance ionization. However, because a single-colour scheme has been used, relatively high irradiances are required to couple with the ionization continuum. The 15 mJ per pulse dye laser output was focused into the source region with a 30 cm focal length bi-convex lens. The laser power was then attenuated with a series of neutral density filters, to examine the relative power dependence of the molecular ion signals and the atomic ion signal, both on and off resonance. Both the molecular and the atomic signals were diminished as laser power was reduced but no improvement was observed in the $U^+:UO^+$ signal ratio.

Conclusions

A number of non-resonant ionization processes, yielding U^+ ions, have been observed that might contribute an unquantifiable degree of isotopic bias to our uranium measurements. One of the more important routes appears to be the photodissociation of the neutral UO species. The use of carbon rich environments has been shown to reduce, and for resin beads eliminate, molecular uranium ions. However, it is clear that the absence of molecular ions does not preclude the formation of U^+ from molecular neutral species which are not themselves observed in the mass spectrum.

The laser power studies show that selectivity is not improved at lower laser powers but that sensitivity is significantly diminished. An alternative ionization scheme, which perhaps uses the Nd:YAG fundamental for the ionization step is likely to reduce the laser power requirements for the resonance steps and offer greater discrimination against molecular interferences.

ACKNOWLEDGEMENTS

The authors would like to thank AWE, Aldermaston and the Royal Society of Chemistry for their support. One of the authors was supported by a Royal Society Research Fellowship and another by a Manchester Metropolitan University, Faculty of Science and Engineering Ph.D. Studentship.

REFERENCES

1. Donohue, D.L., Smith, D.H, Young, J.P., McKown, H.S. and Pritchard, C.A.. *Anal. Chem.* **56**, 379-381, (1984).

2. Fassett, J.D., Travis, J.C., Moore, L.J. *Proc. SPIE.* **482**, 36-43, (1984)

3. Greenwood, N.N. and Earnshaw, A., *Chemistry of the Elements*, Pergamon Press, 1984, p. 1472

Diode-Laser-Based Resonance Ionization Mass Spectrometry of Gadolinium

K. Blaum[†], B. A. Bushaw[*], C. Geppert[†], P. Müller[†],
W. Nörtershäuser[†], A. Schmitt[†], N. Trautmann[‡], K. Wendt[†]

[†]*Institut für Physik*, [‡]*Institut für Kernchemie,
Johannes Gutenberg-Universität, D-55099 Mainz, Germany*
[*]*Pacific Northwest National Laboratory, Richland, WA 99352*

Abstract: A compact diode-laser-based resonance ionization mass spectrometer has been adapted for determination of stable gadolinium isotopes. The experimental system combines narrowband resonance ionization and quadrupole mass spectrometry. It will be primarily used for precise isotope ratio determination for cosmochemical studies. Other possible applications are discussed. Excitation schemes for single- to triple-resonance ionization are presented which can be realized using solid-state diode laser systems. First studies include the isotopic abundance sensitivity of the mass spectrometer and high-resolution isotope shift measurements of all gadolinium isotopes in the $6s^2\ ^9D_2 \rightarrow 6s6p\ J=3$ transition, which are reported for the first time.

INTRODUCTION

Sensitive trace determination of stable and long-lived isotopes of gadolinium has a variety of applications in a number of fields: In cosmochemical studies, the ratio of the minor stable isotopes ^{152}Gd/^{154}Gd reflects the branching ratio R of the s-process (neutron capture at slow rate) path at ^{151}Sm. Thus, isotope anomalies in single grain meteorite inclusions give information about R, leading to a value of the stellar temperature during the s-process (1,2). However, conventional thermal-ionization mass spectrometry is limited by isobaric interferences from ^{152}Sm and only a few limited results are available (3). Furthermore, there is demand for biomedical studies to determine the abundance of Gd in blood and human tissue since Gd-chelate is used as a primary contrast medium for magnetic resonance imaging and little is known about the kinetics and physiological relevance of Gd in the human body. Another application of an efficient excitation scheme for multi-step resonance ionization might be the isotope enrichment of ^{155,157}Gd. Since these isotopes have a very large neutron capture cross-section, they are used as burnable poison in nuclear reactors (4). Due to this wide range of interest, we chose Gd as a test candidate for adapting our diode-laser-based resonance ionization mass spectrometer (5,6) to the detection of rare earth elements.

The ionization potentials of Gd (6.15 eV) and Ca (6.11 eV) are similar. Hence, resonance ionization can be performed using single- to triple-resonance excitation with visible light and subsequent photo-ionization with an appropriate wavelength. Corresponding excitation schemes which can be realized using solid-state diode laser systems are given in Figure 1. The simplest excitation scheme is a single-resonance two-photon ionization (Fig. 1a), which is well suited for precise measurements of isotope ratios and will be addressed in this work. The $6s6p$ $J=3$ state is populated by

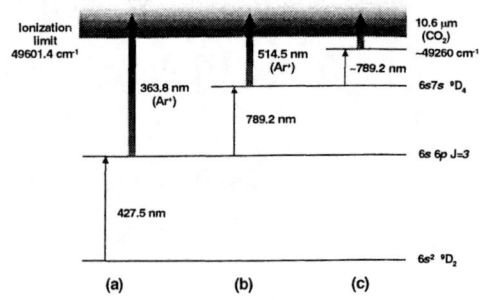

FIGURE 1. Typical single-, double-, and triple-resonance RIS schemes for gadolinium. **a-c** see text

blue light at 427.5 nm, produced by frequency doubling of an extended cavity diode laser. Nonresonant ionization is accomplished with UV radiation from an Ar ion laser. Both, selectivity and sensitivity can be increased using double-resonance excitation (Fig. 1b) and can use nonresonant ionization in the visible or, even better, resonant excitation to autoionizing levels. Both approaches are limited in excitation efficiency since the uppermost state of each excitation ladder is short-lived and will decay into a number of lower lying ("trapped") levels, hence will be lost for further excitation. Therefore, it will be desirable to use a third step to transfer population to a higher-lying long-lived state offering (a) excitation rates more competitive with decay rates into the "trapped" states and (b) the possibility of efficient ionization performed, e.g., with a CO_2 laser.

First studies have been carried out on the single-resonance scheme of Figure 1(a). Overall isotope selectivity that can be achieved depends on the abundance sensitivity of the quadrupole mass spectrometer (QMS) as well as on the isotope shifts in the resonance transition. Here we present results for the performance of the QMS and high-resolution measurements of isotope shifts (IS) and hyperfine structure (HFS) in the $6s^2$ $^9D_2$ → $6s6p$ $J=3$ transition for all naturally occuring isotopes of Gd.

EXPERIMENTAL

The diode-laser-based mass spectrometric trace determination system, as described in (2), can be divided into three major components: A graphite crucible for atomic beam production, the laser system for multi-step resonant excitation and ionization of the neutral atoms, and a quadrupole mass spectrometer (QMS) for mass separation and ion detection.

Gd-nitrate samples placed in a cylindrical graphite crucible and heated to 1200 – 1500°C are efficiently reduced and evaporated forming a collimated atomic beam with full-angle divergence of about 10°, enabling good overlap with laser beams of 3 mm diameter. Ions created in the ionization region are extracted with conventional ion optics and analyzed with a commercial QMS. Ions passing through the mass filter are detected with an off-axis channeltron particle multiplier operated in the ion counting mode. A detailed description of crucible-QMS system has already been given in (7).

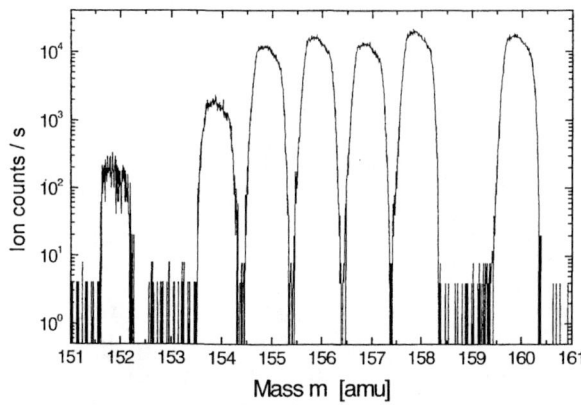

FIGURE 2: QMS mass spectrum in the range of the gadolinium isotopes.

To increase the mass range of the QMS up to 200 amu, it was necessary to decrease the radio-frequency aplied to the electrodes of the mass filter from 2.9 MHz to 1.2 MHz.

MASS SPECTROMETER PERFORMANCE

First studies include the determination of the abundance sensitivity of the QMS, operated at an RF frequency of 1.2 MHz. A typical high-dynamic-range mass spectrum of the Gd isotopes is shown in Figure 2. It was obtained by nonselective thermal surface ionization of a Gd-nitrate sample at 1400°C. The mass spectrum shows all naturally occuring Gd isotopes, with correct relative abundances indicating isobaric or molecular contaminations do not contribute significantly to the Gd peak intensities at these sample loadings. An isotopic abundance sensitivity of about 10^5 is achieved.

ISOTOPE SHIFT MEASUREMENTS

The optical isotope selectivity available in a specific excitation scheme depends on the isotope shifts and hyperfine structure splittings (for isotopes with nuclear spin) between the various isotopes, as well as the experimental lineshapes. Figure 3 shows a typical experimental spectrum of the $6s^2\ ^9D_2 \to 6s6p\ J=3$ transition for naturally occuring Gd isotopes obtained with the single-resonance ionization scheme. DC bias on the mass spectrometer was turned off to transmit all masses while the frequency-doubled diode laser was scanned in 5 MHz steps and ~500 mW of 363.6 nm light was used for the ionization step. The ion count rate was recorded as function of the laser frequency and the sampling time around the least abundant ^{152}Gd was increased by a factor of 10 for better statistics. From the measured spectra, we estimated the isotope shifts between the even Gd isotopes and ^{158}Gd, as given in Table 1.

TABLE 1. Isotope shifts ($v^A - v^{158}$) of even Gd isotopes in $6s^2\ ^9D_2 \to 6s6p\ J=3$. All values in MHz.

^{152}Gd	^{154}Gd	^{156}Gd	^{158}Gd	^{160}Gd
9459(60)	3971(30)	1697(10)	0	-1768(10)

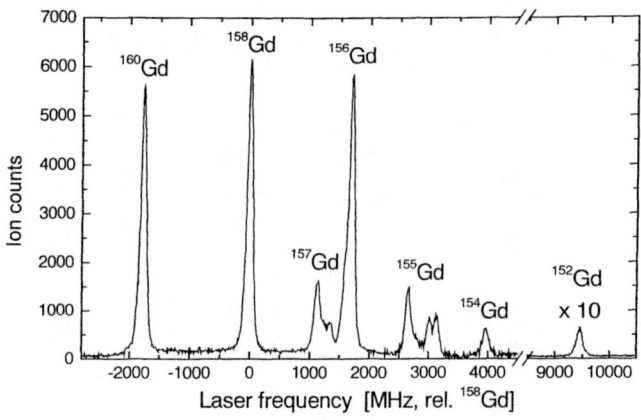

FIGURE 3. Isotope shift measurement for the even gadolinium isotopes in the $6s^2\ {}^9D_2 \rightarrow 6s6p\ J=3$ transition. Laser detuning is given relative to the ^{158}Gd resonance. QMS operated in rf-only mode (no mass selection). In the range around ^{152}Gd, the sampling time is increased by a factor of 10.

CONCLUSION AND OUTLOOK

Diode-laser-based RIMS measurements have been used for spectroscopic studies on gadolinium, with precision IS measurements in the $6s^2\ {}^9D_2 \rightarrow 6s6p\ J=3$ transition for all even stable isotopes. Final data evaluation of IS and HFS for the odd isotopes, which are evaluated with mass-selected spectra because of overlap with the even isotopes, is in progress. Because of the large isotope shifts and an isotopic abundance sensitivity of $>10^5$ achieved by the QMS, excitation with either single-step cw or even broad-band pulsed laser is feasable. To achieve high overall efficiency of $>10^{-5}$, which is necessary for analytical measurements, multi-step resonance ionization, combining cw excitation in the first step and subsequent high repetition rate pulsed ionization with powerful Ti:Sa lasers, might be favorable.

ACKNOWLEDGEMENTS

Funding from the Deutsche Forschungsgemeinschaft and the "Zentrum für Umweltforschung der Johannes Gutenberg-Universität Mainz" is gratefully acknowledged.

REFERENCES

1. Käppeler, F., *et al.*, *Astrophys. J.* **354**, 630-643 (1990).
2. Rolfs, C. E. and Rodney, W. S., *Cauldrons in the Cosmos*, Chicago, Chicago Press, 1988, ch. 9.
3. Hidaka, H., *et al.*, *Anal. Chem.* **67**, 1437-1441 (1995).
4. Santala, M. I. K., *et al.*, *Appl. Phys. B* **64**, 339-347 (1997).
5. Bushaw, B. A., *et al.*, "Multiple Resonance RIMS Measurements of Calcium Isotopes Using Diode Lasers", in *Proceedings of the Conference on Resonance Ionization Spectroscopy*, 1996, pp. 115-118.
6. Wendt, K., *et al.*, *Fresenius J. Anal. Chem.* **359**, 361-363 (1997).
7. Blaum, K., *et al.*, submitted to *Int. J. Mass Spectrom. Ion Process.* (1998).

Trace Analysis of Plutonium by Resonance Ionization Mass Spectroscopy

N. Erdmann, C. Grüning, N. Trautmann, A. Waldek

Institut für Kernchemie, Universität Mainz, D-55099 Mainz (Germany)

G. Huber, P. Kunz, M. Nunnemann, G. Passler

Institut für Physik, Universität Mainz, D-55099 Mainz (Germany)

Abstract. Resonance ionization mass spectroscopy (RIMS) is well suited for trace analysis of long-lived radioisotopes in various types of samples due to its extremely high element and isotope selectivity and its excellent sensitivity. Plutonium atoms evaporated from a filament are excited and ionized in a three-step resonant process. Information about the isotopic composition of a sample is obtained by subsequent mass analysis of the ions in a time-of-flight spectrometer.

A detection limit of $10^6 - 10^7$ atoms of plutonium has been achieved. For most of the long-lived plutonium isotopes, this value is distinctly below the detection limit of α-spectroscopy, which is the standard detection method for plutonium. Different samples (soil, sediment and urine) have been measured with respect to their content and the isotopic composition of plutonium in the ultra-trace regime.

INTRODUCTION

Trace amounts of plutonium are present in the environment mainly as a result of global fallout from nuclear weapons tests, nuclear accidents and releases from nuclear facilities. A sensitive and fast detection method is required for risk assessment, low-level surveillance of the environment, personnel dose monitoring and studies of the migration behavior. The determination of isotopic abundances yields important information about the source and the origin of a contamination as the isotopic ratios depend on the way of production. The standard method for plutonium determination, α-spectroscopy, suffers from difficulties in resolving ^{239}Pu and ^{240}Pu due to their similar α-energies, ^{241}Pu is not detected as it is a β-emitter.

RIMS combines the following requirements for ultra-trace analysis of plutonium: excellent element- and isotope selectivity and a low detection limit, which is independent of the half-live of the isotope. The chemical separation procedure for plutonium from samples can be simplified and information about the total content of plutonium in a sample as well as its isotopic composition is obtained within short measuring time.

SAMPLE PREPARATION

Prior to the sample preparation, tracer isotopes are added to monitor the yield of the chemical separation and of the RIMS measurement. Then, the sample is ashed, fused with Na_2O_2/NaOH and dissolved in water. The residue is dissolved in 8N HCl, and the plutonium is coprecipitated with $Fe(OH)_3$. After dissolution of the precipitate in 4N HNO_3, this solution is passed through a column filled with TEVA•Spec SPS resin (Aliquat 336N). The column is washed with 4N HNO_3 and the plutonium is eluted from the column with 0.5N HCl. The eluate is evaporated to dryness and fumed with concentrated H_2SO_4 to destroy polymeric and colloidal species. Subsequently, plutonium is electrolytically deposited on a tantalum backing from a 20% $(NH_4)_2SO_4$ solution. Finally, it is covered with a thin titanium layer by sputtering (1). By heating such a 'sandwich type' filament in a vacuum chamber, the hydroxide is converted to the oxide. During diffusion through the titanium layer, plutonium oxide is reduced to the metallic state, and an atomic beam is evaporated from the surface of the filament (2).

EXPERIMENTAL SETUP

The experimental setup for RIMS consists of the laser system and the time-of-flight (TOF) mass spectrometer and is described in detail in (3,4). Three tunable dye lasers are simultaneously pumped by two copper vapor lasers at a pulse repetition rate of 6.5 kHz. In order to obtain optimum spatial overlap, the three dye laser beams are guided to the vacuum chamber by a single quartz fiber and focused into the interaction region, where they cross the atomic beam evaporated from the filament. The ions produced via three-step, three-color resonant excitation leading to an autoionizing state (λ_1 = 586.49 nm, λ_2 = 665.57 nm, λ_3 = 577.28 nm for ^{239}Pu) are detected in a reflectron type TOF mass spectrometer to achieve sufficient mass resolution ($m/\Delta m_{FWHM} \approx 600$).

RESULTS

With this facility, a reproducible detection efficiency of $\epsilon = 4\times10^{-5}$ has been determined. ϵ is defined as the ratio of the number of ions detected with RIMS to the number of atoms deposited on the filament. Taking into account the background count rate, a detection limit of 1×10^6 atoms for a single isotope is derived, which is by more than two orders of magnitude better than that for α-spectroscopy

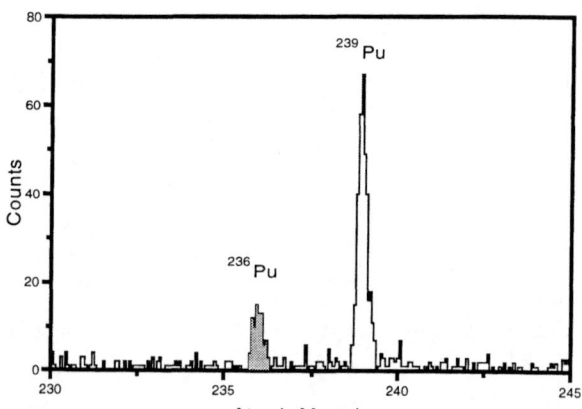

FIGURE 1. RIMS-spectrum of a sample containing 3.6×10^6 atoms of ^{236}Pu and $\approx 2\times10^7$ atoms of ^{239}Pu.

of ^{239}Pu. The high sensitivity is demonstrated by Fig. 1, which shows the RIMS-spectrum of a sample containing 3.6×10^6 atoms of ^{236}Pu and $\approx 2\times10^7$ atoms of ^{239}Pu.

To determine the isotopic composition of plutonium, the first and third laser excitation step must be scanned due to the isotope shifts in the transition wavelengths (5). This leads to small losses in the ionization efficiency, so that 10^7 atoms of the least abundant isotope are necessary to obtain an accuracy of $\approx 10\%$.

The RIMS method has been applied for trace analysis of different environmental samples with respect to the total amount and isotopic composition of plutonium:

Several soil samples from the Chernobyl area were investigated (6). The measured isotopic composition is typical for 'reactor plutonium', and in good agreement with values calculated from reactor parameters given by the management of the Chernobyl nuclear power plant (7).

Also, sea sediment collected at the Mururoa Atoll in the Pacific Ocean (reference material IAEA-368) was analyzed. The isotopic composition was determined to be 97 % ^{239}Pu and 3 % ^{240}Pu which is typical for weapons plutonium. The specified content of 31 mBq/g (^{239}Pu + ^{240}Pu) could be confirmed.

RIMS-measurements of urine samples

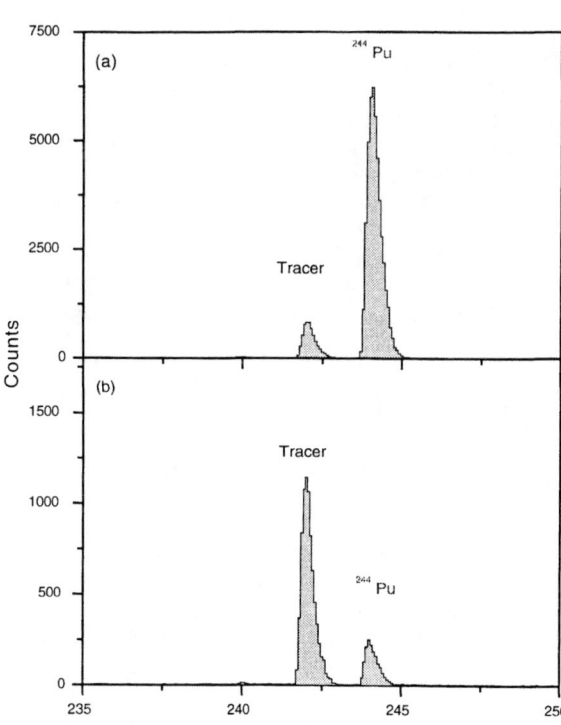

FIGURE 2. RIMS-spectra of plutonium in urine samples (see text for details).

The National Radiological Protection Board (NRPB) in Great Britain is interested in determining the uptake and urinary excretion of plutonium in human volunteers (8). These were injected with $\approx 2\times10^9$ atoms of ^{244}Pu (NBS reference material SRM966); the chemical separation of plutonium from the urine samples after spiking with ^{242}Pu tracer was done at NRPB. A series of 30 urine samples was measured with RIMS. Fig. 2 shows the RIMS-spectra of two urine samples: one taken the first day after injection (a), the other one taken after 30 days; for both samples the same amount of tracer ($\approx 4\times10^9$ atoms of ^{242}Pu) was added.

The amount of plutonium excretion per day in dependence of the time elapsed after injection for all volunteers is shown in figure 3. The RIMS results are in excellent agreement with mass spectrometric data for other volunteer samples.

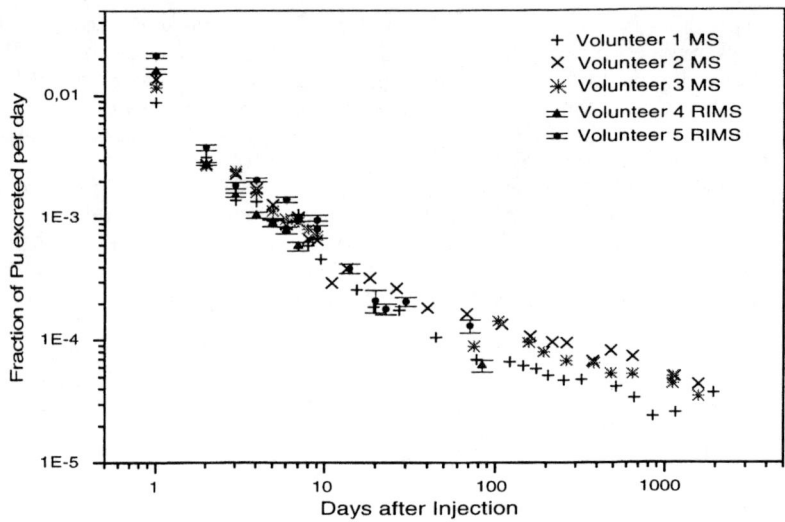

FIGURE 3. Time-dependent urinary excretion of plutonium after injection of $\approx 2\times10^{12}$ atoms of ^{244}Pu.

CONCLUSION AND OUTLOOK

RIMS is a powerful tool for ultra-trace analysis of plutonium in environmental samples due to its excellent detection limit which gives access to the sub femtogram regime, and because of its extreme elemental and isotopic selectivity. The detection limit of this technique is by more than two orders of magnitude better than that of conventional α-spectroscopy for ^{239}Pu. The practicability and reliability of RIMS has been proven for a variety of samples. In the future an all-solid state laser system consisting of three Ti:Sapphire lasers which are pumped by a high repetition rate Nd:YAG laser (9) will be used. A compact system using diode lasers and a quadrupole mass spectrometer is under construction.

REFERENCES

1. K. Eberhardt et al., *Resonance Ionization Spectroscopy 1994*, AIP Conf. Proc. **329** (1995) 503
2. B. Eichler et al., *Radiochim. Acta* **79**, 221 (1997)
3. W. Ruster et al., *Nucl. Instr. Meth.* **A 281** (1989) 547
4. F.-J. Urban et al., Resonance Ionization Spectroscopy 1992, *Inst. Phys. Conf. Ser.* **128** (1992) 233
5. J. Blaise, J.F. Wyart, Energy Levels and Atomic Spectra of Actinides. *Tables internationales de Constantes Selectionées*, Université P. et M. Curie, **Vol. 20**, Paris (1992)
6. N. Erdmann et al., Resonance Ionization Spectroscopy 1996, AIP Conf. Proc. **388** (1997) 205
7. S.N. Begichev, Fuel of the 4th Reactor Unit - Brief Reference Book, Moscow (1990)
8. D.S. Popplewell et al., *Radiat. Prot. Dosim.* **53** (1994) 241
9. C. Grüning et al., these conference proceedings

POSTER SESSION IV
NEW LASER DEVELOPMENTS AND APPLICATIONS

A High Repetition Rate Solid State Laser System for Resonance Ionisation Mass Spectrometry of Actinides

*C. Grüning, *N. Erdmann, †G. Huber, *P. Klopp, *J. V. Kratz, †P. Kunz, †M. Nunnemann, †G. Passler, *O. Stetzer, *A. Waldek and †K. Wendt

*Institut für Kernchemie, Johannes Gutenberg-Universität Mainz,
†Institut für Physik, Johannes Gutenberg-Universität Mainz,
D-55099 Mainz, Germany

Abstract. A new, high repetition rate solid state laser system consisting of three Titanium-Sapphire (Ti:Sa) lasers pumped by one Nd:YAG laser has been set up for resonance ionisation mass spectrometry for routine trace analysis of actinides. Each Ti:Sa laser produces up to 4 W of laser light with a bandwidth of 2-6 GHz continuously tuneable in a range from 725 to 900 nm. Using a three step ionisation scheme with $\lambda_1 = 420.76$ nm, $\lambda_2 = 847.28$ nm and $\lambda_3 < 760$ nm for ^{239}Pu, the overall detection efficiency of the setup has been measured to be $\varepsilon = 8.0 \times 10^{-6}$. Thus a detection limit of 1×10^7 atoms of ^{239}Pu is derived and opens up the way to use a reliable and easy to handle laser system for routine applications of RIMS.

INTRODUCTION

Resonance ionisation mass spectrometry (RIMS) has been extensively applied by our group for ultra-trace analysis and atomic spectrometry of plutonium and other actinides (1,2,3). The laser system used so far consists of three dye lasers pumped by two powerful Cu - vapor lasers. Its important properties are a high repetition rate of 6.6 kHz, a broad tuning range through the usage of suitable dyes and a linewidth of 1 - 4 GHz (1). The drawbacks of this system like its size and high maintenance efforts lead to the development of a new powerful and easy to handle solid state laser system to further facilitate the application of RIMS for routine analysis of actinides. The characteristics of this laser system as well as first measurements carried out on ^{239}Pu at our time-of-flight mass spectrometer (TOF) for the demonstration of the capability and specifications of the system will be presented in this article.

THE NEW LASER SYSTEM

Primary requirements for a pulsed laser system applicable for very efficient resonance ionisation in combination with a TOF - mass spectrometer are high power, a high repetition rate, broad tuning range and narrow linewidth.

A commercially available, Q-switched and intra-cavity doubled Nd:YAG pump laser (Clark-MXR ORC-1000) with a repetition rate of 1 - 25 kHz, a power of up to 50 W at 532 nm and a pulse length of approximately 400 ns is used to simultaneously pump three Titanium-Sapphire (Ti:Sa) lasers. The design of the optical system involves a Z - shaped laser resonator geometry to compensate for the astigmatism of the Ti:Sa - crystal (4,5). With one single set of mirrors, the laser is continuously tuneable from 725 to 900 nm. For wavelength selection, a three plate birefringent filter and a thin etalon of d = 0,33 mm, corresponding to a free spectral range of 300 GHz are installed inside the resonator. This arrangement reduces the linewidth to 2 - 6 GHz. The efficiency of the Ti:Sa lasers over their tuning ranges is shown in figure 1. The drop in the conversion efficiency around 820 nm is caused by the tendency of the laser to oscillate simultaneously at two different orders of the birefringent filter.

The temporal length of the Ti:Sa pulses is 60 - 150 ns. An intracavity Pockels cell is used as a Q - switch to optimize the temporal overlap of the pulses of all three lasers. It allows to shift the Ti:Sa pulses for 3.2 μs relative to the pump pulse, corresponding to the fluorescence lifetime of the excited state of the Ti:Sa crystal.

As the ionisation energies of the actinides are approximately 6 eV, it is necessary to frequency double one laser in order to use a three step ionisation scheme. This is done with an external, single pass setup applying a BBO crystal. Conversion efficiencies of up to 7% are reached. All three laser beams are coupled into an optical fibre and finally focused into the source region of the TOF for optimum spatial overlap with the atomic beam.

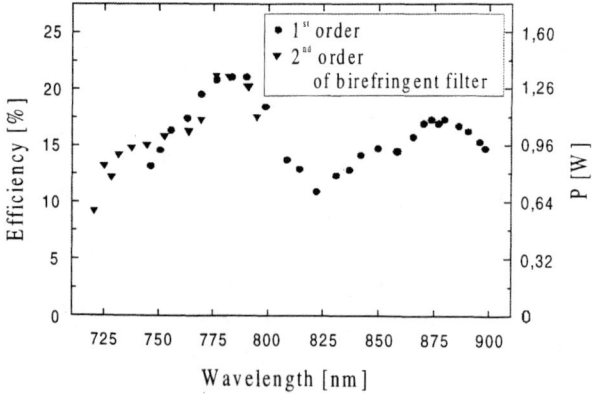

FIGURE 1. Tuning range of the Ti:Sa laser, repetition rate 5 kHz, P_{Pump} = 6,4 W.

MEASUREMENTS

After a number of principal tests of the performance of the laser system regarding:

- laser power
- tuning range
- spectral structure and bandwidth and
- temporal pulse structure,

first measurements of the system on ^{239}Pu were carried out at our standard TOF mass spectrometer (6) with the following procedure: An atomic beam of plutonium is produced by evaporation from a sandwich filament. The atoms are ionised by two step resonant excitation and non-resonant and resonant ionisation. The photoions are subsequently mass analysed in a reflectron time-of-flight mass spectrometer with a mass resolution of $m/\Delta m \approx 600$. It serves for isotope selectivity and efficient background suppression.

Excitation Scheme for Plutonium

Spectroscopic investigations in the spectrum of Pu I have been carried out to search for an efficient three step ionisation scheme using wavelengths of 725 - 900 nm and 365 - 450 nm available from the new laser system.

Starting from the $5f^6\ 7s^2\ ^7F_0$ ground state, a favourable excitation scheme uses a wavelength of $\lambda_1 = 420{,}76$ nm in the first step to the $5f^6\ 7s\ 7p\ ^7D_1$ state. The second step leads to a $J = 2$ state with $\lambda_2 = 847{,}28$ nm. Scans of the ionisation laser showed several autoionising states in the range from $\lambda_3 = 750$ to 758 nm. Those with the highest resonance enhancements R were at $\lambda_3 = 756{,}27$ nm with $R \approx 5$ and $\lambda_3 = 751{,}72$ nm with $R \approx 6$, respectively.

Detection Efficiency with the New Laser System

After exploring the most promising excitation scheme, measurements of the overall efficiency of the new laser system combined with the TOF mass spectrometer for plutonium have been carried out. The repetition rate of the lasers has been set to 3.3 kHz. With laser powers of 20 mW at λ_1, 260 mW at λ_2 and 465 mW at λ_3 in the interaction region of the laser with the atomic beam, it could be demonstrated that the transitions of all three steps were saturated. Figure 2 shows a time-of-flight spectrum of a ^{239}Pu sample, where the filament contained 3.25×10^{11} atoms of ^{239}Pu (notice the logarithmic scale). Pu$^+$ can be clearly seen. Its oxides PuO$^+$ and PuO$_2^+$ are suppressed by more than three orders of magnitude.

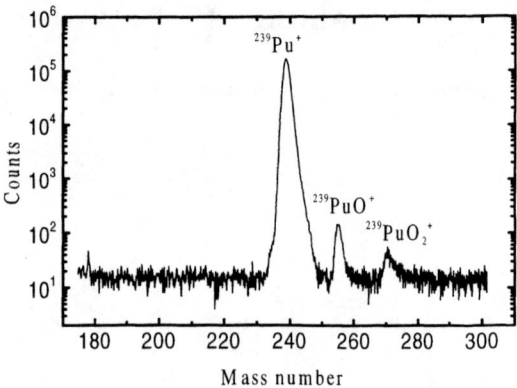

FIGURE 2. Time-of-flight spectrum of ^{239}Pu. Mass peaks of plutonium and its oxides can be seen.

A total number of 2.6 x 10^6 ions of ^{239}Pu$^+$ have been detected during a measuring time of 2.5 hours. Thus, an overall efficiency of ε = 8.0 x 10^{-6} is obtained. The background was less than 30 counts per mass - channel. Taking the background into account, we derive a detection limit of approximately 1 x 10^7 atoms of ^{239}Pu at a confidence level of 99,7 % (3σ).

CONCLUSION

It has been demonstrated that the new Nd:YAG pumped Titanium-Sapphire laser system is well suited for ultra-trace analysis of plutonium. The detection limit of 1 x 10^7 atoms is comparable to the one obtained with the dye - / Cu - vapor laser system (6). The routine operation of this laser system for ultra-trace analysis and atomic spectroscopy of actinides is planned.

REFERENCES

1. Passler, G. et al., *Kerntechnik* **62,** pp. 85 (1997).
2. Erdmann, N. et al., *Fresenius J. Anal. Chem.* **359**, pp.378 (1997).
3. Köhler, S. et al., *Spectrochimica Acta* **B52**, pp. 717 (1997).
4. Knowles, M. R. H. et al., *Optics Communications* **89**, pp. 493 (1992).
5. Kogelnik, H.W. et al., *IEEE Journal of Quantum Electronics* **QE-8**, pp. 373 (1972).
6. Erdmann, N. et al, "Trace Analysis of Plutonium by Resonance Ionization Mass Spectroscopy (RIMS)", *this volume*.

CW Laser-initiated Resonance Ionization Mass Spectrometry of Gd Atoms

Do-Young Jeong, Yongjoo Rhee, Jongmin Lee, and Bum Ku Rhee*

Laboratory for Quantum Optics, Korea Atomic Energy Research Institute, P. O. Box 105, Yusong, Taejon 305-600, Korea
* Department of Physics, Sogang University, 1 Sinsoo-dong, Mapo-ku, Seoul 121-742, Korea

Abstract. The cw laser-initiated resonance ionization mass spectrometrry (RIMS) of Gd atoms has been investigated. Three-color three-photon ionization of Gd atoms was achieved by using a cw ring dye laser for the first transition (999 cm^{-1} - 18084 cm^{-1}) and two pulsed lasers for the second transition (18084 cm^{-1} - 34755 cm^{-1}) followed by an autoionization transition. The spectrum consisting of three peaks as a function of the cw laser detuning was observed for a strongly driven second transition. A narrow dip at the resonance frequency was also observed, which was ascribed to the population loss from the first excited state.

Resonance ionization spectroscopy (RIS) combined with mass spectrometers has been rapidly developed during past decades not only for atomic spectroscopy but also for mass spectrometric analysis. The selectivity inherent in this processes can eliminate the problem of isobaric interferences, allowing the precise determination of isotopic ratios in mixed elemental samples. The sensitivity allowing single atom detection under controlled conditions has also been demonstrated. The methods to excite and photoionize atoms in RIS can be classified as follows : (a) excitation and ionization with pulsed lasers, (b) excitation and ionization with cw lasers, and (c) excitation with a cw laser and ionization with pulsed lasers. Most studies of RIS have been carried out with pulsed lasers. A pulsed laser has in general high peak power, which makes it possible to achieve high ionization efficiency nearly to unity, and wide tunability which makes nearly all of the elements amenable to RIS. On the contrary, high selectivity under certain conditions, which is essential in rare isotope detection, can be achieved in RIS with narrow bandwidth cw lasers.

The advantages of cw laser-initiated RIS are its higher selectivity than in RIS with pulsed lasers and higher ionization efficiency than in RIS with cw lasers. However, it has not been extensively studied up to now, compared with the other methods. In this paper, we report the cw laser-initiated resonance ionization mass spectrometry (RIMS) in a Gd atomic beam. Our interests were focused not only on the spectral resolution achieved in this method but also on the photoionization lineshapes depending both on the intensity of a cw laser and on the energy of a pulse laser. We have observed the triplet spectrum for a strongly driven transition (18084 - 34755 cm^{-1}) of Gd atoms and a narrow dip at the resonance frequency under certain conditions, when a cw laser is scanned across the resonance.

EXPERIMENTAL SETUP

The natural solid Gd sample placed in tungsten crucible was heated by electron bombardment in a vacuum chamber. The atomic beam was collimated and directed perpendicular to laser beam propagation direction. The temperature of the crucible was kept at 1750 °C within +/- 20 °C. The atomic beam velocity was 510 m/s and the Doppler width for this source in the crossed beam configuration was measured to be 70 MHz. A cw ring dye laser (Coherent 899-29) which has typically 1 MHz bandwidth was used as a probe laser to excite Gd atoms at the metastable state 999 cm^{-1} (J=5) to 18084 cm^{-1} (J=5). A pulsed dye laser as a pump laser induced a strongly driven transition of 18084 - 34755 cm^{-1} (J=6) and an additional pulsed laser was introduced to photoionize Gd atoms via the autoionization state 51241 cm^{-1}. Both pulsed dye lasers (Lumonics HD-500) pumped by a excimer laser have 1.5 GHz linewidth and 10 ns pulsewidth in FWHM. A cw laser beam was focused by a convex lens with 100 cm focal length and arranged to counterpropagate against pulsed beams. After spatial filtering by lenses and an aperture, pulsed lasers were aligned to copropagate simultaneously through the Gd atomic beam with uniform intensity distribution. The diameter of the two pulsed lasers was 2.0 mm in the interaction region. All lasers were linearly polarized along the same direction. The frequencies of a cw laser was monitored by recording the fringe from a Fabry-Perot interferometer (free spectral range : 300 MHz). The time-of-flight mass spectrometer was used to detect a single kind of Gd isotope ions (^{160}Gd in this experiment). The spectrometer was operated in a pulse-mode in which the electric field for ion acceleration was biased after the ionizing laser pulse, in order to avoid the spectral broadening as well as the spectral splitting induced by the electric field.

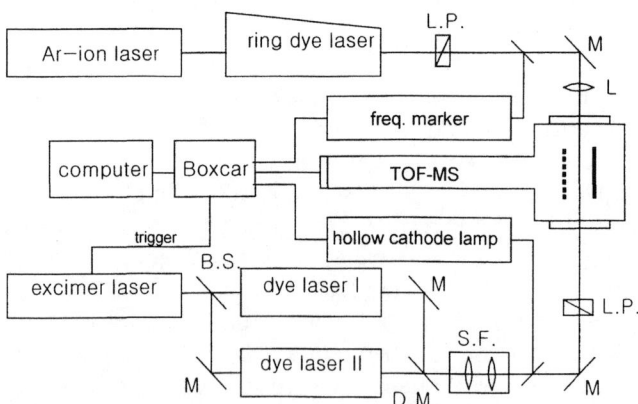

FIGURE 1. Experimental setup for the cw laser-initiated resonance ionization mass spectrometry

RESULTS AND DISCUSSION

Fig. 2 (a) shows the typical photoionization spectrum of ^{160}Gd as a function of the probe laser detuning when the power of the probe laser was 630 mW and the pump laser has the detuning of -3 GHz. The spectrum shows the triplet structure consisting of three peaks, of which the central peak has the relatively narrower linewidth and two side peaks have the broader one. The separation between the two side peaks was about 7 GHz. This value is comparable to the peak Rabi frequency of the pump transition. Fig. 2 (b) shows changes of the linewidth and shape of the central peak as the probe laser intensity was varied, while the detuning of the pump laser was fixed at 3 GHz., When the laser power is 30 mW, the symmetric spectrum and the narrow linewidth could be obtained. The minimum linewidth of the center peak was measured to be 70 MHz with the Gaussian fit, which corresponds to Doppler width of an atomic beam. As the power increases, the spectrum becomes asymmetric and broadened.

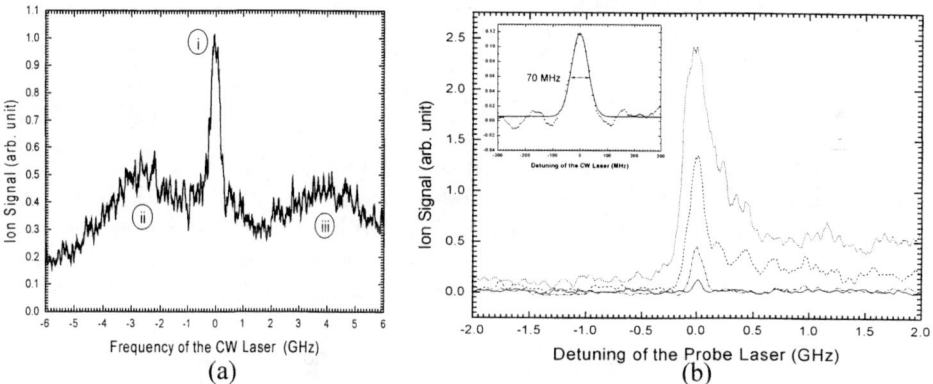

FIGURE 2. (a) The typical photoionization spectrum as a function of the probe laser detuning
(b) Changes of the linewidth and shape of the center peak as the probe laser intensity was varied.

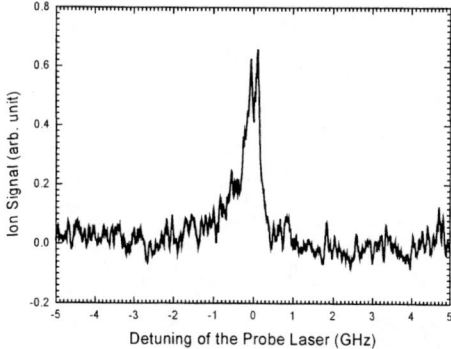

FIGURE 3. The observed dip at the resonant frequency, which is originated from the population loss from the first excited state.

When the pulse beams were focused by a convex lens with 50 cm focal length and the power of the probe laser was 630 mW, only the center peak could be obtained and the dip at the resonant frequency appeared as shown in Fig. 3. The dip is originated from the population loss from the first excited state (18084 cm^{-1}) and the small diameter of focused pulse lasers enhanced the population loss in the photoionization process. Note that the branching ratio of the transition from 18084 cm^{-1} to 999 cm^{-1} is 0.13.

CONCLUSION

We studied experimentally the cw laser-initiated resonance ionization of Gd atoms. We have observed the triplet spectrum for a strongly driven transition of Gd atoms and a dip at the resonance frequency under a certain condition, when a cw laser is scanned around the resonance. It is also found that the observed dip is originated from the population loss from the excited state and the ultimate spectral resolution was limited by the Doppler broadening of an atomic beam in the crossed beam configuration.

REFERENCES

[1] B.D. Cannon, B.A. Bushaw, and T.J. Whitaker, J. Opt. Soc. Am B 2, 1542 (1985)
[2] C.M. Miller, R. Engleman, Jr., and R.A. Keller, J. Opt. Soc. Am 2, 1503 (1985)
[3] B.L. Fearey, D.C. Parent, R.A. Keller, and C.M. Miller, J. Opt. Soc. Am B 7, 3 (1990)
[4] T.W. Ducas, M.G. Littman, R.R. Freeman, and D. Kleppner, Phys. Rev. Lett. 35, 366 (1975)
[5] R.W. Shaw, J.P. Young, D.H. Smith, A.S. Bonanno, and J.M. Dale, Phys. Rev. A 41, 2566 (1990)
[6] Jianan Qu, Fucheng Lin, Zhiyap Zhou, Lizhou Zhu, and Caiyan Luo, Opt. Comm. 78, 153 (1990)

Lasing threshold reduction and the enhancement of mode selection in a novel Littrow-type coupled cavity

Do-Kyeong Ko, Sung-Ho Kim, Byung Heon Cha, and Jongmin Lee

Laboratory for Quantum Optics, Korea Atomic Energy Research Institute, P. O. Box 105, Yusong, Taejon 305-600, Korea

D. J. Binks, L. A. W. Gloster, and T. A. King

Laser Photonics Group, Department of Physics and Astronomy, University of Manchester, Manchester, M13 9PL, United Kingdom

Abstract. We demonstrate a new grating-tuned coupled cavity which enhances mode selection and reduces the lasing threshold. This cavity consists of a Littrow cavity and an additional mirror to feedback the light, which is normally lost through grating reflection, into the cavity. This cavity corresponds to a three-mirror cavity with a spectral gate. We compare the overall performance of this cavity in terms of mode selectivity and threshold with those of the standard grazing incidence cavity(GIC) or with three- and four-arm GICs. Threshold gain and cavity losses are calculated and the predicted mode spacing and complex reflectance are experimentally verified.

INTRODUCTION

Single longitudinal mode(SLM) operation of pulsed tunable laser oscillators typically requires SLM master oscillator as injection seeder. The authors formed a dual-cavity configuration by adding an additional feedback mirror to the grazing-incidence cavity(GIC) and produced narrow linewidth pulses without external injection seeding(1,2). The conventional GIC is an open cavity and has very low efficiency. Feedback occurs by the first-order diffracted beam which has very low diffraction coefficient and the zeroth order diffracted beam which has higher diffraction coefficient is used as the output. By putting an additional mirror at the output of the GIC, we could construct a Michelson interferometric(MI) cavity. The MI cavity consists of a standard GIC and a linear cavity(LC) and they are coupled to each other through the grating. By increasing the cavity length and decreasing the mirror reflectivity of the LC, the GIC pulse builds up more quickly than the LC pulse. In this case, the GIC and the LC act as a seeder and a slave oscillator, respectively. Meriam and Yin have obtained a SLM pulse with a threshold of 20mJ, and a slope efficiency greater than 40% in a Ti:sapphire laser by this technique(2). A similar injection-seeding experiment with a Fizeau interferometer was demonstrated by Peshev *et al.*(3).

The second type of three arm GIC was configured to form a Fox Smith(FS) interferometer which has much greater mode selectivity than that of the Michelson interferometer. The degree to which the GIC mode selectivity is enhanced by the interferometer configuration is determined by the loss modulation produced. The Michelson interferometer modulates the loss sinusoidally which produces moderate additional selection. The Fox-Smith interferometer, however, modulates the loss in an etalon-like manner which is more favourable to mode discrimination. We readily observed single mode operation with a bandwidth of <200 MHz, in comparison, the conventional GIC which operated on 5 modes under identical conditions(4). Recently, Zhang and Tokaryk reported lasing threshold reduction in similar Fox-Smith configurations(5). Although the Fox Smith configuration produces superior mode selection and reduces lasing threshold to some extent, it does not appreciably decrease the threshold of the GIC because it is an open cavity, unlike the Michelson configuration. By adding both the extra tuning mirror and the output coupler in the standard GIC, we could construct a four arm GIC which can be regarded as a superposition of the Michelson and Fox-Smith GICs(6).

In this paper, we report a novel Littrow-type coupled cavity. This cavity is a kind of a three-mirror cavity and has an effective etalon inside the cavity similar to the Fox-

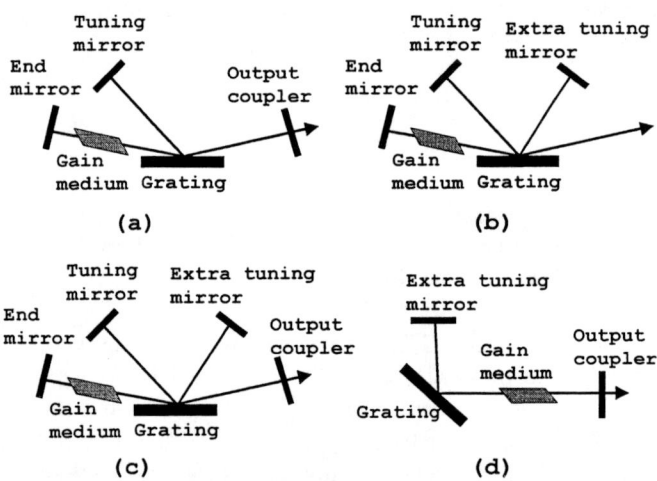

FIGURE 1. Schematic diagrams of various kinds of coupled cavities. (a) MI cavity (b) FS cavity (c) four-arm grazing incidence cavity (d) Littrow-type coupled cavity

Smith GIC. All four cavities are shown shematically in figure 1. We have compared overall performance of these cavities in terms of mode selectivity and the threshold.

LITTROW-TYPE COUPLED CAVITY

The Littrow-type coupled cavity is shown in figure 1(d). A 2400 lines/mm holographic grating 20mm long was positioned at an Littrow reflection angle. The gain medium used was a 0.5mM solution of Rh6G in ethanol. The 2nd harmonics of a Q-switched Nd:YAG laser (Continuum, NY81C-10) with 12 ns duration and the repetition rate of 10 Hz was used as a pump source. The length of the Littrow cavity was 22cm. The output spectrum was measured by a wavemeter(Burleigh WA-4500). The measured linewidth(FWHM) of the standard Littrow cavity was 4.6pm at 588nm. The threshold energy was reduced from 0.4mJ to 0.32mJ when an extra tuning mirror(ETM) was installed to feedback the 0th order (specular) diffracted beam. The distance between the ETM and the grating was 14cm. When the ETM is too close to the grating, broadband emission occurred. By replacing the ETM by a grating, we could enhance the mode selectivity further. The distance between the two grating was about 2.5cm. The linewidth was reduced to 1.87pm as shown in figure 2. In spite of the linewidth reduction, the cavity was still operated in multimode.

GRAZING-INCIDENCE COUPLED CAVITIES

The oscillation condition of the coupled cavities is given by $r_{eff} g_{th} r_o \exp(i2kl) = 1$, where r_o, gth, and $reff$ are the amplitude reflectivity of the output coupler(OC), threshold gain, and the complex amplitude reflectivity due to the grating and mirrors except the OC and l is the cavity length between the grating and the OC. We have measured the effective reflectances(R_{eff}) of the GI coupled cavities. An tapered doide laser (SDL-TC30) seeded by an external cavity diode laser (New Focus 6202) was used

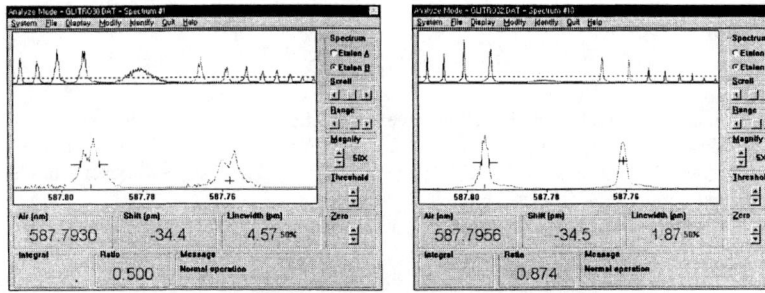

FIGURE 2. Output spectra of the Littrow cavity(left) and the Littrow-type coupled cavity

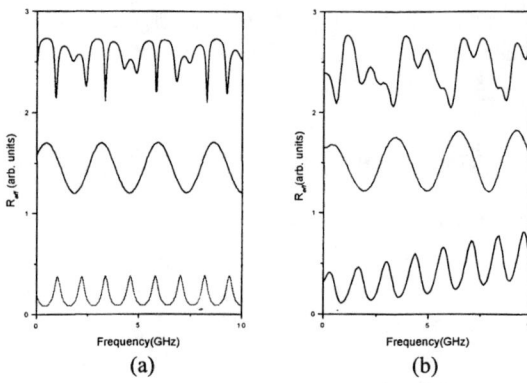

FIGURE 3. Effective reflectances of the Fox-Smith(bottom), the Michelson(middle), and the four-arm GI coupled cavity(up). (a) and (b) are theoretical prediction and the experimental results, respectively.

as a tunable single mode laser source. In order to prevent optical feedback from causing multi-mode operation of the diode, it was isolated from the complex reflector by a Faraday isolator.

Figure 3 shows the theoretical predictions (left) and the experimental results (right) of the reflectances of the Fox-Smith(bottom), the Michelson(middle), and the four-arm GI coupled cavity(up). For the four-arm GI coupled cavity, there are some discrepancies in shape. But the positions of the valleys and peaks are almost the same. By selecting the appropriate lengths of the arms, we can suppress the adjacent modes and get the stable single mode. The R_{eff} of the Littrow-type coupled cavity which was not shown in the figure, on the other hand, has the reversed shape of the R_{eff} of the Fox-Smith coupled cavity. This is a reason why it was difficult to obtain stable single mode although we could get the narrower linewidth.

In summary, we have demonstrated three types of grazing-incidence coupled cavities and a Littrow-type coupled cavity. By the technique of 'self-seeding', the threshold reduction could be achieved and the mode selectivity could be enhanced by interference effect. And the effective reflectances of the coupled cavities were measured and compared with the theoretical predictions.

REFERENCES

1. Ko, D. K., Lim, G., Kim, S. H., Cha, B. H., and Lee, J. Opt. Lett. **20**, 710-712 (1995).
2. Binks, D. J., Gloster, L. A. W., King, T. A., and McKinnie, Appl. Opt. 36, 9371-9377 (1997).
3. Merriam, A. J., and Yin, G. Y., "Efficient self-seeding of a pulsed Ti:sapphire laser," in CLEO'98, Technical Digest Series, 255-256 (1998).
4. Peshev, Z. Y., Deleva, A. D., and Aneva, Z. I., Appl. Phys. Lett. **69**, 2000-2002 (1996).
5. Binks, D. J., Ko, D. K., Gloster, L. A. W., and King, T. A., Opt. Commun. **146**, 173-176 (1998).
6. Zhang, G. Z., and Tokaryk, D. W., Appl. Opt. **36**, 5855-5858 (1997).
7. Binks, D. J., Ko, D. K., Gloster, L. A. W., and King, T. A., J. Mod. Opt. **45**, 1249-1258 (1998).

High-Efficiency Parametric Oscillation in Beta-Barium Borate with Pump Reflection

Jongmin Lee, Sung-Woo Lee, Do-Kyeong Ko, Sung-Ho Kim,
Jae-Min Han, and Byung Heon Cha

Laboratory for Quantum Optics, Korea Atomic Energy Research Institute, P. O. Box 105, Yusong, Taejon 305-600, Korea

Abstract. Beta-barium borate (β-BBO) cut for type I phase matching is pumped by the second harmonics of a Q-switched Nd:YAG laser. In a noncollinear phase matching configuration with pump reflection, the maximum output efficiency of 22.9% was obtained at the wavelength of 756 nm. The threshold was 4.9 mJ and the measured linewidth was 1.5 nm. In collinear phase matching with pump and idler reflection, walkoff compensation was achieved. In this scheme, the maximum output efficiency was 11.5% at 758.5 nm and the threshold was 1.4 mJ.

The optical parametric oscillator (OPO) has been widely used as a wavelength tunable coherent radiation source, since it offers a broad spectrum bandwidth and some of its wavelength is not easily accessible to other tunable laser systems. But, a large birefringence of the nonlinear medium causes the walkoff which seriously deteriorate efficiency of the system by reducing the effective interaction length. To compensate the walkoff, several types of configurations have been proposed. Noncollinear phase matching in which the optical paths between the wave vector of the resonant wave and the Poynting vector of the pump wave can be overlapped inside the medium has proved to be effective way of compensating walkoff.[1,2] Two β-BBO crystals were employed so that each crystal could compensate for the walkoff of the other.[3]

In this manuscript, we present the characteristics of noncollinear phase matching configuration with pump reflection and walkoff compensation by reflecting the pump and the idler wave.

NONCOLLINEAR PHASE MATCHING CONFIGURATION WITH PUMP REFLECTION

β-BaB_2O_4 crystal (CASIX) has an aperture of 4×4 mm^2 with an interaction length of 12 mm and is cut for the type I phase matching (e → o + o) with the cut angle of 21°. The second harmonics of the multimode Q-switched Nd:YAG laser (Lumonics,

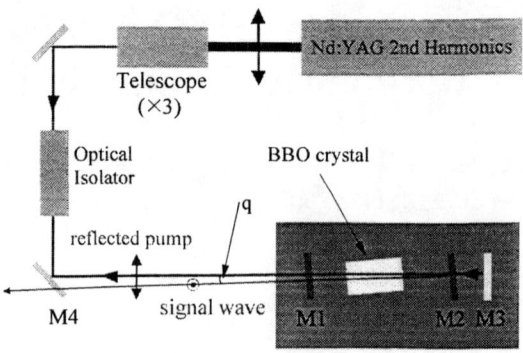

FIGURE 1. A scheme of the noncollinear phase matching configuration with pump reflection

HY750) with 20 ns duration and the repetition rate of 10 Hz was used as a pump source. A schematic diagram of the noncollinear OPO with the pump reflection configuration is shown in Fig. 1. A telescope (×3) is used to reduce the pump diameter to 2 mm. After reflected at the dichroic mirror M4 which has a high reflectance at 532 nm and a high transparency over 680 ~ 780 nm, the pump beam is introduced to the OPO cavity which consists of two flat mirrors M1 and M2 with T > 97% at 532 nm. These two mirrors are separated by 24 mm and tilted by 3° from the pump direction to compensate for the spatial walkoff. The pump beam gets into the cavity through the output mirror M1 which has a dielectric coating of R > 90% over 680 ~ 780 nm. This transmitted pump beam is then reflected back to the cavity by the flat mirror M3. β-BBO crystal is mounted on a rotation stage so that wavelength could be tuned by rotating the crystal. The complete walkoff compensation is achieved at $q = 5.7°$. But, this condition was unable to be attained, because of the crystal entrance aperture. The experiment has been performed at $q = 3°$ at which the angle between the Poynting vector of the pump beam and the wave vector of the signal wave is 1.4°. In collinear phase matching geometry, the angle is 3.1°.[1]

Fig. 2 shows the output energy of the signal wave against the input pump energy at the wavelength of 756 nm in the double-pass noncollinear (■), double-pass collinear (◆), and single-pass noncollinear phase matched geometry (●). The solid line represents a fitting curve. In the double-pass noncollinear phase matched geometry, the maximum output energy of 7.86 mJ corresponding to the efficiency of 22.9% was obtained The maximum output energy of 6.5 mJ corresponding to the efficiency of 18.9% was obtained in the double-pass collinear phase matched geometry, and the maximum output energy of 4.8 mJ which is the efficiency of 16.4% was measured in the noncollinear phase matched geometry. The threshold energy was 4.8 mJ, 4.9 mJ, and 8.6 mJ, respectively.

The spectrum profile of the output signal measured by using Optical Multichannel Analyzer (Oriel, Instaspec III). The linewidth of 1.5 nm is measured in the double-pass

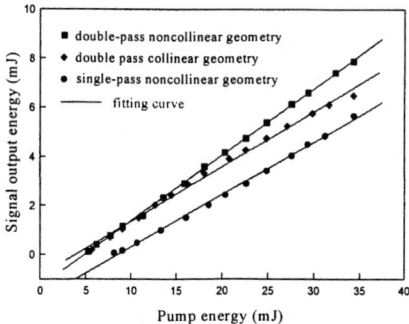

FIGURE 2. Output energy versus input pump energy at the signal wavelength of 756 nm. The solid line represents a fitting curve.

noncollinear configuration, which is much smaller than the linewidth of 7.2 nm in the single-pass noncollinear configuration. In the double-pass geometry, the interaction length is considered to be extended, which causes the linewidth to decrease.

ONE CRYSTAL WALKOFF COMPENSATION GEOMETRY

Figure 3 (a) shows the schematic diagram of the one crystal walkoff compensation configuration. β-BBO (CASIX) was pumped by the second harmonics of the multimode Q-switched Nd:YAG laser (Continuum, NY81C) with 12 ns duration. β-

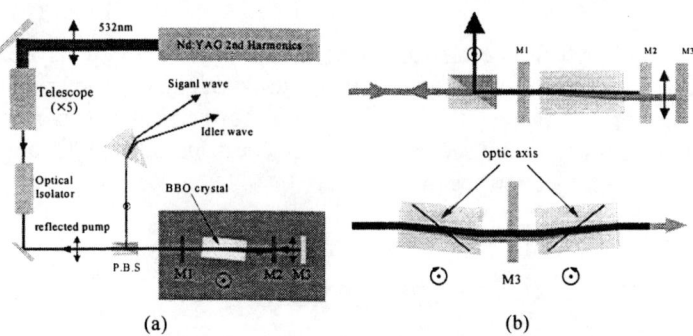

FIGURE 3.(a) Schematic diagram of the one crystal walkoff compensation. (b) detailed description of the cavity (upper), two crystal walkoff compensation equivalent to the upper scheme (below).

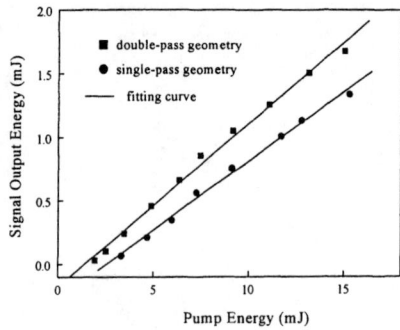

FIGURE 4. Output energy response versus pump energy at 785.5 nm.

BBO is cut for type-I phase matching with 21° cut angle and has the dimension of 4 × 4 × 12 mm. Telescope (×5) reduces the pump beam diameter to 1.5 mm. Polarization beam splitter separates the pump beam from the generated signal and idler wave. The pump beam is reflected at M3 (R > 99% at 532 nm), and the idler wave is reflected at M2 (R > 99% at 758.5 nm and 1781 nm, T = 85% at 532 nm) and pass out the cavity through M1 (R = 90% at 758.5 nm, T > 99% at 1781 nm and 532 nm). The OPO cavity is composed of two flat mirrors of M1 and M2 which are separated by 24 mm, and the distance between M2 and M3 is 4 mm. The detailed description of the cavity is depicted in Figure 3 (b), which shows that our configuration is the same as the two crystal walkoff compensation geometry with its optic axis inverted.

Figure 4 shows the output energy according to pump energy at 758.5 nm. In double-pass configuration, the maximum output efficiency of 11.5% was obtained at 13.1 mJ and the threshold was measured to be 1.4 mJ. In single-pass configuration, the maximum output efficiency of 8.8% at 12.7 mJ and the threshold of 2.3 mJ was obtained.

In this scheme, alignment is easier than that of two crystals, and the synchronization of the angle of each crystal when you change the wavelength is not needed. Due to surface damage of the crystal, the efficiency was lower than we expect. But, this scheme could be applied to the second harmonic generation, CW OPO, amplifier, and the slave oscillator of injection seeding system.

REFERENCES

1. Lee, S. W., Kim, S. H., Ko, D.-K., Han, J. M., and Lee, J., *Opt. Comm.*, **144**, 241- 244 (1997).
2. Gloster, L. A. W., Mckinnie, I. T., King, T. A., *IEEE J. Quantum Electron.*, **30**, 2961-2969 (1994).
3. Bosenberg, W. R., Pelouch, W. S., Tang, C. L., *Appl. Phys. Lett.* **55**, 1952-1954 (1989).

Probing the Response Mechanism of the Thermionic Detector by Resonance Enhanced Ionization Spectroscopy

A.W. McMahon and P.A. Schofield

Department of Chemistry and Materials, Manchester Metropolitan University, Chester St., Manchester M1 5GD

Abstract. The thermionic detector is a sensitive, selective device used in gas chromatography. It responds selectively towards nitrogen- and phosphorus-containing organic compounds with detection limits in the picogram range. The detector is of great importance for the measurement of trace levels of drugs, pesticides and herbicides in biological matrices and the environment. There is, however, some dispute in the literature regarding the detector's response mechanism (1,2,3). The detector is based on a hydrogen-air diffusion flame. Two electrodes polarise the flame with a potential difference of about 200 V and the current through the flame is measured using an electrometer amplifier. The selectivity of the system relies on the presence of a ceramic bead in the flame doped with an alkali metal, usually Rb. In the presence of nitrogen- and phosphorus-containing organics, CN⁻ and PO⁻ anions are formed, yielding a current which is the measured response. It has been suggested (1) that this selective response arises from a charge transfer reaction between the Rb excited states and CN• or PO• and PO$_2$• radicals. Using an AlGaAs diode laser, the Rb excited state population can be modulated and the influence on detector current monitored. The Rb resonance-enhanced ionization signal, laser-induced fluorescence and emission signals have all been used to probe the response mechanism of the detector. The results demonstrate that in the gas phase, the above charge transfer reaction plays little if any role in detector response.

INTRODUCTION

The thermionic detector is a selective ionization detector used in gas chromatography for the trace determination of nitrogen- and phosphorus-containing organic substances. The device's sensitivity and selectivity are of great importance in the analysis of environmental samples for a variety of pesticides and herbicides and for the detection of drugs and toxins in biological systems. There is however some dispute in the literature as to the response mechanism of the device. The detector is a small hydrogen-air

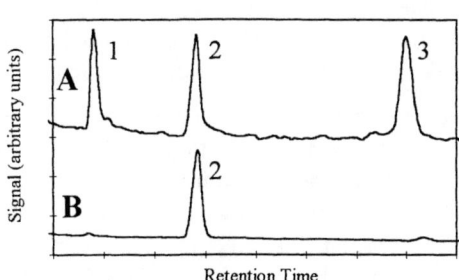

Figure 1. Chromatograms showing Thermionic Detector response to 1) azobenzene, 2) octadecane and 3) malathion, with the Rb source A) hot and B) cold.

CP454, *Resonance Ionization Spectroscopy*
edited by J. C. Vickerman, I. Lyon, N. P. Lockyer, and J. E. Parks
© 1998 The American Institute of Physics   1-56396-810-X/98/$15.00

diffusion flame, through which an ionisation current is measured at a potential difference of a few hundred V/cm. Such an arrangement responds non-selectively to the vast majority of volatile organic chemicals that combust to generate CHO^+ ions in the flame. A sensitive and highly selective component of response is introduced when a Rb-doped ceramic bead comes into contact with the flame. In early designs of detector, the bead was heated only through contact with the flame, however modern detectors use an electrically heated bead.

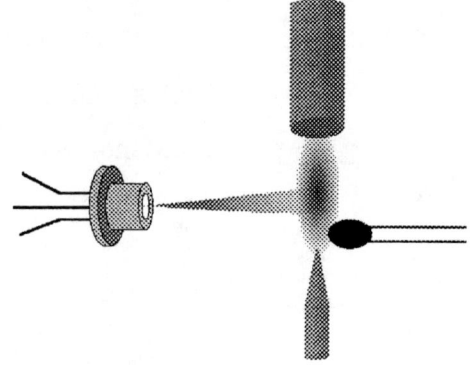

Figure 2. Schematic diagram of the thermionic detector - diode laser system.

The unselective response can be suppressed by adjusting the flame gas ratio, in some systems to a point where combustion is not supported. It has been suggested that Rb is evaporated from the bead and electronically excited in the flame. The $5^2P_{3/2,1/2}$ excited states of Rb (~2.5 eV ionization potential) can then take part in a thermodynamically favourable electron transfer to $CN^{\bullet}$ radicals (3.821 eV electron affinity), yielding ionic products and therefore a signal.

$$Rb^* + CN^{\bullet} \longrightarrow Rb^+ + CN^-$$

The $CN^{\bullet}$ radical is formed by combustion of nitrogen-containing organic chemicals in the flame. The reaction is not energetically favourable for ground state Rb, which has an ionisation potential of 4.177 eV (see Figure 3). A similar mechanism has been proposed for phosphorus-containing organic chemicals which combust to form $PO^{\bullet}$, $PO_2^{\bullet}$ and $HPO^{\bullet}$ radicals. Of these, $PO_2^{\bullet}$ has the necessary electron affinity to allow electron transfer, however $HPO^{\bullet}$ is important as it provides an intense emission, allowing the presence of phosphorus species in the flame to be readily observed.

It is serendipitous that the Rb $5^2P_{3/2}$ state can be excited at 780.0 nm by inexpensive, commercially available AlGaAs, semiconductor lasers. Excitation to this state also populates the $5^2P_{1/2}$ state via collisional energy loss in the flame. The possibility of modulating the excited-state population and thereby modulating the above charge-transfer reaction was investigated. If the reaction could be modulated by simply modulating the excitation laser, then it should be possible to decouple signals arising from the charge transfer reaction from background ionisation signals based on CHO^+ formation. This would give rise to a greater degree of selectivity and even sensitivity. The charge transfer reaction also suggests a route to the enhancement of REIS signals, by charge transfer to high electron affinity neutrals doped into the flame, rather than collisional ionisation.

EXPERIMENTAL

The detector was constructed from a Swagelok cross into which a Hewlett Packard flame ionization detector jet and collecting electrode were inserted, held by electrically insulating Vespel ferrules. Holes were drilled in the sides of the cross for optical access through glass windows held in place against PTFE o-rings. The horizontal arms of the cross held the ceramic bead and a window to allow flame emissions monitoring. The semiconductor laser, a Sharp LT02MD0 50 mW, AlGaAs laser, was used without an external cavity and was therefore susceptible to mode hops within its tuning range. However, it was relatively straightforward to find and maintain resonance. The laser temperature and current were controlled using a Seastar TC 5100 thermoelectric controller and a Seastar LD 23100 laser diode driver.

RESULTS AND DISCUSSION

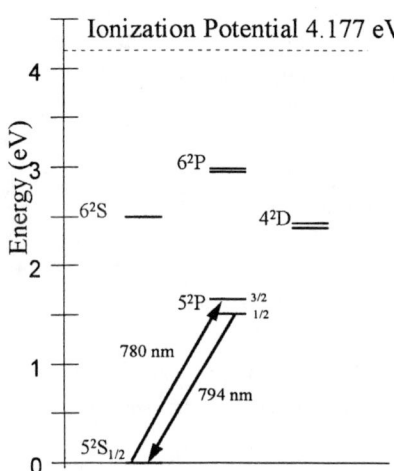

Figure 3. Partial Grottrian diagram for Rb, showing diode laser excitation at 780 nm and non-resonant fluorescence at 794 nm.

Figure 2 shows a schematic of the experimental configuration. Observation of an REIS signal clearly indicates that the $5^2P_{3/2}$ excited state population can be modulated by the laser. However, none of the selective nitrogen or phosphorus responses exhibit a modulated component. The laser beam position has been moved within the flame to examine different flame regions and the beam has been both focused and defocused to achieve high irradiances and high beam overlap with the flame. Despite observing clear REIS signals, no modulation of response has been observed. Before concluding that the gas-phase electron transfer mechanism described above, does not play a role in response, it is necessary to consider other reasons why excited state population modulation might not influence response.

The first possibility is that the degree of excitation is small and any modulation signal is lost in the noise. This argument can be countered by monitoring laser induced fluorescence at 794 nm (see Figure 3). Comparison of the emission and laser-induced fluorescence signals at 794 nm indicates that the $5^2P_{1/2}$ and therefore also the $5^2P_{3/2}$ excited state concentrations have been more than doubled by laser excitation (see Figure 4).

A second possibility is that the Rb $5^2P_{3/2}$ concentration is in vast excess and the CN• radical concentration is immediately consumed on generation and is rate-limiting. This is not supported by the fact that CN• radical emission at 388 nm is readily observed when nitrogen-containing organics enter the flame and that this emission is not influenced by modulation of the Rb excited state concentration. Whilst PO_2• could not be directly observed by optical emission, HPO• emission, at 526 nm, was not influenced by Rb excited state modulation. It is also possible that the 1 kHz laser modulation mostly used in this work was faster than the time constant of the overall detection mechanism. However, at lower frequencies and even an unmodulated laser showed no influence on any of these signals.

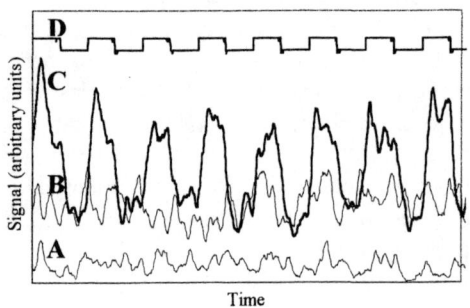

Figure 4. Rb emission and LIF signals at 794 nm. A) emission with ceramic bead cold, B) emission with ceramic bead hot and C) LIF with ceramic bead hot. D) Laser output, to indicate frequency and phase.

CONCLUSIONS.

On the basis of our observations, we conclude that the gas phase charge transfer mechanism involving Rb $5^2P_{3/2,1/2}$ states does not play a significant role in the response mechanism of the detector. The previously reported mass spectrometric observation of enhanced Rb^+ and CN^- ions in the presence of analyte (4), suggest that the reaction is important however our results imply that the reaction occurs at the bead surface.

ACKNOWLEDGEMENTS

The authors would like to thank AEA Technology for their support and the Faculty of Science and Engineering, Manchester Metropolitan University for a PhD studentship.

REFERENCES

1) Kolb, B. and Bischoff, J., *J. Chromatogr. Sci.* **12**, 625-629, (1974)
2) Olah, K., Szoke, A. and Vajta Zs., *J. Chromatogr. Sci.* **17**, 497-502, (1979)
3) Van der Weijer, P., *Anal. Chem.* **60**, 1380-87, (1988)
4) Bombick, D.D. and Allison, J., *J. Chromatogr. Sci.*, **27**, 612-619, (1989)

On-line Monitoring of Trace Compounds in the Flue Gas of an Incineration Pilot Plant: Formation of Polycyclic Aromatic Hydrocarbons

H. J. Heger[1,2], R. Zimmermann[2,*], R. Dorfner[2], A. Kettrup[2], U. Boesl[1]

[1] *Technische Universität München, Institut für Physikalische und Theoretische Chemie, Lichtenbergstr. 4, D-85747 Garching, Germany*
[2] *GSF Forschungszentrum, Institut für Ökologische Chemie, Ingolstädter Landstraße. 1, D-86764 Neuherberg, and TUM, Lehrstuhl f. ökol. Chemie u. Umweltanalytik, D- 85350 Freising, Germany*

Abstract: Laser mass spectrometry is applied for on-line analysis of PAHs from a complex flue gas matrix in the combustion chamber of an incineration plant. Process monitoring of industrial processes can be performed. New insights into the formation of toxic combustion byproducts are possible.

INTRODUCTION

Resonance-enhanced multiphoton ionization (REMPI) combined with Time-of-Flight Mass Spectrometry (TOFMS) is a two dimensional analytical method suitable for direct on-line detection of trace amounts of target components from complex matrices without any cleanup steps. These properties predestine laser mass spectrometers for on-line process analysis of industrial processes. In this contribution the incineration process is used as an example for complex industrial processes. The major source of toxic combustion byproducts is incomplete combustion. Especially waste incineration has to deal with that problem due to inhomogeneous feed composition. One important class of toxic byproducts are the Polycyclic Aromatic Hydrocarbons. With REMPI these substance class can be selectively ionized from the flue gas matrix and on-line monitoring of the toxic products itself can be performed.

EXPERIMENTAL

The mobile laser mass spectrometer for substance class specific PAH analysis is equipped with a compact laser, either Nd:YAG (266nm) or KrF excimer (248nm), and a small linear TOF (50 cm flight tube). The whole instrument is setup in a rack with 80x100 cm dimensions. Data acquisition was performed using a 500 MS/s transient recorder PC Card that enables the storage of every single mass spectrum to hard disk with a repetition rate of 5 Hz. This allows extensive post-acquisition data analysis including unlimited integrators or addition of mass spectra for reduction of the detection limit. Quantification of the monitored species concentration is performed using an external calibration standard based on permeation tubes, providing e.g. naphthalene at ppb level.

---

* corresponding author

CP454, *Resonance Ionization Spectroscopy*
edited by J. C. Vickerman, I. Lyon, N. P. Lockyer, and J. E. Parks
© 1998 The American Institute of Physics   1-56396-810-X/98/$15.00

The measurements[1] were performed at the waste incineration pilot plant of the CUTEC institute GmbH in Clausthal-Zellerfeld[2]. The sampling point was directly in the post combustion chamber with a flue gas temperature of 600 - 800°C.

RESULTS AND DISCUSSION

Typical results showing the time resolution and sensitivity of the REMPI-TOFMS method are given in Fig. 1. The ionization is performed with a KrF-excimer laser (248 nm). Dynamic processes in the combustion chamber can be observed. The left picture shows three short peaks and the respective conventional CO measurement, demonstrating that even in large scale industrial processes rapid, process related effects can be observed. In the right picture a step of about 100 ppt in the naphthalene trace is depicted, showing the start of a change of the combustion conditions.

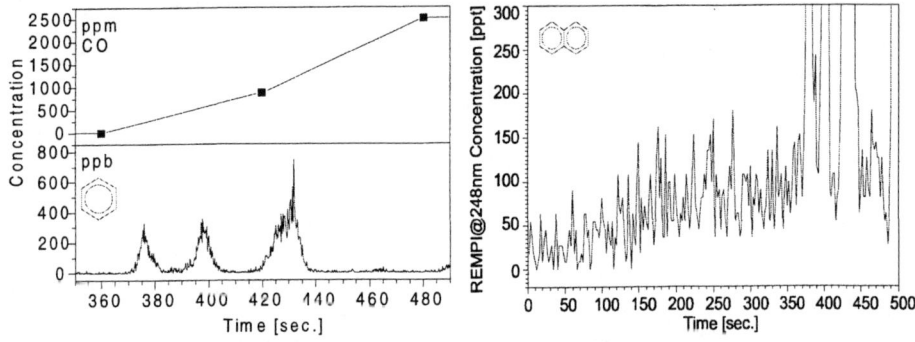

FIGURE 1. Time resolution and sensitivity of the REMPI-TOFMS method.

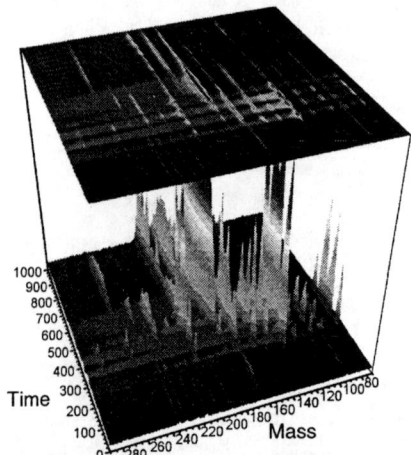

FIGURE 2. Complete Data Set of a REMPI@266nm measurement sequence.

With substance class selective ionization (effusive molecular beam inlet) and acquisition of complete TOF mass spectra a variety of PAHs can be observed in a single laser shot mass spectrum. The data set recorded during a measurement sequence is depicted in Fig. 2. Short Peaks can be observed for PAHs up to 276 amu, which shows that sampling and inlet system are transferring the concentration changes in the flue gas into the mass spectrometer without memory effects.

PAHs are growing[3] by acetylene addition and ring closure. Successive addition of acetylene leads to the PAHs depicted in row I of Fig. 3. Alternatively rings with 5 carbon atoms can be build (row II). Furthermore addition of two PAHs can occur, leading to e.g. biphenyl (row III). With REMPI-TOFMS additionally the alcylated PAHs can be observed (row IV: methylated PAHs).

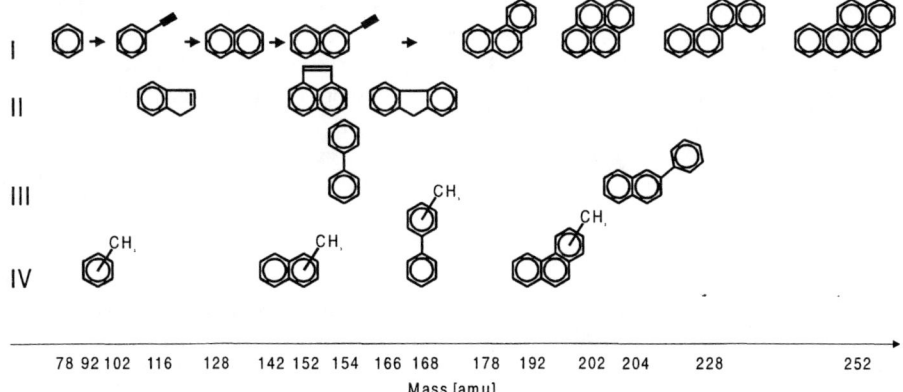

FIGURE 3. Growing of PAHs from benzene.

The formation of the different types of PAHs depends on the conditions during the formation reaction. Basic rules (4) can give hints for an interpretation of the different concentration profiles of the PAHs observed in the combustion chamber. For example, the degree of alkylation is dependent on the temperature: the lower the temperature the higher alkylated the resulting PAHs. Furthermore, PAHs with five rings are formed preferably at lower temperatures. Clustered PAHs (e.g. pyrene) are preferred against chained PAHs (e.g. chrysene) at high temperatures. Although it is not shown yet, whether these simple rules apply for the conditions present in an incinerator or not, they give a first idea for interpretations.

With the above mentioned rules in mind, a contour plot of the flue gas concentrations recorded in a REMPI-TOFMS measurement sequence (Fig. 4) can be seen as a mirror of the underlying combustion conditions. The PAHs and intermediates depicted in row I are simultaneously rising. Fluorene, as an example for PAHs with 5 member rings (row II), shows a nearly phase shifted time behavior. Biphenyl (row II) acts again similar to the PAHs in row I while the methylated PAHs (row III) show slightly delayed concentration peaks.

After the phase with short concentration peaks from 78 amu (benzene) up to 276 amu (e.g. benzo[ghi]perylene) broad peaks of e.g. 178 amu (phenanthrene) and 202 amu (pyrene) are

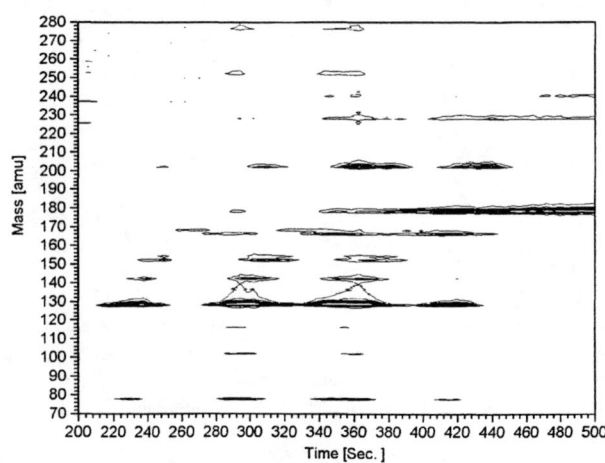

FIGURE 4. REMPI@266nm plot.

observed without simultaneous occurrence of precursor PAHs. This suggests that these emissions are not due to gas phase reactions as before, but maybe a memory effect in the plant due to adsorption and subsequent reaction/desorption from a surface. The PAH emissions can be devided into process related emissions (gas phase reaction) and plant construction related emissions (e.g. memory effects).

CONCLUSION

An on-line monitoring of the formation of PAHs is now possible with REMPI-TOFMS and with the help of experiments performed under well defined combustion conditions in e.g. a flow reactor a detailed understanding of the formation of PAHs in incinerators may be possible.

The variety of concentration to time profiles found for different PAHs indicates that the state of the plant is mirrored by the relation of the PAHs observed by REMPI-TOFMS on-line monitoring. A description of the respective state of the plant should be possible using e.g. pattern recognition methods.

Application of REMPI-TOFMS can lead to an improvement in the plant layout reducing the plant construction related emissions. REMPI-TOFMS experiments in combination with recent combustion simulation methods would be useful for optimizing combustion chamber layout.

Using REMPI-TOFMS for on-line monitoring in combination with feedback steering of the combustion process can reduce the process related emissions. This may decrease the effort of flue gas cleaning and thus decrease the residual waste flow and the costs of building and running an incineration plant.

ACKNOWLEDGMENT

The project is financially supported by the "Deutsche Bundesstiftung Umwelt", Osnabrück. HJH wishes to thank the Max-Buchner-Stiftung der DECHEMA for a research grant. We are indebted to Prof. E. W. Schlag, TU München, Prof. Leschonsky and Dr. M. Beckmann, CUTEC GmbH and Prof. Scholz, TU Clausthal for support of this work.

REFERENCES

1. a) Zimmermann, R.; Heger, H. J.; Kettrup, A.; Boesl, U. *Rapid Comm. Mass. Spectrom.* **1997**, *11*, 1095-1102. b) Heger, H. J.; Zimmermann, R.; Dorfner, R.; Beckmann, M.; Griebel, H.; Kettrup, A.; Boesl, U. *submitted for publication* **1998**
2. Beckmann, M.; Scholz, R.; Wiese, C.; Davidovic, M. *International Incineration Conference* **1997**.
3. a) Bittner, J. D.; Howard, J. B. In *Eighteenth Symposium (International) on Combustion*; The Combustion Institute: Pittsburgh, **1981**, pp 1105-1116. b) Frenklach, M. In *Twenty-Second Symposium (International) on Combustion*; The Combustion Institute: Pittsburgh, **1988**, pp 1075-1082. c) Badger, G. M.; Kimber, R. L. W.; Spotswood, T. M. *Nature* **1960**, *187*, 663-665. d) Siegmann, K.; Hepp, H.; Sattler, K. *Combust. Sci. and Tech.* **1995**, *109*, 165-181.
4. Blumer, M. *Scientific American* **1976**, 234(3), 35-45

Laser Ionisation Mass Spectrometry (REMPI-TOFMS) for On-line Analysis of Volatiles in Food Science: Coffee-Roasting and Headspace Experiments

Ralph Dorfner[1], Ralf Zimmermann[1*], Chahan Yeretzian[2], Antonius Kettrup[1]

[1] *GSF Forschungszentrum Umwelt und Gesundheit GmbH, Institut für Ökologische Chemie, Ingolstädter Landstraße. 1, D-85764 Neuherberg, Germany and Technische Universität München, Lehrstuhl für Ökologische Chemie und Umweltanalytik, D-85748 Freising, Germany*

[2] *Nestlé Research Center, Nestec Ltd., Vers-chez-les-Blanc, P.O. Box 44, CH-1000 Lausanne 26, Switzerland*

* *corresponding author*

Abstract. Using REMPI@266 nm or 248 nm several phenolic compounds, such as Phenol or the flavour-active 4-vinylguaiacol, can be detected in the roast off-gas.

INTRODUCTION

Coffee represents the second most important food in world trade. The world production of coffee is about 5.7 Mio. tons, which corresponds to a proceeds of exportation of 12 billion US $[1]. Taking the high value of coffee into account, the large commercial importance of coffee-processing is obvious. This is particularly true for the roast process, as the typical coffee flavour mostly arises during the roasting of the green coffee beans. Standard roast processes last from 1.5 up to 10 min. Most roast processes start at a roast temperature of 100 °C and end at a roast temperature of 260 °C[1]. Afterwards the cooling of the roasted coffee beans by air or water is necessary to stop the roasting process suddenly. During the roast process, hundreds of different chemical compounds are produce by pyrolysis of the organic material of the green beans. About 1000 volatile compounds have been identified and many of them contribute to the coffee flavour.[2]

The analysis of coffee volatiles is still a challenge for modern trace analysis. High sophisticated analytical techniques and time consuming procedures are required for identification of the compounds that are decisive for the unique taste and flavour of coffee.

Due to the fundamental importance of the roast process for the aroma and quality of the final coffee product, extensive research activities are carried out to get a better insight into the formation of the roasting products and the influence of the process parameters onto the final chemical composition. In the context, the development of on-line monitoring devices to assess time-concentration profiles of specific volatiles during coffee processing is very important for fundamental studies on the generation of flavour-active compounds as well as for process control and optimisation in industrial applications. For monitoring and a future steering of e.g. the roast process, a fast and selective, real-time on-line measurement technique is needed.

EXPERIMENTAL

One approach for on-line analysis of coffee roasting gases is the combination of resonance-enhanced multi-photon ionisation and time-of-flight mass spectrometry (REMPI-TOFMS) which presents a two-dimensional technique, that combines selective ionisation with mass-selective detection. With a suitable fixed-wavelength laser and an effusive gas inlet (warm sample gas) it is possible to achieve a reasonable species-selectivity. For investigation of coffee roasting we used a fixed-wavelength laser @ 266nm (Nd:YAG) or @ 248nm (KrF-Excimer). With these wavelength several phenole derivates and some nitrogen-heterocyclic compounds like indole (m/z 117) or caffeine (m/z 194) are detectable.[3] The multitude of the other roast products is effectively suppressed.

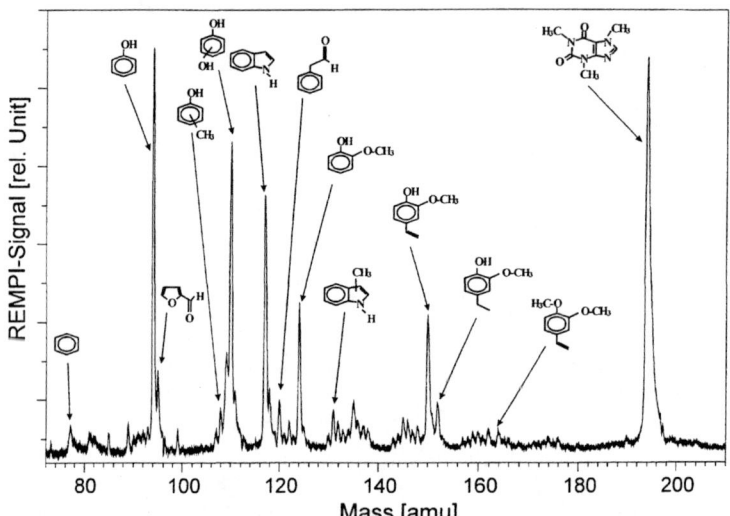

FIGURE 1. On-line REMPI@266nm – TOF mass spectrum of coffee roasting off-gas taken with an effusive sample inlet.

In the laboratory setup the roasting process simulation was performed in a quartz glass tube which is heated by a laboratory heating system. The heating system allows to maintain the roasting temperatures up to 250 °C. The roasting gases are transferred into the ion source of the TOFMS via a heated sampling system and a heated transfer line which contains a deactivated fused silica capillary. The design of the transfer line and sampling system avoids memory effects.

The coffee roasting was performed as follows: The heater was adjusted at a temperature of 250 °C. 1-20 beans green arabica coffee beans were placed into the cold glass tube. Then the glass tube was located into the preheated heating system. At the top of the tube the sampling system was fixed.

RESULTS AND DISCUSSION

A typical REMPI@266 nm mass spectrum of arabica coffee roasting gas is shown in Fig. 1. As mentioned above, the phenolic compounds are very prominent in the REMPI@266 nm TOF mass spectra. Several phenolic compounds as well as nitrogen or oxygen containing heterocyclic compounds are identified (e.g. phenol [m/z 94], cresols [m/z 108], dihydroxyphenol [m/z 110], guaiacol [m/z 124], 4-vinylguaiacol [m/z 150], dimethoxystyrene [m/z 164], indole [m/z 117], methylindoles [m/z 131], caffeine [m/z 194], furfural [m/z 96]), using the molecular mass and the REMPI wavelength as parameters for identification. The peaks at m/z 135 or 109 are due to the weak fragmentation of two phenolic derivates with methoxy groups.

In Fig. 2 REMPI@266nm-TOFMS time-concentration profiles of benzene, phenol, 4-vinylguaiacol and 4-ethylguaiacol, registered during the coffee roasting process, are shown. The different time-intensity behaviour of these compounds suggest an application of the technique for real-time control of technical coffee roasting processes. The relation between

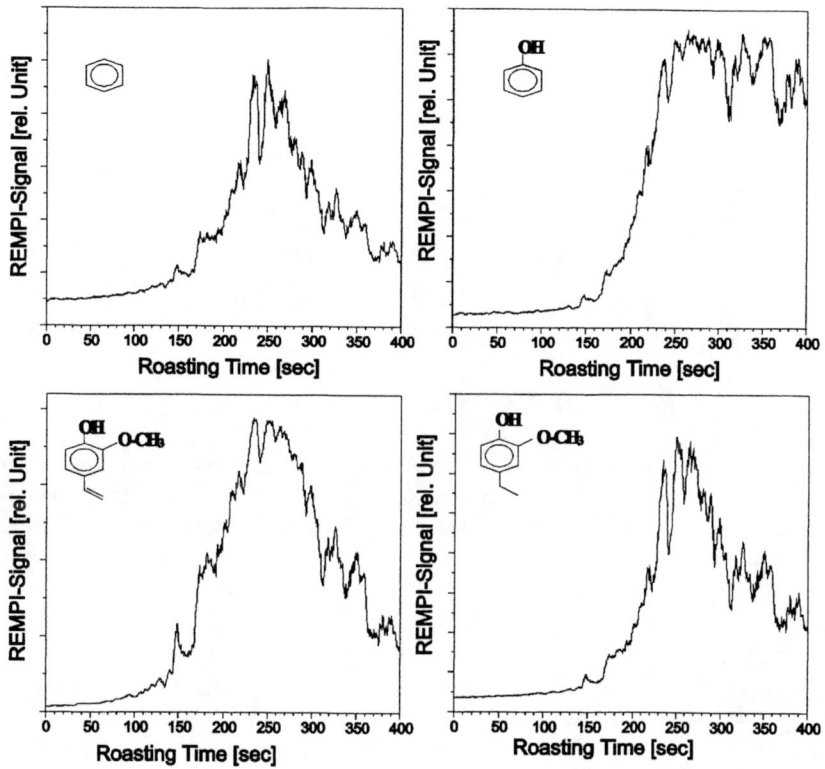

FIGURE 2. REMPI@266nm-TOFMS time-concentration profiles of volatile compounds registered during a single bean roasting process.

phenol and 4-vinylguaiacol is already known as an indicator for the roast degree[4]. So it may be possible to achieve a steady high quality and flavour of the coffee products.

A time-concentration profile of caffeine measured with REMPI@248 nm-TOFMS at very high temperature (400°C) is shown in Fig. 3. After a short latency period, the beans start to emit cracking sounds, indicating "popping-effects" due to carbon dioxide, mostly formed according to the pyrolysis of organic acids. During each "popping" caffeine enriched carbon dioxide gas is emitted. These highly dynamic caffeine eruption are clearly visible in the time-intensity profile, demonstrating the high time resolution and the appropriate design of our

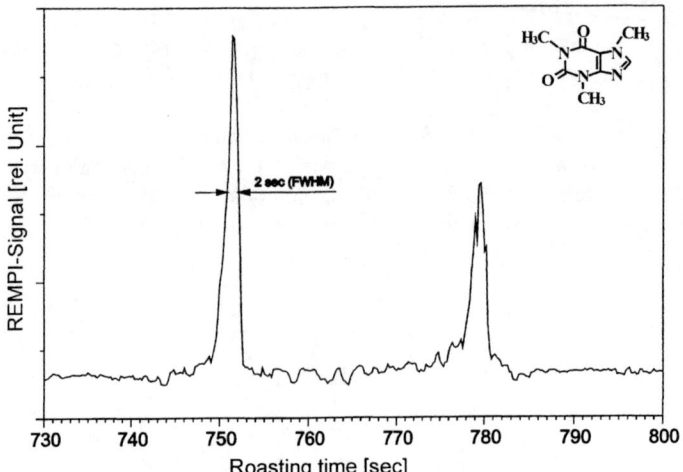

FIGURE 3. On-line registered REMPI@248nm – TOFMS time-concentration profile of caffeine registered during a coffee bean roasting experiment at very high temperatures.

heatable inlet system (no memory effects are observed). Some other compounds like indole show similar behaviour.

CONCLUSION

The REMPI-TOFMS technique can be applied to on-line trace analysis and control in several fields of public or commercial interest (e.g. process and quality control, investigation of the formation and degeneration of the roast products). An on-line, real-time steering of industrial processes is achievable.

ACKNOWLEDGMENT

Financial support from the Nestlé Research Center Lausanne (Nestec Ltd.) is gratefully acknowledged. Further on the author thanks H. J. Heger for support and motivating discussions.

REFERENCES

1. F. Rotzoll, H. G. Müller-Henniges; *Kaffee-Bibliothek*, Hamburg: Deutscher Kaffee-Verband, 1997
2. W. Grosch; Chemie in unserer Zeit, Vol. 30, 126, (1996)
3. R. Zimmermann, H.J. Heger, C. Yeretzian, H. Nagel, U. Boesl; *RCM*, Vol. **10**, 1975-1979, (1996)
4. R. Tressel; *Terminal Generation of Aromas*, T. H. Parliment, R. J. Mc Gorrin and C.-T. Ho (Eds), ACS Symposium Series 409, American Chemical Society, Washington, DS (1989), pp. 285-301

POSTER SESSION V
PHOTOELECTRONS AND DISSOCIATION

Investigation of the multiphoton ionization of Ba atoms with photoelectron spectroscopy.

A. Yu. Elizarov

A.F.Ioffe Physical-Technical Institute, Russian Academy of Sciences, 194021 Politehnichescaja 26, St Peterburg, Russia.

Abstract. An experimental and theoretical study is performed of the angular photoelectron distribution for three-photon ionization of Ba atoms through the 2ω-excited intermediate state $6p^2(^1S_0)$ and the auto-ionized state $6p8s(^3P_1)$. Rotation of the polarization plane of dye-laser radiation allowed us to investigate the photoelectron angular distribution. Electrons were counted with the help of a time-of-flight electron spectrometer. Employing the density-matrix formalism, expressions have been obtained for angular dependence of the differential ionization probability. Possible experiments are discussed.

INTRODUCTION

The study of the photoelectron angular distribution for stepwise ionization of atoms by polarized laser radiation from an excited oriented state is a branch of polarization laser spectroscopy and has been pursued experimentally and theoretically for more than 20 years. This method allows one to obtain unique information about various (including autoionized) states of atoms. The use of tunable lasers and synchrotron radiation opens up new experimental possibilities, for example, the possibility of studying auto-ionized states lying in the VUV region of the spectrum. [1–3]

Excitation of autoionized states by laser radiation in the visible region of the spectrum can be realized with the help of multiphoton resonant or nonresonant excitation processes. An additional advantage of two-photon processes over two-step ones arises for intermediate states with nonzero total angular momentum, when the orientation of atoms in intermediate states can be destroyed as a result of processes of radiation effusion and collisional depolarization. Thus, for example, depolarization of atoms in the intermediate state $6s6p(^1P_1)$ during stepwise excitation by polarized radiation of Ba atoms was observed for atom concentrations in the beam greater than $10^{11} cm^{-3}$ [4].

The influence of depolarization processes is greatly diminished if a two-photon process is used to excite the intermediate state. In this case radiative relaxation to the ground state is forbidden by the selection rules, as a result of which radiation effusion and collisional depolarization have practically no effect on the orientation of the atoms in the intermediate state for atom concentrations in the beam up to $10^{13} cm^{-3}$ ([5]), which provides experimental advantages in high-resolution studies of angular distributions.

A theoretical description of excitation of autoionized states by polarized radiation was first made in [6], where the wave-function formalism was used. The use of density-matrix methods has made it possible to substantially simplify the summation over unobservable projections of the angular momenta. This method has proved to be especially convenient for the description of excitation of autoionized states of polarized atoms. [7–10]. Preliminary polarization of the target atom opens up additional possibilities in the setup of a complete quantum-mechanical photoionization experiment [9].

In this work we consider the photoelectron angular distribution for resonant two-step ionization of auto-ionized states of the Ba atom with the configuration $6p8s(^3P_1)$ through the the 2ω-excited intermediate state $6p^2(^1S_0)$.

I THEORY

For two-step ionization by polarized radiation the photoelectron angular distribution can be represented analytically in the form

$$\frac{dW}{d\Omega} \propto Tr(R\rho_a\rho^\gamma \varepsilon_i \varepsilon_e R^*) \qquad (1)$$

where dW is the probability of emission of a photoelectron within the solid angle $d\Omega$, ρ_a is the density matrix of the atoms in the intermediate state, ρ^γ is the density matrix of the ionizing radiation, ε_i and ε_e are the detection efficiency matrices of the ions and electrons, R is the radiative interaction operator describing the transition in the presence of electromagnetic radiation between the intermediate and final state.

The multipole of the 2ω-excited intermediate state can be represented as a superposition of excitation channels through the intermediate states with total angular momentum J_a [12]:

$$\rho^{(2)}_{K_1 Q_1} = \sum_{J_a} \langle J_a \| R \| J_0 \rangle \langle J_1 \| R \| J_a \rangle \rho_{K_0 Q_0} \rho_{k_1 q_1} \cdot$$

$$\sum_{J_a} \langle J_a \| R \| J_0 \rangle^* \langle J_1 \| R \| J_a \rangle^* (E - E_{J_a} - \hbar\omega)^{-1} \qquad (2)$$

where $|J_0\rangle$ and E are the initial state and its energy, $|J_a\rangle$ and E_{J_a} are the intermediate state of two-photon excitation and its energy, $|J_1\rangle$ is the final state of the two-photon excitation process, $\langle J_a \| R \| J_0 \rangle$ is the reduced matrix element of the radiative interaction, $\rho_{K_0 Q_0}$ is the density matrix of the ground state of the atom, $\rho_{k_1 q_1}$ is the density matrix of the electromagnetic radiation of the first step, and $\hbar\omega$ is the energy of the electromagnetic radiation. The sum over J_a follows from the well-known formula of vector manifolds:

$$(\vec{e}_s, \vec{e}_r) = \sum_m (\vec{e}_s, \vec{e}_m)(\vec{e}_m, \vec{e}_r) \qquad (3)$$

In the description of the auto-ionization process we will employ the formalism of irreducible tensor operators. Following [11,13,14] we write the state multipoles of the photons and atoms in the ground state, intermediate state, and final state.

a) The polarization density-matrix of the electromagnetic radiation was obtained in [13]. Employing the explicit expression for it, we write the state multipole of the dipole radiation

$$\rho^\gamma_{k_i q_i}(1,1) = \sum_{\lambda,\lambda'} (-1)^{1-\lambda'} \hat{k} \begin{pmatrix} 1 & 1 & k_1 \\ \lambda & \lambda' & -q_1 \end{pmatrix} (\lambda \mid \rho^\gamma \mid \lambda') \qquad (4)$$

where i is the number of the excitation step, $\begin{pmatrix} 1 & 1 & k_1 \\ \lambda & \lambda' & -q_1 \end{pmatrix}$ is the 3j-symbol, $(\lambda|\rho|\lambda')$ is the photon density matrix, expressed in its usual form in terms of the Stokes parameters ξ_j:

$$(\lambda \mid \rho^\gamma \mid \lambda') = (1/2) \begin{pmatrix} 1+\xi_2 & -\xi_3 + i\xi_1 \\ -\xi_3 - i\xi_1 & 1-\xi_2 \end{pmatrix} \qquad (5)$$

b) For atoms in the isotropically oriented ground state the state multipole has the following form: [11]

$$\rho_{K_0 Q_0}(\gamma_0 J_0, \gamma'_0 J'_0) = \hat{J}_0^{-1} \delta_{K_0 0} \delta_{Q_0 0} \delta_{J_0 J'_0} \qquad (6)$$

where $\hat{J} \equiv (2J+1)^{1/2}$ for arbitrary angular momentum, J_0 is the total angular momentum of the initial state, and $\gamma_{(j)}$ are the remaining quantum numbers needed to describe the jth state of the atom.

c) Employing expression (2), we write an expression for the state multipole of the atom in the oriented intermediate state $|\gamma_1 J_1\rangle$:

$$\rho_{(2)K_1Q_1}(\gamma_1 J_1, \gamma_1' J_1') = \sum_{J_a}(E - E_{J_a} - \hbar\omega)^{-1} \cdot$$

$$\langle \gamma_0 J_0, \gamma_a J_a \parallel R \parallel \gamma_a J_a \rangle \langle \gamma_0 J_0, \gamma_a J_a \parallel R \parallel \gamma_a J_a \rangle^* \cdot$$

$$\langle \gamma_a J_a, \gamma_1 J_1 \parallel R \parallel \gamma_1 J_1 \rangle \langle \gamma_a J_a, \gamma_1 J_1 \parallel R \parallel \gamma_1 J_1 \rangle^* \cdot$$

$$\sum_{K_0 k_1} \hat{J}_a \hat{k}_1 \hat{K}_0 \, (K_0 Q_0 k_1 q_1 \mid K_a Q_a) \begin{Bmatrix} J_0 & J_0 & K_0 \\ 1 & 1 & k_1 \\ J_a & J_a & K_a \end{Bmatrix} \cdot$$

$$\rho_{K_0 Q_0}(\gamma_0 J_0, \gamma_0' J_0') D_{Q_0 q_1}^{k_1 *} \rho_{k_1 q_1}^{\gamma}(1,1) \cdot$$

$$\sum_{K_a k_1'} \hat{J}_1 \hat{k}_1' \hat{K}_a \, (K_a Q_a k_1' q_1' \mid K_1 Q_1) \begin{Bmatrix} J_1 & J_1 & K_1 \\ 1 & 1 & k_1' \\ J_a & J_a & K_a \end{Bmatrix} \cdot$$

$$\rho_{k_1' q_1'}^{\gamma}(1,1) D_{Q_a q_1'}^{k_1' *}$$

(7)

where $(K_a Q_a | K_0 Q_0 k_1 q_1)$ is the Clebsch–Gordon coefficient, $\{\ldots\}$ is the $9j$-symbol, and $D_{Q_0 q_1}^{k_1' *}(\hat{A})$ is the matrix of finite rotations [15] translating the treatment of the excitation process from the laboratory coordinate system to the atomic coordinate system defined by the polarization vector of the atom, in which the density matrix is diagonal.

d) We represent the multipole of the autoionized state of the atom in the form

$$\rho_{K_2 Q_2}(\gamma_2 J_2, \gamma_2' J_2') =$$

$$\langle \gamma_1 J_1, \gamma_2 J_2 \parallel R \parallel \gamma_2 J_2 \rangle \langle \gamma_1 J_1, \gamma_2 J_2 \parallel R \parallel \gamma_2 J_2 \rangle^* \cdot$$

$$\rho'_{K_2 Q_2}(J_2, J_2) \rho'_{K_2 Q_2}(J_2, J_2) =$$

$$\sum_{K_1 k_2} \hat{j}_2^2 \hat{k}_2 \hat{K}_1 \, (K_1 Q_1 k_2 q_2 \mid K_2 Q_2) \begin{Bmatrix} J_1 & J_1 & K_1 \\ 1 & 1 & k_2 \\ J_2 & J_2 & K_2 \end{Bmatrix} \cdot$$

(8)

$$\rho_{(2)K_1 Q_1}(\gamma_1 J_1, \gamma_1' J_1') \rho_{k_2 q_2}^{\gamma}(1,1) D_{Q_1 q_2}^{k_2 *}$$

where $|\gamma_2 J_2\rangle$ is the auto-ionized state with total angular momentum J_2 and $\rho_{k_2 q_2}^{\gamma}(1,1)$ is the state multipole of the electromagnetic radiation of the second step.

Since in the nonrelativistic approximation the ionization probability does not depend on the electron spin and the polarization characteristics of the states of the electron and ion are not fixed, the expression for the product of the state multipoles of the ion and photoelectron, arising as a result of decay of the autoionized state, can be represented in the form

$$\rho_{K,Q}(\gamma_i J_i, \gamma_i J_i) \rho_{k_e q_e}(lj, l'j') =$$

$$(\frac{1}{4\pi}) \langle lj, \gamma_i J_i \parallel R \parallel \gamma_2 J_2 \rangle \langle l'j', \gamma_i J_i \parallel R \parallel \gamma_2 J_2 \rangle^* \cdot$$

$$\sum_{K_2} \rho_{K_2 Q_2}(\gamma_2 J_2, \gamma_2' J_2')(-1)^{J_i + J_2 + K_2 + j'} \hat{j}_2^2 \hat{j}_i^{(-1)} \begin{Bmatrix} J_2 & j & J_i \\ j' & J_2 & K_2 \end{Bmatrix},$$

(9)

where $|\gamma_i J_i\rangle$ is the state of the atom with one electron knocked out, V is the Coulomb interaction operator, and $\{\ldots\}$ is the $6j$-symbol.

When using the state multipole formalism, expression (1) transforms to

$$\frac{dW}{d\Omega} = \tag{10}$$

$$\sum \rho_{K,Q,}(\gamma_i J_i, \gamma_i J_i) \rho_{(2)k_e q_e}(lj, l'j') \varepsilon^*_{K,Q,}(\gamma_i J_i, \gamma_i J_i) \varepsilon^*_{k_e q_e}(lj, l'j')$$

where the sum is over all repeated indices, $\varepsilon^*_{K,Q,}(\gamma_i J_i, \gamma_i J_i)$ and $\varepsilon^*_{k_e q_e}(lj, l'j')$ are the detection efficiency tensors of the ions and electrons, respectively. In the case when the polarization characteristics of the state of the ion are not fixed, $\varepsilon_{K,Q,}(\gamma_1 J_1, \gamma_1 J_1) = \hat{J}_i \delta_{K,0} \delta_{Q,0}$ (Ref. 14) and the expression for the product of the detection efficiency tensors of the ions and the electrons has the form [14]

$$\varepsilon_{K,Q,}(\gamma_i J_i, \gamma_i J_i) \varepsilon_{k_e q_e}(lj, l'j') =$$

$$\frac{1}{4\pi} \sum_{k_e} \hat{J}_i (-1)^{1/2-j} \bar{Z}(lj, l'j'; 1/2 k_e) D^{k_e}_{q_e 0}(R) \tag{11}$$

where $\bar{Z}(lj, l'j'; 1/2 k_e)$ is the Huby function [14].

Substituting Eqs. (8), (9), and (11) in Eq.(10), we obtain an expression for $dW/d\Omega$ (here we use the atomic system of units)

where the sum is over $l, l', j, j', J_2, J_i, J_a$; P is the photon flux density, α is the fine structure constant, and

$$\frac{dW}{d\Omega} = 2\pi (P 2\pi \alpha \omega)^3 \cdot$$

$$\sum \langle \gamma_1 J_1, \gamma_2 J_2 \| R \| \gamma_2 J_2 \rangle \langle \gamma_1 J_1, \gamma_2 J_2 \| R \| \gamma_2 J_2 \rangle^* \cdot \tag{12}$$

$$(1/4\pi) \langle lj, \gamma_i J_i \| R \| \gamma_2 J_2 \rangle \langle l'j', \gamma_i J_i \| R \| \gamma_2 J_2 \rangle^* \cdot$$

$$K(l, l', j, j' J_2 J_i J_a K_2, k_e)$$

The necessity of summing over K_2 in the case of excitation of an auto-ionized state was first shown in [16]. For excitation of such states from an isotropically oriented intermediate state, let us consider the case in which the coordinate system is related to the polarization plane such that $Q_2 = q_e = 0$. By virtue of the properties of the 6j-symbol, expression (12) is nonzero for $K_2 = 0, 2$ and transforms to [18]

$$K(l, l', j, j' J_2 J_i J_a K_2, k_e) \equiv$$

$$\equiv \hat{J}_2^2 \sum_{K_2} \rho'_{K_2 0}(J_2 J_2)(-1)^{J_i + J_2 + K_2 + j'} \left\{ \begin{array}{ccc} J_2 & j & J_i \\ j' & J_2 & K_2 \end{array} \right\} \cdot \tag{13}$$

$$\sum_{k_e} (-1)^{1/2-j} \bar{Z}(lj, l'j'; 1/2 k_e) D^{k_e}_{00}(\hat{R}^{-1} \hat{A})$$

where σ_s is the total cross section, β is the anisotropy parameter of the angular distribution, $\langle \cos \theta_e \rangle$ is the Legendre polynomial of degree 2, and θ_e is the angle between the axis of the electron energy analyzer and the direction of polarization of the laser radiation.

On the same way the expression of the double ionization:

$$\hat{A}(\gamma_0 J_0) + \hbar_1 \longrightarrow \left[\hat{A}^+ (\gamma_{i'} J_{i'}) + \hat{e}_1 (l_1 j_1) \right]^{J_1}_{(\gamma_1 J_1)} + \tag{14}$$

$$\hbar \nu_2 \longrightarrow \left[\hat{A}^{2+} (\gamma_{i''} J_{i''}) + \hat{e}_2 (l_2 j_2) \right]^{J_2}_{(\gamma_2 J_2)},$$

can be represented in the follows form [17]:

$$\frac{dW}{d\Omega_1 d\Omega_2} =$$
$$= const \cdot \text{Tr} \sum_{k_2,q_2,\gamma_2,\gamma'_2,J_2,J'_2,} \rho^{f_2}_{k_2 q_2}(\gamma_2 J_2, \gamma'_2 J'_2) \varepsilon^{f_2*}_{k_2 q_2}(\gamma_2 J_2, \gamma'_2 J'_2) =$$
$$\rho^{f_2}_{k_2 q_2}(\gamma J_2, \gamma' J'_2) \cdot \varepsilon^{f_2*}_{k_2 q_2}(\gamma_2 J_2, \gamma'_2 J'_2) =$$

$$\langle \gamma_{J_2} J_{f_2}; l_2 j_2, \gamma_{i''} J_{i''} \| R \| \gamma_{f_1} J_{f_1}\rangle \langle \gamma_{f_2} J_{f_2}; l_2 j_2, \gamma_{i''} J_{i''} \| R \| \gamma_{f_1} J_{f_1}\rangle^* \times$$
$$\langle \gamma_{f_1} J_{f_1}; l_1 j_1, \gamma_{i'} J_{i'} \| R \| \gamma_0 J_0\rangle \langle \gamma_{f_1} J_{f_1}; l_1 j_1, \gamma_{i'} J_{i'} \| R \| \gamma_0 J_0\rangle^* \times$$
$$\sum_{k_1,k'_e k''_\gamma} \rho^\gamma_{k'_\gamma q'_\gamma}(1,1) \rho^\gamma_{k''_\gamma q''_\gamma}(1,1)(-1)^{J_0 + J_i - j'_2} \hat{k}_1^2 \hat{k}''_\gamma \hat{k}'_e (\hat{J}_2 \hat{J}'_2)^2 (\hat{J}_1)^4 \hat{J}_0^{-2} \times$$ (15)
$$(k_1 0, k''_\gamma 0 \mid k_2 0)(k'_e 0, k_1 0 \mid k_2 0) W(1 J_1 1 J_1; J_0 k_1) \times$$
$$\bar{Z}\left(l_1 j_1 l'_1 j'_1; \tfrac{1}{2} k_1\right) \bar{Z}\left(l_2 j_2 l'_2 j'_2; \tfrac{1}{2} k_2\right) \times$$
$$\left\{\begin{array}{ccc} j_1 & J_1 & J_i \\ J_1 & j'_1 & k_1 \end{array}\right\} \left\{\begin{array}{ccc} J_1 & 1 & J_2 \\ J_1 & 1 & J'_2 \\ k_1 & k''_\gamma & k_2 \end{array}\right\} \left\{\begin{array}{ccc} l_2 & J_1 & J_2 \\ l'_2 & J_1 & J'_2 \\ k'_e & k_1 & k_2 \end{array}\right\} \times$$
$$D^{k_1}_{00}(\Omega_1) \otimes D^{k_2}_{00}(\Omega_2),$$

where

$$D^{k_1}_{00}(\Omega_1) \otimes D^{k_2}_{00}(\Omega_2) = 4\pi \hat{k}_1^{-2} \left[Y^*_{k_1}(\phi_1,\theta_1) \otimes Y^*_{k_2}(\phi_2,\theta_2)\right]_{00} \delta_{k_1 k_2} =$$
$$(-1)^{q_1} 4\pi \hat{k}_1^{-3} \sum_{k_1=-q_1}^{q_1} Y_{k_1-q_1}(\phi_1,\theta_1) Y_{k_1 q_1}(\phi_2,\theta_2), \quad 0 \leq q_1 \leq k_1$$ (16)

II EXPERIMENTAL SETUP

The time-of-flight electron energy analyzer was specially developed and prepared for recording the electron angular distribution and discriminating their energies in multiphoton ionization of atoms. Details of the experimental setup are given in [19]; here we describe briefly its operation. The energy of the recorded electrons is less than 3eV. The electron-optical circuit of the spectrometer employs a weak accelerating electrostatic field following a short interval of field-free electron drift during which ionization takes place. The choice of this design is explained by the fact that in the interaction region of the radiation and atoms electrostatic fields distorting the angular dependence of the electron distribution are absent. The Ba atom concentration in the beam was varied in the range 10^{10}–$10^{12} at/cm^3$.

Excitation and ionization of Ba atoms were realized by pulsed dye-laser radiation oriented perpendicular to the direction of the Ba atomic beam. The dye-laser had the following parameters: spectral width of the lasing line $2cm^{-1}$, frequency scanning region 570–590nm, pulse energy 1.5mJ, pulse duration 20ns. The degree of linear polarization was 98%. The interaction region of the atoms with the laser radiation was screened from the Earth's magnetic field with the help of a triple permalloy screen, where the width of each screen was 2mm. The residual magnetic field in the electron drift region was 2mG.

III DISCUSSION

Three-photon ionization of the Ba atoms was realized according to the following scheme: [19]

$$6s^2(^1S_0) \longrightarrow 6p^2(^1S_0) \longrightarrow 6p8s(^3P_1) \longrightarrow [6s(^2S_{1/2} + e^{(-)}(\varepsilon_{1/2,3/2})]$$

The energy of the continuum corresponding to three-photon ionization coincided with the position of the auto-ionized state $6p8s(^3P_1)$.

By rotating the polarization plane of the laser radiation we measured the distribution of the photoelectrons corresponding to the $6s$ state of the ion. By virtue of the isotropic orientation of the $6p^2(^1S_0)$, whose choice was dictated by the experimental capabilities of the setup, the analytical expression for the electron angular distribution has the form (13).

For the electrons corresponding to the $6s$ state of Ba^+ we recorded the photoelectron angular distribution for different Ba atom concentrations in the interaction region with the laser radiation. Expression (13) was used to fit the experimental dependence of the electron signal intensity on the rotation angle of the polarization plane of the

TABLE 1.

j, j'	$\mathcal{K}(j, j', 0)$	$\mathcal{K}(j, j', 2)$
1/2, 1/2	−1.197	0
1/2, 3/2	0	−0.219
3/2, 3/2	−0.423	0.155

radiation. The fit yielded the parameter value $\beta_2 = 0.74 \pm 0.1$, which remains within the limits of experimental error up to Ba atom concentrations of $5.0 \times 10^{-12}\,cm^{-3}$.

The main contribution to the probability of two-photon excitation comes from the excitation channel with participation of the state that is closest in energy to the virtual level. In [5] it was shown that for 2ω-excitation of the $6p^2(^1S_0)$ state the main contribution to the probability of two-photon excitation comes from the channel with participation of the resonant state $6s6p(^1P_1)$ standing off from the virtual level of two-photon excitation by $808\,cm^{-1}$. The following selection-rule allowed state stands off $13695 cm^{-1}$ from the virtual level. Therefore it is possible to limit the sum over J_a to the one term corresponding to the $6s6p(^1P_1)$ state.

For the case of the intermediate state $6p^2(^1S_0)$ ($J_2 = 0$) expression (12) simplifies substantially: $K_2 = k_e$ and the numerical values for $\mathcal{K}(1, 1, j, j', 1, 1, 1/2, 1, K_2, K_2) \equiv \mathcal{K}(j, j', K_2)$ are nonzero only for $K_2 = 0$ and 2 (see Table I). Comparison with experiment is hindered by the fact that we have not calculated the Coulomb and dipole matrix elements.

IV CONCLUSION

We have considered the problem of the electron angular distribution in three-photon ionization of Ba atoms. As the intermediate states we used a discrete state and an auto-ionized state. Excitation of an even state with the help of a two-photon process ensures minimal depolarization of the atoms in the intermediate state. This fact is especially important for the accuracy of the results of the complete quantum-mechanical experiment on photoionization of polarized atoms, which records the angular distribution of the electrons. From the point of view of experimental realization this variant of the setup of the complete experiment is the most convenient since it does not require a determination of the spin state of the photoelectrons, which is a complicated experimental task. We hope that the method proposed here of decreasing the influence of depolarization of the atoms in the intermediate state on the recording of the photoelectron angular distribution will be of interest to experimentalists.

ACKNOWLEDGMENTS

This work was carried out with the support of the program "Optics: Laser Physics."

REFERENCES

1. Meyer M., Muller B., Nunneman A., Prescher Tn., von Raven E., Richter M., Schmidt M., Sontag B. and Zimmermann P., *Phys. Rev. Lett.* **59**, 2963 (1987).
2. Fedotov A. B., Ilyasov O. S., Koroteev N. I., and Zheltikov A. M., *Nuovo Cimento* **14**, 1003 (1992).
3. Baier S., Grum-Grzhimailo A. N., and Kabachnik N. M., *J. Phys.B: At. Mol. Opt. Phys.* **27**, 3363 (1994).
4. Elizarov A. Yu. and Cherepkov N. A., *Zh. Eksp. Teor. Fiz.* **96** 1224 (1989) [*Sov. Phys. JETP* **69**, 695 (1989)].
5. Cherepkov N. A. and Elizarov A. Yu., *J. Phys. B: At. Mol. Opt. Phys.* **24**, 4169 (1991).
6. Cleff B. and Mehlhorn W., *J.Phys. B: At. Mol. Opt. Phys.* **7**, 593 (1974).
7. Kabachnik N. M. and Sazhina I. P., *J. Phys. B: At. Mol. Opt. Phys.* **9**, 1681 (1976).
8. Bobashev S. V., Elizarov A. Yu., Prilipko V. K., and Cherepkov N. A., *Laser Phys.* **3**, 751 (1993).
9. Klar H., *J.Phys. B: At. Mol. Opt. Phys.* **13**, 4741 (1980).
10. Klar H. and Kleinpoppen H., *J.Phys. B: At. Mol. Opt. Phys.* **15**, 933 (1982).
11. Devons S. and Goldfarb L. J., *Handbuch der Physik* (Springer-Verlag, Berlin, 1957), p.362.
12. Goeppert-Mayer M., *Ann. Phys.* **9**, 273 (1931).
13. Peshkin M., *Adv. Chem. Phys.* **18**, 1 (1971).
14. Ferguson A. J., *Angular Momentum Methods in Gamma-Ray Spectroscopy* (North-Holland, Amsterdam, 1965), p.246.
15. Varshalovich D. A., Moskalev A. N., and Khersonskii V. K., *Quantum Theory of Angular Momentum* (World Scientific, Singapore, 1988).
16. Berezhko E. G. and Kabachnik N. M., *J.Phys. B: At. Mol. Opt. Phys.* **10**, 2467 (1977).

17. Elizarov A. Yu., Laser Phys. in press.
18. Yang C. N., Phys. Rev. **74**, 764 (1948).
19. Elizarov A. Yu., JETP Lett. **62**, 23 (1996).

POSTER SESSION VI
MOLECULAR RIS

Higher Rydberg States of C_6H_6, C_6D_6 and C_6H_5F Studied by Two-photon Resonance Ionization Spectroscopy

S. Wang, R.J. Donovan, T. Ridley and K.P. Lawley

Department of Chemistry, The University of Edinburgh, King's Buildings, West Mains Road, Edinburgh EH9 3JJ.

Abstract: The higher Rydberg states of jet-cooled C_6H_6, C_6D_6 (n≤30) and C_6H_5F (n≤13) have been studied using two-photon resonance-enhanced multiphoton ionization (REMPI) spectroscopy. In C_6H_6 and C_6D_6 long series of three gerade nd Rydberg series converging on the first ionization potential, previously observed up to n=8, are extended up to n=30. The spectra are dominated by origin transitions with only weak vibrational structure being observed. The series with A_{1g} and E_{1g} or E_{2g} symmetry were distinguished by polarization experiments. Similar long series of three nd Rydberg states are also observed for C_6H_5F and are reported here for the first time. In contrast to benzene, more vibrational activity, especially in v_{6a} was observed, and thus Rydberg series can only be resolved up to n=12 for C_6H_5F.

INTRODUCTION

The Rydberg states of benzene, one of the most widely investigated molecules of the class of polyatomic molecules, continue to attract particular interest. The gerade Rydberg states of benzene up to n=8 have been studied Grant and co-workers (1) using (2+1) REMPI. Very recently, Neusser and co-workers (2) reported a double resonance study of high Rydberg states (n=48-110) of C_6H_6 and C_6D_6 using two Fourier-transform limited nanosecond laser pulses and a full review of previous studies can be found therein.

The Rydberg states of monohalogenated benzenes, however, have not been studied very thoroughly. Price and co-workers (3) found two Rydberg series of fluorobenzene with quantum defects of 0.49 and 0.03 in the far ultraviolet absorption spectrum. They found a slight difference between the limits of convergence of the two series. A later study by Gilbert and Sandorfy (4) recorded the VUV spectrum of fluorobenzene with a resolution superior to that of the early study and found both series (n=3-8) converge to 9.20 eV.

The first REMPI study on the Rydberg states of fluorobenzene was reported by Goodman and co-workers using (2+1) REMPI (5). In a later study (6), the same group observed the n=4 and possibly n=5 members of the d series of fluorobenzene with resolved vibrational structure using (3+1) REMPI.

In this work we present a detailed study of the higher Rydberg states of C_6H_6, C_6D_6 and C_6H_5F by mass-resolved (2+1) REMPI. The goal of this study is to investigate the higher Rydberg states of benzene in the energy region which Grant and co-workers (1) were unable to access and below the energy region studied by Neusser and co-workers (2). A comparison study for C_6H_5F was also performed.

EXPERIMENTAL

The molecular beam was generated by pulsing a mixture of benzene or fluorobenzene at its vapour pressure and one atmosphere of He, through a General Valve pulsed nozzle with a 250 μm diameter aperture, into the ionization region of a linear time-of-flight mass spectrometer. Ions were collected at 90° to both the molecular and laser beams. Mass-resolved ion signals were processed by a Stanford Research SR250 boxcar integrator and stored on a PC. Radiation was provided by a Lamdba Physik 3002 dye laser pumped by a Lambda Physik EMG 201 MSC excimer laser. The frequency doubled laser beam was focused and directed into the chamber by a 5cm focal length lens. The laser pulse energy delivered in each laser pulse after frequency doubling is 2-5 mJ. Circular polarization was achieved by passing the beam through a Soleil Babinet prism. All spectra were recorded in the C^+ and C_2^+ mass channels at the same time with the power curve and all spectra presented were normalised to the square of the laser power.

RESULTS AND DISCUSSION

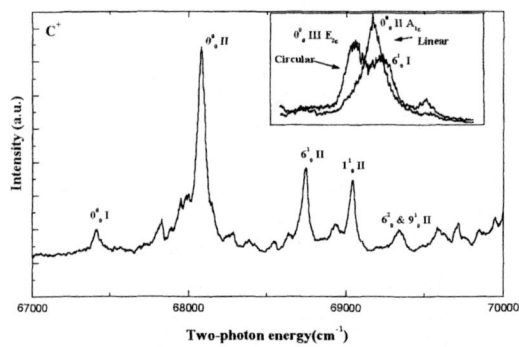

FIGURE 1. The linearly polarized two-photon spectrum of C_6H_6 in the 67000-70000 cm^{-1} region. The insert shows a high resolution scan of the strong peak under linear and circular polarization.

The mass resolved (2+1) REMPI spectra of jet-cooled C_6H_6 and C_6D_6 were recorded using both linearly and circularly polarized light in the energy region from 60000 cm^{-1} to the first IP (~74600 cm^{-1}). In the low energy region up to 73000 cm^{-1} studied by Grant and co-workers (1) by collecting all ions rather than using C^+ and C_2^+, our spectra reproduce theirs. For reference, the two-photon REMPI spectrum of benzene(C_6H_6) recorded in the C^+ mass channel in the 4d region is shown in Fig.1. It can be seen that the strongest feature in this spectrum is the origin of n=4 at 68074 cm^{-1} with a quantum defect of -0.12 and A_{1g} symmetry. Built upon the origin are the weaker vibrational levels $6a^1_0$ and 1^1_0 which have about 1/3 intensity of the origin. This relatively weak vibrational activity makes the origin of higher Rydberg states easy to resolve and recognize. Because this origin is weaker in the spectrum recorded with circularly polarized light a second underlying origin can be observed at 68051 cm^{-1} which corresponds to an E_{2g} origin. Further to the red of the A_{1g} origin there is another origin at 67411 cm^{-1} with E symmetry which has small near-zero quantum defect (0.08). Another Rydberg series starting at n=5 (70170 cm^{-1}) with quantum defect (0.00) appears under circular polarization.

Fig.2 presents the Rydberg series from n=9 to 30 of C_6H_6 recorded in the C^+ mass channel with linearly and circularly polarized light. The linear spectrum consists of a series of sharp distinct features with a quantum defect of −0.13 which is obviously the extension of the strong d Rydberg series observed strongly at the n=4 level. The circular spectrum consists of the extension of the two very closely spaced Rydberg series (with quantum defects of 0.04 and 0.00) whose low members can be resolved but merge as n increase. All of the same features are observed in the equivalent spectrum of C_6D_6.

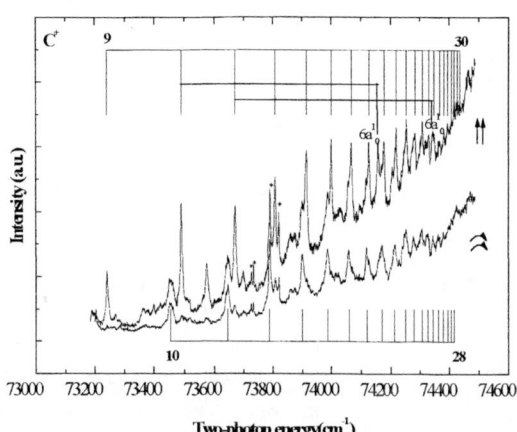

FIGURE 2. The linearly and circularly polarized two-photon spectra of C_6H_6 in the 73000-74600 cm^{-1} region. The peaks marked with * are atomic carbon resonance

The two-photon REMPI spectrum of fluorobenzene in the region 66000-69500 cm^{-1} is shown in Fig.3. The features appearing in Fig.3 are very similar to Fig.1 and correspond to 4d Rydberg states based on the ground state of the ion. The strong sharp feature at 67828 cm^{-1} has almost the same quantum defect (-0.14) as the A_{1g} origin of benzene and the small feature at 67331 cm^{-1} has almost zero quantum defect (0.03) as does the E_{1g} origin of benzene. An interesting difference is the prominent intensity of the $6a^1_0$ transition compared with the spectrum of benzene(Fig.1). One of the possible reasons is that in C_{2v} symmetry the 6a mode is a totally symmetric vibration while in the case of benzene v_6 can only gain intensity by Jahn-Teller distortion. Just as in benzene, another near-zero quantum defect (0.00) series starts at n=5 (69848 cm^{-1}) as shown in Fig.4 and it can be seen more obviously under circular polarization. Rydberg series can only be resolved up to n=13 due to the increased activity of mode 6a, with a resultant increase in spectral congestion.

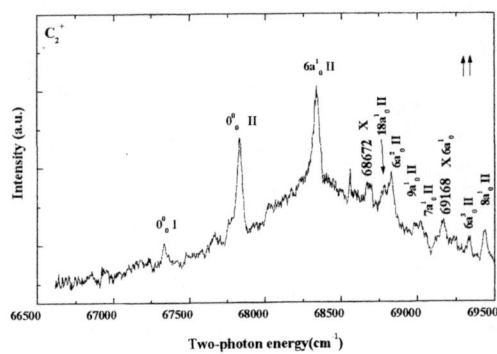

FIGURE 3. The linearly polarized two-photon spectrum of C_6H_5F in the 66500-69500 cm^{-1} region.

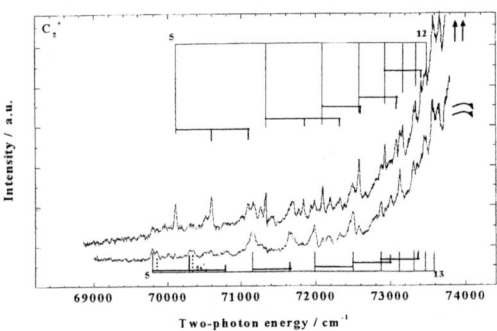

FIGURE 4. The linearly and circularly polarized two-photon spectra of C_6H_5F in the 69000-74000 cm^{-1} region.

The frequencies of the totally symmetric modes of the strongest A_1-component at n=4 are (in cm^{-1}): v_{6a} (506), v_{7a} (1188), v_{8a} (1606), v_{9a} (1168). The peak at 68825 cm^{-1}, 997 cm^{-1} from origin at 67829 cm^{-1} could corresponding to $6a^2_0$ or 1^1_0. The peak at 69330 could be assigned as $6a^3_0$ or $1^1_0 6a^1_0$. The features at 68672 and 69168 cm^{-1} could not assigned to symmetric vibration built on the observed origin and possibly correspond to $5p$ origin with a quantum defect 0.56 and one quanta of 6a built on it.

For substituted benzenes, two-photon transitions from the HOMO ($^1A_1$) to ns(1), np(3), nd(5), nf (7) states are all allowed and thus more Rydberg series should be observed. The number of Rydberg series of flurobenzene observed in this work, however, is essentially the same as in benzene. The relatively intensity and quantum defects of these series of these two molecule are also very similar. This indicates that there is very little influence of the substituent (F) on the electron excitation to Rydberg states converging to first IP of fluorobenzene. Therefore these three series should still be assigned to nd series. Further study on heavier halobenzenes such as C_6H_5Cl and C_6H_5Br, and $C_6H_4X_2$ to investigate the effect of the departure from D_{6h} symmetry is underway.

REFERENCE

1. Whetten, R.L., Grubb, S.G., Otis, C.E., Albrecht, A.C., and Grant, E.R., J. Chem. Phys. **82**,1115(1984).
2. Neuhauser, R.G., Siglow, K. and Neusser, H.J., J. Chem. Phys. **106**, 896(1997).
3. Hammond, V.J., Price, W.C., Teegan, J.P. and Walsh, A.D., Faraday Discuss. Soc. **9**, 53(1950).
4. Gilbert, R.and Sandorfy, C., Chem. Phys. Lett. **9**, 121(1971).
5. Krogh-Jespersen, K., Rava, R.P.and Goodman, L., Chem. Phys. Lett. **64**, 413(1979).
6. Krogh-Jespersen, K., Rava, R.P.and Goodman, L., Chem. Phys. **47**, 321(1979).

POSTER SESSION VII
FEMTO-SECOND AND HIGH INTENSITY RIS

Ionization and Fragmentation of Small Molecules under psec and fsec Laser Excitation

S. Couris, E. Koudoumas and C. Fotakis

*Foundation for Research and Technology-Hellas,
Institute of Electronic Structure and Laser,
P.O. Box 1527, 71110 Heraklion, Crete, Greece*

Abstract. Experimental results are presented concerning the interaction of small molecules with high intensity short pulse laser light. The time-of-flight technique was used in order to study the ionization and fragmentation processes taking place. Laser pulses of different wavelengths and pulse durations were employed and preliminary results will be presented. Evidence for strong alignment of the multiply charged fragments along the electric field of the laser light will be discussed.

INTRODUCTION

The study of ionization, dissociation and/or fragmentation of molecules in the gas phase under irradiation with psec and fsec laser pulses has received a lot of interest during the last few years (1-5), because of the significant information that they can offer in the understanding of elementary and fundamental physical and chemical processes. In this work, the interaction of small molecules such as CS_2, OCS, CO_2 and CH_3I with intense short pulse laser radiation was investigated experimentally using time of flight mass spectroscopy. Mass spectra were recorded under several experimental conditions such as radiation intensity, wavelength, pulse duration and polarization of the laser beam.

2. EXPERIMENTAL

2.1 Laser systems

Three different laser systems have been employed in order to study the dependence of the molecular fragmentation and/or ionization upon wavelength and pulse duration. Fisrt, a hybrid excimer-dye laser system (Lambda Physik EMG 150 MSC) based on the concept of distributed feedback dye laser (DFDL) was employed as a 5 psec and 0.5 psec laser source at 248 and 497 nm. In short, the 308 nm output of a double-cavity excimer laser pumps a cascade of dye laser modules to produce 0.5 psec pulses at 497 nm with an energy of 100 μJ. The output is frequency doubled in a BBO crystal and then amplified in the second cavity of the excimer laser. The final output is a UV laser beam at 248 nm with a pulse duration of 0.5 psec and an energy of 20 mJ. In addition, by inserting a bandwidth limiting etalon, pulses of 5 psec duration at both 497 and 248 nm are obtained.

The two other laser systems were both Ti:Sapphire lasers operating at 800 nm. The first system consists of a Spectra Physics Ti:Sapphire oscillator and a BMI regenerative Ti:Sapphire amplifier. The oscillator (a Tsumani Ti:Sapphire laser pumped by a Beamlock Ar^+ laser) delivers pulses of 100 fsec duration, 0.8 W power at a wavelength of 800 nm and a repetition rate of 82 MHz. The oscillator output is stretched and then used as seeding for the regenerative Ti:Sapphire amplifier (ALPHA 1000S) which is pumped by the second harmonic of a BMI Nd:YLF laser operating at a repetition rate of 1 kHz. The amplified pulse is then compressed to get the output of 200 fsec.

The other laser system consists of a Coherent Ti:Sapphire oscillator and a BMI regenerative Ti:Sapphire amplifier. The oscillator (a Mira Ti:Sapphire laser pumped by an Innova Ar^+ laser) delivers pulses of 50 fsec duration, 0.7 W power at a wavelength of 800 nm and a repetition rate of 76 MHz. The oscillator output is also stretched and then used as seeding for the regenerative Ti:Sapphire amplifier (ALPHA 1000US) which is also pumped by the second harmonic of a Nd:YLF laser operating at 1 kHz. The amplified pulse is then compressed to get the output of 50 fsec.

2.2 The Time-of -Flight mass spectrometer

A home made time-of-flight mass spectrometer of a 60 cm long field-free drift tube has been used for the present study. An appropriate mixture of the gas phase molecule under study seeded in 200 mbars of Ar 50 mbars CS_2 (or the other molecules under study) in 200 mbars argon was expanded through a 0.3 mm nozzle into the TOF. The ions were accelerated between a repeller plate (+1900 V) and a grounded electrode separated by 1.5 cm. Signals were detected by a pair of 1 inch diameter channel plates and recorded with a LeCroy 9414 digital oscilloscope. The pressure in the chamber was always kept below 10^{-6} mbars during the recording of the mass spectra in order to avoid saturation effects. Spectroscopic grade samples were used after several repeated freeze-thaw degassing cycles under vacuum.

3. RESULTS AND DISCUSSION

Although several molecules have been studied in our laboratory (4,5) and experimental work is continuing, in this presentation we will report only on experimental results concerning small molecules like CS_2, OCS, CO_2 and CH_3I, remaining focused mainly on the carbon disulfide molecule. In Fig. 1 the mass spectra of the above mentioned molecules are presented at low incident energies as were observed under 200 fsec excitation at 800 nm. As can be seen at low incident intensities only the parent ion was present in the mass spectra recorded. The intensity required to get an observable signal for each molecule was found to increase with increasing ionization potential of the molecule. Thus, in order to ionize the CO_2 molecule, which has an ionization potential of 13.769 eV, six times higher intensity is required in order to get an observable signal compared with the case of CH_3I which has an ionization potential of only 9.538 eV.

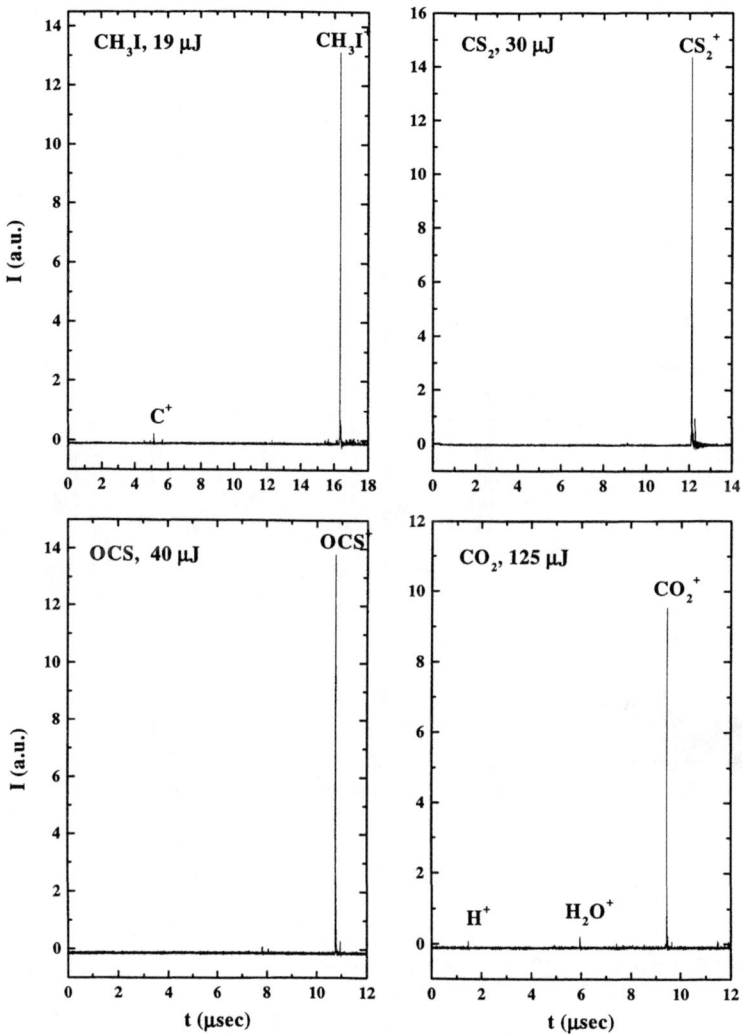

FIFURE 1. Ionization/fragmentation of some small molecules at 800 nm, 200 fsec excitation at low incident energies:

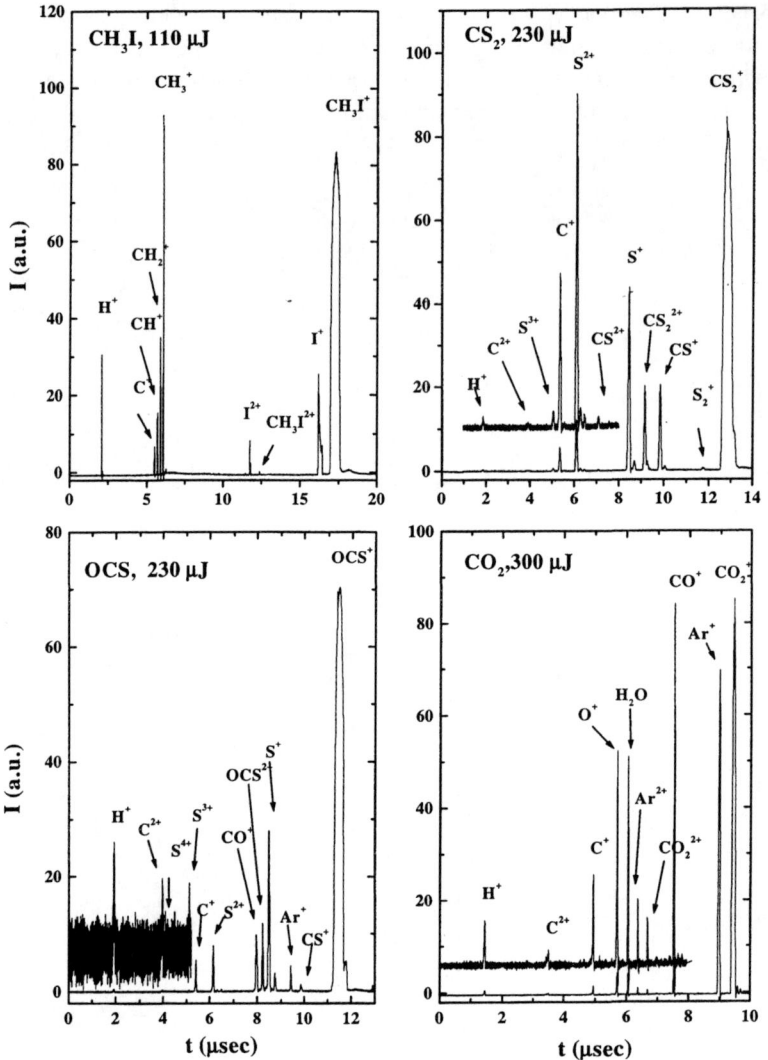

FIFURE 2. Ionization/fragmentation of some small molecules at 800 nm, 200 fsec excitation at high incident energies.

When higher intensities are used, fragmentation becomes important for all molecules and multiply charged parent ions and fragments (S^{2+}, S^{3+}, S^{4+}, I^{2+}, C^{2+})

were observed as depicted by Fig. 2. In order to show the extended ionization/fragmentation in the mass spectra presented, the parent ion peaks were saturated due to the high intensity used. As can be seen, in all cases the doubly charged parent ions were present. Moreover other multiply charged fragments appeared easily due probably to the fragmentation after subsequent absorption of even more photons by the parent molecular ion and/or the doubly charged parent ion. Although the responsible mechanisms are not yet clear to us, in order to clarify the observed behavior, more experimental work is actively going on in our laboratory using measurements of the kinetic energies of the emitted photoelectrons and pump-probe experiments.

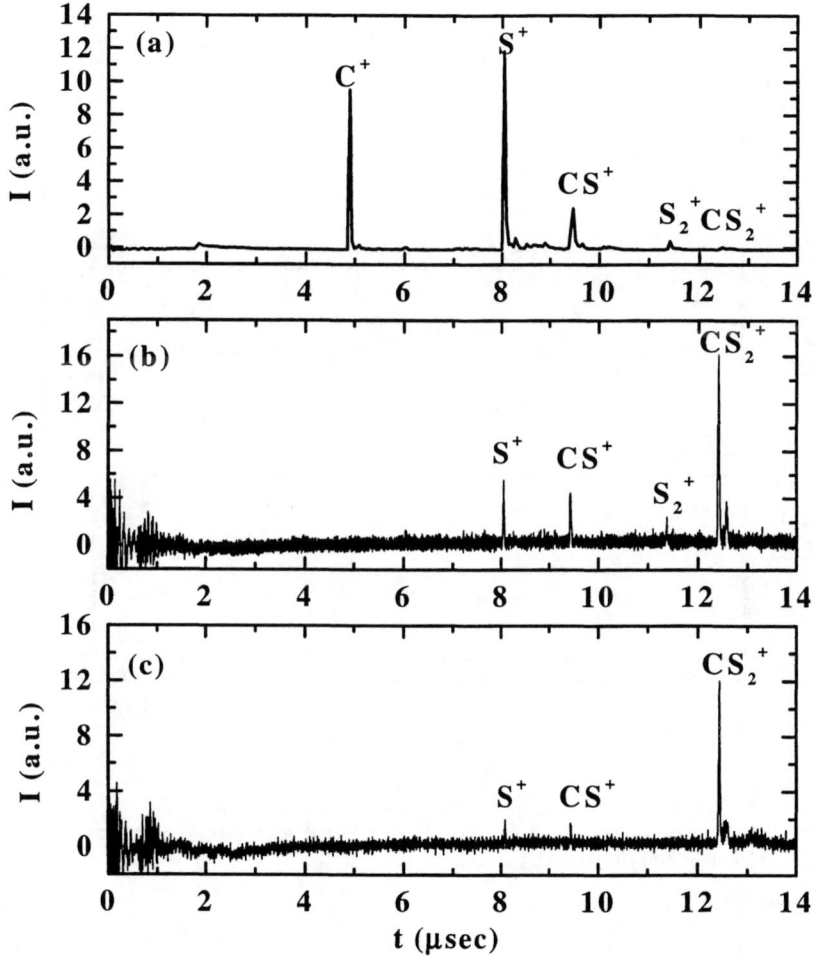

FIFURE 3. Ionization/fragmentation of CS_2 at 248 nm for: a) 15 nsec, b) 5 psec and c) 0.5 psec.

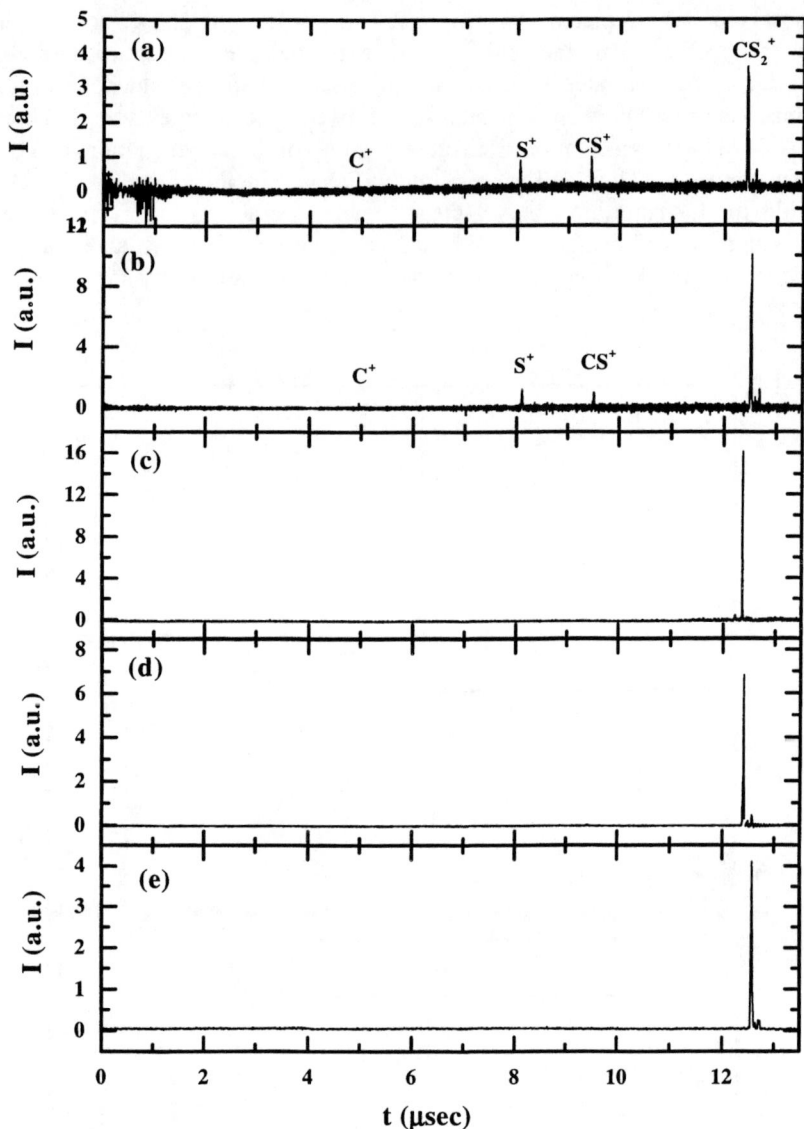

FIFURE 4. Ionization/fragmentation of CS2 for very low incident energy at: a) 5 psec 497 nm, b) 0.5 psec 497 nm, c) 200 fsec 400 nm, d) 200 fsec 800 nm and e) 50 fsec 800 nm.

Particularly for the CS$_2$ molecule, mass spectra were recorded under several experimental conditions, at different wavelengths and for various pulse durations. The first and very important observation was that while for nsec laser excitation the fragmentation of the molecule dominates and the parent ion peak can be observed only for very high intensities, as the pulse duration becomes shorter, the ionization of the parent molecule becomes more efficient. This is presented in Fig. 3 where the mass spectra of CS$_2$ at 248 nm for three different laser pulse durations (15 nsec, 5 psec and 0.5 psec) are shown. As can be seen, the parent molecular ion is almost absent for the case of 15 nsec and increases rapidly with the shortening of the pulse duration. At 248 nm (6,7), CS$_2$ has to absorb at least three photons to be ionized, which brings the molecule to an energy of 15 eV, slightly above the dissociation limit of the molecular ion CS$_2^+$, which can subsequently dissociate to S$^+$(^{4}S) and CS($^1\Sigma^+$) (with a threshold limit at 14.81 eV). This means that the S$^+$ will be produced inevitably during the CS$_2$ ionization process. The produced neutral CS($^1\Sigma^+$) can also absorb three photons (CS ionization potential is 11.33 eV) and be ionized within the duration of the laser pulse. Two photon processes are also possible, that excite the molecule to an energy of about 10 eV, slightly below its ionization potential, where it is known that a great number of high lying Rydberg states are present, generally of strong dissociative character. The products of such a dissociation are excited CS (mainly at the A$^1\Pi$ state) and S (at the ^{3}P$_{J=0,1,2}$ or ^{1}D$_2$ levels). Then subsequent ionization of the products is possible. Moreover, increasing the intensity of the incoming laser radiation, other channels can become active either through the absorption of more photons and subsequent fragmentation of the so produced molecular ions or during the fragmentation of the multiply charged parent molecule.

In figure 4, we present a number of CS$_2$ mass spectra obtained at the lowest possible laser energy for the signal to be observed at five different laser pulses. In all cases the most prominent peak is the molecular parent ion peak. In particular, for the 200 and 50 fsec laser pulses used, only the CS$_2^+$ mass peak was present, independently of the wavelength used (400 or 800 nm). It is worthy to mention that the molecule has to absorb 4 and 7 photons respectively, in order to be ionized. For the longer laser pulses used (0.5 and 5 psec at 497 nm) the appearance of the S$^+$, the CS$^+$ and the C$^+$ mass peaks is evident, even for the lowest intensities used. These are significantly smaller than the molecular parent ion signal corresponding to about 12-15% of the parent ion peak intensity.

With increasing incident intensity, the presence of molecular fragments starts becoming important and even dominant. This is depicted in figure 5 which presents the CS$_2$ mass spectra obtained for high input intensities at various pulse durations. As can be seen, for longer wavelengths, the multiply charged fragments and the doubly charged parent molecule are more efficient generated. In contrast, for short wavelengths, significant generation of S$_2^+$ appeared.

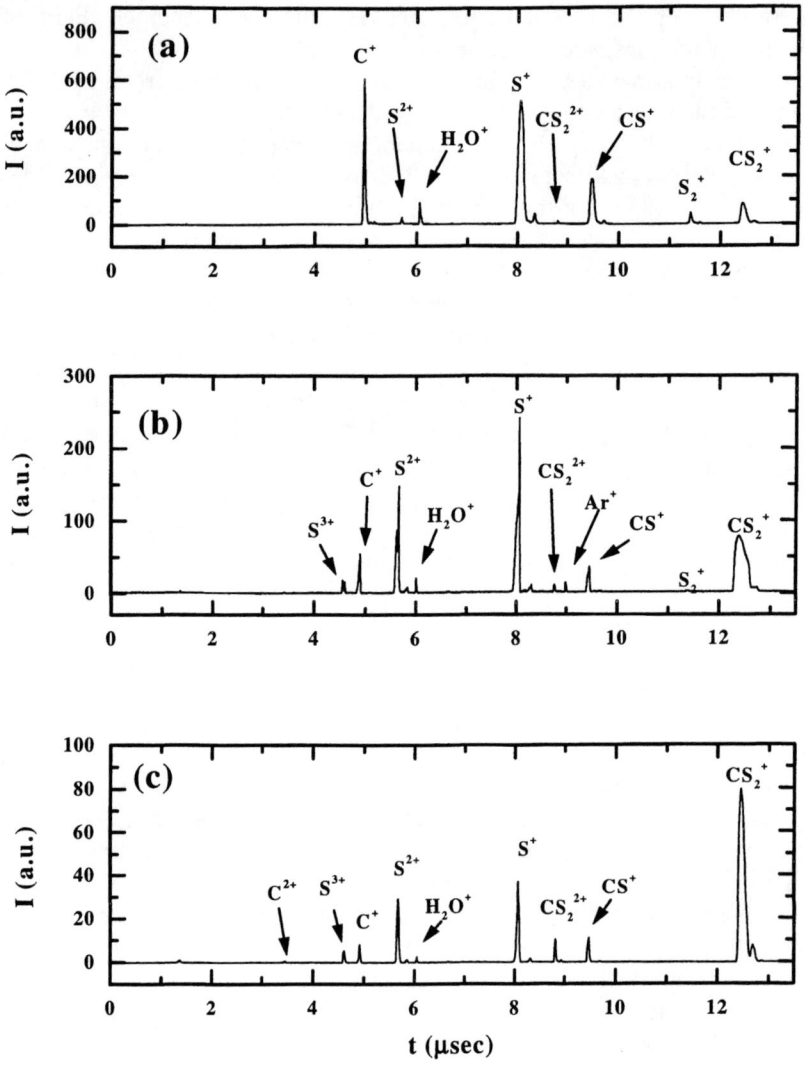

FIGURE 5. Enhanced molecular fragmentation and generation of multiply charged ions at high incident intensities for: a) 248 nm 0.5 psec, b) 497 nm 0.5 psec and c) 800 nm 200 fsec

The polarization of the laser electric field was found to be of great importance for the mass spectra characteristics. This effect was examined in detail for all the laser pulses used with CS_2 as the prototype molecule. With increasing incident intensity where fragmentation is comparable to the ionization and multiply charge fragments are observed, these fragments were found to be preferential aligned along the laser electric field. This resulted in the absence of any detectable signal when the laser polarization was perpendicular to the TOF axis.

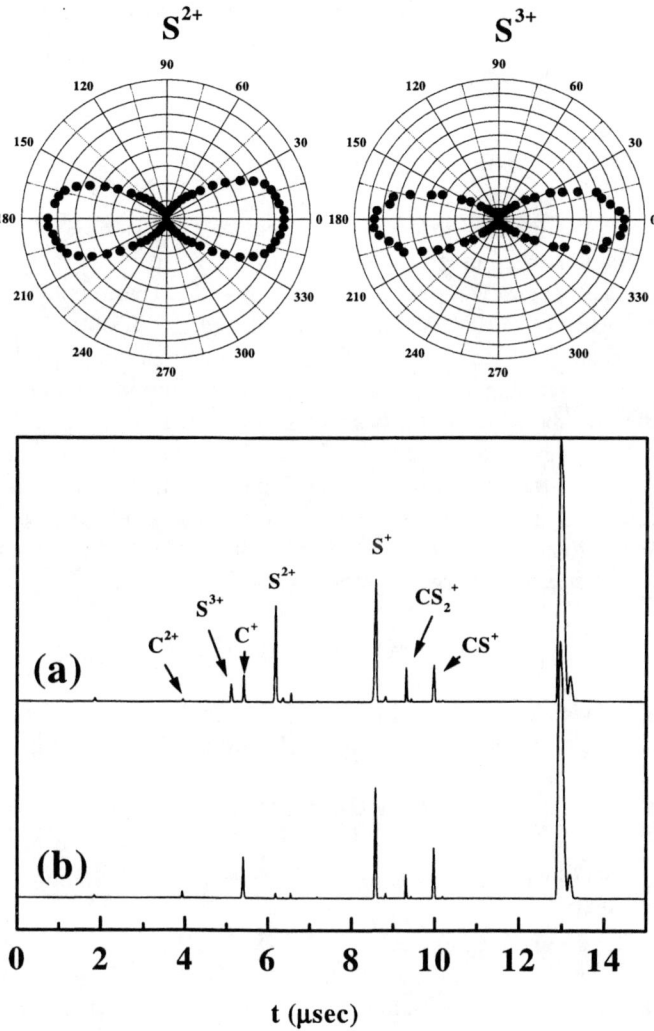

FIGURE 6. Preferential alignment of S^{2+} and S^{3+} along the laser electric field axis.

This is clearly shown in figure 6 where two CS_2 mass spectra are presented for two orthogonal polarizations of the 800 nm, 200 fsec laser beam. As can be seen, the S^{2+} and S^{3+} peaks appear only when the laser polarization is parallel to the TOF axis (p-polarization), the peaks being either very small or even not observable for s-polarization. Although the spectra shown in Fig. 6 were recorded using an extraction field of about 1250 V/cm similar results were obtained for various values of the extraction field ranging from 30 up 1250 V/cm. Moreover, similar polarization effect was observed for the other pulses used.

4. CONCLUSIONS

In this work, the interaction of small molecules such as CS_2 with intense short pulse laser radiation was investigated experimentally using time of flight mass spectroscopy. Mass spectra were recorded under several experimental conditions such as radiation intensity (10^9-10^{13} Watt/cm^2), wavelength (800, 400, 496 and 248 nm), pulse duration (5 and 0.5 psec, 200 and 50 fsec) and polarization of the laser beam. For the shortest pulses used and at low incident intensities, the molecular parent ion was found to be the only peak in the mass spectra recorded. For longer pulses (0.5 and 5 psec), molecular fragmentation appeared even for the lowest intensities used. Increasing incident intensity resulted in an enhancement of the fragmentation and multiply charged ions and fragments were observed. The multiply charged fragments were found to be aligned parallel to the laser electric field, the alignment being more pronounced with increasing charge for each fragment.

ACKNOWLEDGMENT

The authors wish to acknowledge support from the Ultraviolet Laser Facility operating at FORTH-IESL within the Large Installations Plan of the EEC.

REFERENCES

1. R. Weinkauf, P. Aicher, G. Wesley, J. Grotemeyer and E.W. Schlag, J. Phys. Chem. 98, p. 8381 (1994).
2. DeWitt and R.J. Levis, J. Chem. Phys. 102, p. 8670 (1995).
3. DeWitt, D.W. Peters and R.J. Levis, Chem. Phys. 218, p. 211 (1997).
4. M. Castillejo, S. Couris, E. Koudoumas and M. Martin, Chem. Phys. Lett. (1998) (in press).
5. K.W.D. Ledingham, C. Kosmidis, S. Georgiou, S. Couris and R.P. Singhal, Chem. Phys. Lett. 247, p. 555 (1995).
6. J.-P. Berger, S. Couris and D. Gauyacq, J. Chem. Phys. 107(21), p.8866 (1997).
7. C. Fotakis, D. Zevgolis, T. Efthimiopoulos and E. Patsilinakou, Chem. Phys. Lett. 110(1), p.73 (1984).

Dissociative Ionization and Angular Distributions of CS_2 and its Ions

P. Graham[1], K.W.D. Ledingham[1], R.P. Singhal[1], D.J. Smith[1], S.L. Wang[1], T. McCanny[1], H.S Kilic[1], A.J. Langley[2], P.F. Taday[2], C.Kosmidis[3].

1 *Department of Physics & Astronomy, University of Glasgow, Glasgow, G12 8QQ, Scotland, UK.*
2 *Central Laser Facility, Rutherford Appleton Laboratory, Didcot, Oxon., OX11 0QX., England, UK.*
[3]Department of Physics, University of Ioannina, Ioannina, GR-45110, Greece.

Abstract. The dissociative ionization of CS_2 has been investigated at several different wavelengths (375, 750, 395, and 790nm), for pulse-widths as short as 50fs, and laser intensities in the range of $(2.2 \times 10^{13} - 3 \times 10^{16})$ W/cm^2. It is found from the various mass spectra that fragmentation is relatively more pronounced at shorter wavelengths, whilst at longer wavelengths, the parent dominates. Another interesting feature, occurring in the data and literature, is the presence of an S_2^+ ion at wavelengths less than about 532nm, and near total absence at wavelengths longer than this. Angular distributions are presented for both 395 and 790nm. S^{n+}-ions fragment along the polarization direction, while the C^{n+}-ions fragment perpendicularly to it, providing a way of distinguishing the different fragments produced in the coulomb explosion process. Finally, distribution widths decrease with higher charge states of the ions, implying that they are more aligned with the field.

INTRODUCTION

From developments in laser generation of light pulses, very intense (>10^{16} W/cm^2), ultrashort (< 150fs) pulses, made possible by the impact of CPA (Chirped Pulse Amplification) techniques (1), allows the experimental investigation of non-linear phenomena via "dressed" states induced by the laser-molecule interaction.

One of these is the alignment and subsequent ionization of a molecule. A very large electric field associated with the laser light, which can be used to induce dipole moments, aligns or orientates the ions along the polarization vector of the electric field. The degree of orientation depends upon the strength of the electric field, and the magnitude of the polarizability of the ion. Studies of the angular distributions of CS_2 at intensities on the order of 10^{12}-10^{14} W/cm^2 have been previously undertaken (2), and it was now sought to extend this for intensities of the order 10^{15}-10^{16} W/cm^2.

A peak at m/z=64, corresponding to S_2^+, has also been previously identified in several experiments (3), and again presently, and it is thought to arise from the triatomic CS_2 molecule in a bent configuration.

EXPERIMENTAL

The TOF apparatus has been described in detail elsewhere,(4). Briefly, CS_2 was admitted effusively from an inlet system into a high-vacuum chamber, pumped to a base pressure of 1×10^{-8} Torr. CS_2 pressure was of order 1×10^{-5} Torr. The TOF system (fig.1) is a conventional linear system, the field-free length being 1.2m, and an Einzel lens fitted to increase ion extraction. Ions are detected by an electron multiplier connected to a LeCroy 9304 digital oscilloscope.

The laser system used is based on a Spectra Physics Ti:S oscillator pumped by an argon ion laser,(5). It produces 50fs pulses at 790nm, and energy/pulse of 2mJ, after being stretched, amplified in a 7mm long Ti:S rod (Crystal Physics), which is pumped by a Nd:YAG (Spectra Physics) laser, before being re-compressed to 50 fs. The repetition rate was 10Hz. Pulse widths for 750nm were 50fs, whilst that for 375nm were 90fs.

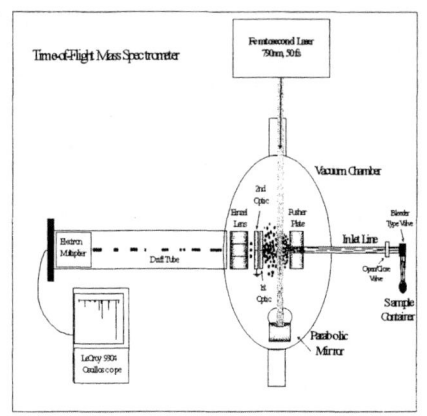

FIGURE 1 TOF System

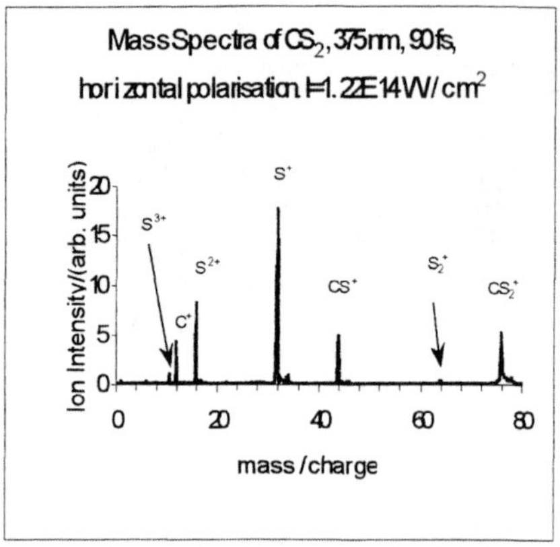

FIGURE 2 Mass spectra at 375nm

RESULTS AND DISCUSSION

Typical mass spectra of CS_2 at 375nm is shown, (fig.2). It can be seen that there is considerable fragmentation, as the parent ion is not the dominant peak, even at these short pulse widths (fragmentation was pronounced at ns time-scales, but reduced for fs studies (6)). It was noted that the intensities are not the same for horizontal and vertical polarization, a consequence of the molecular alignment. The S^+ peak is the largest in both polarisations. It is also possible that post dissociative ionization could occur for some fragments, although it may be a weak process given the short pulse durations.

The dominant peaks for the horizontal case, are: CS_2^+, CS^+, S^{2+}, and C^+, all of approximately equal intensity. There are also S^{3+} and S^{4+} peaks present. Other dominant peaks in the vertical spectra are: C^+, CS_2^+, and CS^+, in decreasing order and a significant C^{2+} peak, as well as a S^{2+}, CS^{2+}, S^{3+}, and a tentative CS_2^{3+} peak. The different sulphur isotopes are present with the correct isotopic abundances. For both polarizations, at 375nm, it has been observed that there is a sizeable S_2^+ peak occurring at m/z=64. This may come about, by transitions to states with bent geometrical configurations, (which cannot be as easily accessed at 750nm).

Typical mass spectra for 750nm, relative to that for 375nm, have less pronounced fragmentation. The parent ion is dominant for both polarization orientations. The strongest fragment ions present in the vertical case are: CS^+, CS_2^{2+}, S^+, and C^+ ions. There are also smaller CS^{2+}, S^{2+}, and C^{2+} ion peaks. In the horizontal spectra, there are: CS^+, CS_2^{2+}, S^+, S^{2+}, and C^+ primary fragment ions, with smaller CS^{2+}, S^{3+}, and C^{2+} ions present.

CS_2 is believed to be linear under 750 nm irradiation, due to the near absence of any peak at m/z = 64, i.e., S_2^+. This moiety is thought to be produced when CS_2 has a bent structure, and thus the 2 peripheral S-atoms are sufficiently close to bond with each other, and break from the C-atom. The lack of the S_2^+ ion is consistent with the idea that no bent geometrical states of the CS_2 molecule, or its ion, can be excited above 532nm, and also the observation of Mathur et al.(3), who noticed that no S_2^+ ions were produced when irradiated by 532nm, 35ps pulses.

Angular distributions for S^{2+}, and C^{2+} are presented, (figs.3-4). It has been seen that all fragments are anisotropic, even the C-atom at 790nm, where an isotropic distribution would be expected from this 'stationary' central atom. The C-distributions are, however, perpendicular to the polarization-vector, for 395 and 790nm. Bhardwaj et al. (7), when considering bent triatomics such as NO_2, explain the perpendicular NO distribution by the directional components of the induced dipole moment of NO_2, whose parallel component is bigger than the orthogonal component, even for small fields, produced in the rising-edge of the pulse. In linear triatomics the fragments should all align along the field direction. The CS distributions are similar to the S distributions.

Since the majority of carbon ions can be detected at angles orthogonal to the axis of polarization, this presents a way of distinguishing the fragments coming from the explosive dissociative ionization of CS_2.

Finally, angular distributions were carried out for up to C^{3+} and S^{5+}, possible with the higher intensities used. It was noticed that they were narrower for each successive charge-state, illustrating that the degree of alignment is stronger, as the electric field needed to produce the higher charge states will cause more of an alignment, and also due to the higher polarization value of the ions. The results are consistent with earlier distributions of CS_2 obtained by Kumar et al, (2).

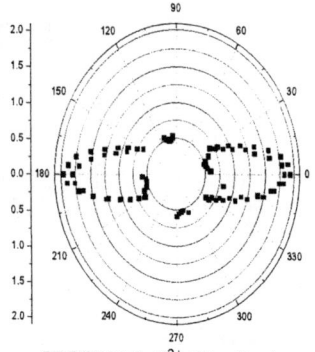

FIGURE 3 S^{2+} Distribution

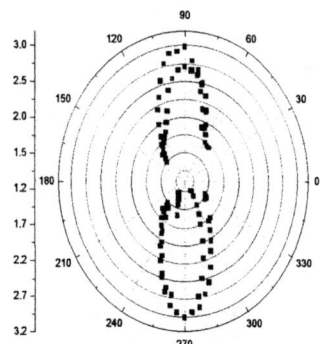

FIGURE 4 C^{2+} Distribution

CONCLUSIONS

The results herein support the view that one charge-state starts when the previous one has saturated. Increasing the laser intensity merely increases the volume of production of an ion. Angular distributions were worked out by varying the polarization direction. It is noted that higher charge state ions have narrower distributions, as expected, and the results indicate alignment of the molecule. Finally, mass spectra were obtained at 375/395nm and 750/790nm, and fragmentation was found to be relatively increased at the shorter wavelength. This has also been observed for C_7H_6O (8), and C_4H_6 (9).

Production of the S_2^+ ion, indicative of exciting the parent to states with a bent geometrical structure, was also increased at 375nm. For certain conditions at 790nm, it was found that the ionization behaved very atomic-like, i.e., little fragmentation and multiple-charged parent ions (10). This phenomenon has also been observed for both CH_3I, and 1-3 Butadiene (9).

REFERENCES

(1). Strickland D and Mourou G, *Opt. Commun.*, **56**, 219 (1985)

(2). Kumar G R, Gross P, Safvan C P, Rajgara F A, and Mathur D, *J. Phys. B: At. Mol. Opt. Phys.*, **29**, L95-L103 (1996)

(3). Mathur D, Kumar G R, Safvan C P and Rajgara F A, *J. Phys. B: At. Mol. Opt. Phys.*, **27**, L603-L610 (1994)

(4). Ledingham K W D, Kilic H S, Kosmidis C, Deas R M, Marshall A, McCanny T, Singhal R P, Langley A J and Shaikh W, *Rapid Commun. Mass Spectrom.*, **9**, 1522 (1995)

(5). Taday P F, Mohammed I, Langley A J, and Ross I N, to be published

(6). Yang J J, Gobeli D A, El-Sayed M A, *J. Phys. Chem.*, **89**, 3426 (1985)

(7). Bhardwaj V R, Safvan C P, Vijayalakshmi K and Mathur D, *J. Phys. B: At. Mol. Opt. Phys.*, **30**, 3821-3831 (1997)

(8). Smith D J, Ledingham K W D, Kilic H S, McCanny T, Peng W X, Singhal R P, Langley A J, Taday P F and Kosmidis C, *J. Phys. Chem. A*, to be published

(9). Graham P, Ledingham K W D, Singhal R P, Smith D J, McCanny T, Langley A J and Taday P F, to be published

(10). Ledingham K W D, Singhal R P, Smith D J, McCanny T, Graham P, Kilic H S, Peng W X, Wang S L, Langley A J, Taday P F, and Kosmidis C, *J. Phys. Chem. A*, **102**, 3002-3005 (1998)

POSTER SESSION VIII
SURFACE APPLICATIONS AND SPUTTERING

Desorption Dynamics Below and Above the Ablation Threshold of van der Waals Films

Maria Lassithiotaki, Antonis Koubenakis, John Labrakis, Savas Georgiou[*]

Institute of Electronic Structure and Laser
Foundation for Research and Technology - Hellas
71110 Heraklion, Crete, Greece

Abstract The 248 nm ablation of organic van der Waals films is studied as function of the laser fluence via time-of-flight quadrupole mass spectrometry. At low fluences, we find desorption of only highly volatile species. In contrast, above the ablation threshold, ejection of even heavy species is highly efficient. The results are in very good correspondance with the delineation drawn by a recent molecular dynamics simulation [L. Zhigilei et al, J. Phys. Chem. B 101 (1997) 2028] for surface vaporization at fluences below the ablation threshold and ejection as a result of pressure buildup above it.

INTRODUCTION

Recently, Zhigilei et al. (1) employed molecular dynamics simulations based on a breathing sphere model for modeling the laser ablation of organic solids. Without any other assumptions, the simulations predicted a well-defined fluence threshold, with distinctly different desorption mechanisms operating below and above it. At low fluences, desorption was shown to result from surface vaporization. In contrast, above the threshold, the heating rate was found to be faster than that for the mechanical relaxation of the material, thus resulting in the buildup of high pressure. This induces a phase - transition explosion from the solid directly to the gaseous phase.

For testing the proposed picture, we turn to the study of the ablation of van der Waals films formed by the condensation of vapours of simple compounds on low temperature substrates. The physicochemical simplicity of these films ensures that theoretical predictions can be tested free from the complications encountered in the study of more complex substrates. Herein, we study the ablation of C_6H_5Cl films at 248 nm. Upon excitation at 248 nm, C_6H_5Cl dissociates exclusively by C-Cl bond scission (2) to yield C_6H_5 and Cl fragments, which can react further to form a number of different products (3). Thus, the suggestion for the operation of different ejection mechanisms above and below the threshold can be tested by comparing the desorption efficiencies of the volatile vs that of the "heavy" photoproducts in the corresponding fluence ranges.

EXPERIMENTAL

The experimental setup is described in detail elsewhere (4). Briefly, experiments are performed on a vacuum system consisting of two, differentially pumped, chambers. In the first cell, a liquid N_2 dewar supports a suprasil substrate for the deposition of the films. Gases are introduced via a stainless needle to form $\approx$ 30-50 µm thick films. Subsequently, the system is pumped to its base pressure (5×10^{-6} Pa) before irradiation commences. The second chamber ($P < 10^{-6}$ Pa) houses the mass spectrometer (Balzers QMG 412) which views the irradiated area in the plane defined by the irradiation axis and the normal to the substrate under a solid angle of $\approx 7.5 \times 10^{-5}$ π sr. The signals are amplified, recorded on a LeCroy LC9400 and sent to a computer.

Desorption is effected by irradiation with the 248 nm output of Lambda-Physik EMG150 excimer laser. The laser incident and detection angles are fixed at ~75° and ~15°, respectively, with respect to the normal to the substrate. Reported fluences are uncorrected for scattering effects at film interface.

Typical TOF spectra recorded for C_6H_5Cl and for the HCl and $(C_6H_5)_2$ photoproducts are depicted in Fig. 1. The TOF curves are corrected for the transit time through the mass spectrometer, and for the inverse dependence of the ionization efficiency of the neutrals on their velocity. Intensities are determined by integration over time of the flux distributions.

C_6H_5Cl (Aldrich) is of high purity (99.5% or better) and is subjected further to careful trap-to-trap distillation and repeated freeze-pump-thaw cycles.

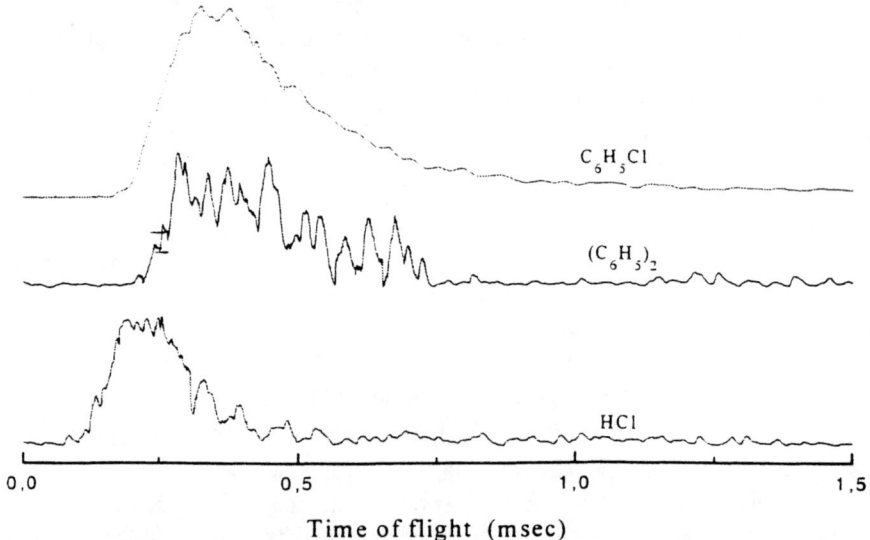

FIGURE 1 Time-of-flight spectra of the C_6H_5Cl, $(C_6H_5)_2$ and HCl recorded upon irradiation of "fresh" films with $F_{LASER} \approx 130$ mJ/cm^2.

RESULTS AND DISCUSSION

Upon irradiation of freshly deposited films of the compound at $F_{LASER} \leq 50$ mJ/cm^2, the intensity of C_6H_5Cl is initially quite low. It rises only with continuing irradiation (Fig. 2), which, as shown below, is due to the accumulation of strongly absorbing photoproducts and the consequent enhancement of the laser energy deposition in the film. In contrast, above 50 mJ/cm^2, desorption of C_6H_5Cl from freshly deposited films is intense from the very first laser pulse (Fig. 2). From the plot of the intensity that is recorded in the first laser pulse as function of the laser fluence, it is apparent that the ablation threshold for neat C_6H_5Cl films is at ≈ 50 mJ/cm^2.

Not only there is an abrupt change in the desorption efficiency of C_6H_5Cl, but additionally different products are observed above the threshold from what below it. Thus, above the ablation threshold, four new major species, namely HCl, $(C_6H_5)_2$, $C_6H_4Cl_2$ and $C_{12}H_9Cl$, are detected in the mass spectrometer. The reactivity responsible for the formation of these species is discussed elsewhere (5). Importantly, the ratio of intensities of the phenyl photoproducts vs. that of the parent peak is found to grow with number of successive laser pulses until reaching a plateau. Since the relative intensities of the phenyl products do not change with successive laser pulses, the pulse evolution is not due to successive reactions i.e., from decomposition of others. Consequently, the pulse dependence reflects the stepwise accumulation of these species in the film. In all, we can view the process as the ablation of a mixture consisting of the indicated photoproducts dissolved in a C_6H_5Cl matrix.

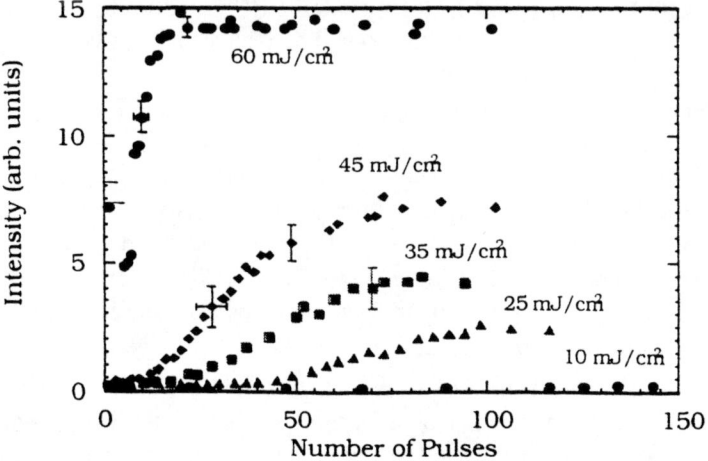

FIGURE 2 Pulse evolution of the C_6H_5Cl intensity in the irradiation of freshly deposited films. The origin of the pulse evolution that is observed at the higher fluence is discussed elsewhere.

In contrast at $F_{LASER} \approx 50$ mJ/cm^2, only HCl is initially observed from freshly deposited films, while the phenyl photoproducts start being observed only as parent signal induction becomes significant. Although the phenyl species are not observed, they are nevertheless formed. Thus, White et al. employed (6) Thermal Desorption Spectroscopy (TDS) and High Resolution Electron Energy Loss Spectroscopy (HREELS) to examine the photolysis of multilayer C_6H_5Cl films with a mercury lamp. The HREELS examination demonstrated photolysis of C_6H_5Cl to occur, while in the postirradiation TDS analysis, $C_{12}H_9Cl$ was found to desorb at ≥ 250 K. Photoproduct formation occurs during the irradiation, as demonstrated by the fact that in the irradiation of thin (2-4 μm thick) C_6H_5Cl films with 1-20 mJ/cm^2, film transmission decreases with successive laser pulses. This strongly suggests the accummulation of species that absorb stronger than the parent molecule. Indeed, biphenyl species are known (7) to absorb at 248 nm much stronger than C_6H_5Cl. Furthermore, we have performed experiments in which freshly deposited films were first irradiated with ≈30 pulses at $F_{LASER} \approx 30$ mJ/cm^2 and then signals for the biphenyl products were recorded by irradiating the same spot at fluences above the ablation threshold. Importantly, the signal recorded in the first few pulses at the monitoring fluence is much higher than that observed in the irradiation exclusively at this higher fluence. The difference is evidently due to biphenyl products that accummulated in the film during the irradiation at ≈ 30 mJ/cm^2. In all, even below 50 mJ/cm^2, formation of phenyl photoproducts does occur and the exclusive reason for failing to observe them in the mass spectrometer is their inefficiency of desorption. Evidently, below the threshold, only the highly volatile HCl can desorb.

The difference between HCl and the phenyl derivatives is fully consistent with the suggested by the simulations thermal nature of desorption below the threshold. The film temperature for $F_{LASER} = 50$ mJ/cm^2 is estimated to be $T = \alpha F_{LASER}/\rho C_p \approx 175$ K (where α is the extinction coefficient of C_6H_5Cl, ρ its density (in gr/cm^3) and Cp its heat capacity (in J/gr K)) (9), which is sufficiently high for HCl desorption, but still quite lower than the T=250 K necessary for biphenyl derivatives evaporation.

Further support for the thermal nature of desorption in this fluence range derives from the examination of desorption from C_6H_5Cl/C_6H_{12} mixtures. C_6H_{12} is almost transparent at 248 nm and thus its molecular photodesorption is negligible (4). Nevertheless, upon irradiation of C_6H_5Cl/C_6H_{12} mixtures, C_6H_{12} desorption is found to be intense, even higher than that of C_6H_5Cl for high dilutions in the nonabsorbing component. The high efficiency of C_6H_{12} desorption is in line with the high volatility of this compound (8), exactly as expected for a thermal process.

On the other hand, the efficient ejection of the "heavy" species above the threshold can not be ascribed to a thermal process. From transmission measurements, we do not find any evidence for the contribution of higher-order absorption process above threshold. In that case, the attainment of $T \approx 250$ K as necessary for the evaporation of the biphenyl derivatives would require fluences of 200 mJ/cm^2, which is well above the experimental threshold, even if no consideration is given to the fact that the experimental values are uncorrected for scattering losses at the film interface. Thus, the desorption of these species is strongly indicative of a nonthermal ejection mechanism in the ablative regime.

In considering the nature of the nonthermal mechanism, there are two ejection characteristics of the biphenyl products that strongly indicate the suggested the photomechanical nature of ejection:

First (Fig. 3), their efficiency of ejection in the gas phase closely corresponds to that of C_6H_5Cl. This correspondence strongly indicates that the driving force for their ejection is their entrainment into the desorbing C_6H_5Cl "jet". Their close correspondence sharply contrasts the nearly complete lack of correlation observed for the highly volatile HCl photoproduct. Second, the velocity distributions of the phenyl photoproducts are nearly identical with that of the parent peak (Fig. 1). This result fully conforms with the picture of the simulations, according to which the "heavy" analyte is entrapped in the desorbing jet of the matrix molecules and are accelerated to nearly the same radial velocity.

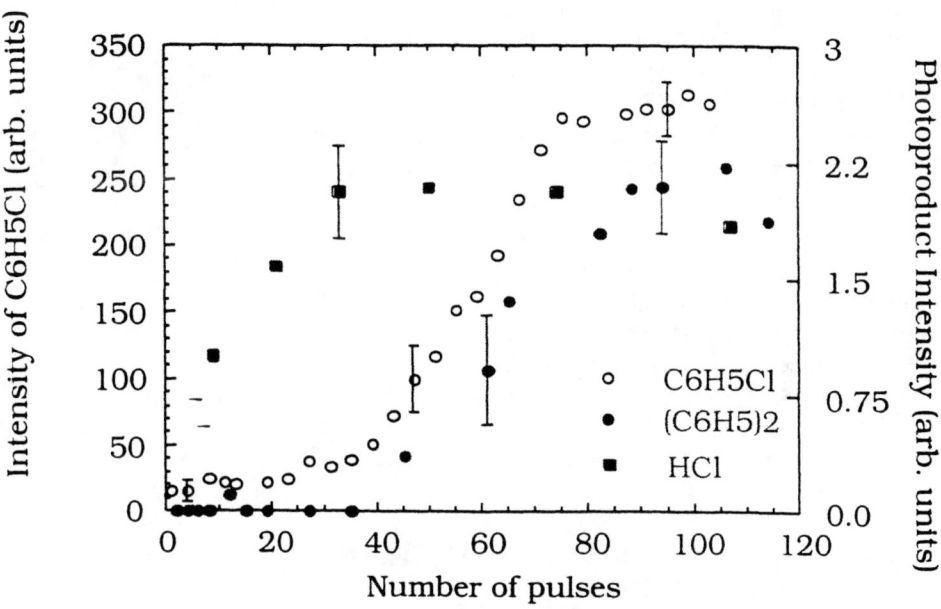

FIGURE 3 Pulse evolution of the intensities of C_6H_5Cl, HCl and $(C_6H_5)_2$ in the irradiation of freshly deposited films with $\approx$30 mJ/cm^2. Essentially the same pulse dependence as that for $(C_6H_5)_2$ is also observed for $C_6H_4Cl_2$ and $C_{12}H_9Cl$.

In all, we find desorption dynamics in the subthreshold range to differ distinctly from that in the ablative regime. The delineation appears to be in very good correspondence with the results of the MD simulations for the thermal nature of desorption below threshold and for photomechanical character of ejection above it.

ACKNOWLEDGEMENTS

The work was supported by the Large Installations Plan DGXII (Project G/89100086/GEP).

REFERENCES

1. (a) Zhigilei, L. V., Kodali, P. B. S., and Garrison, B. J., *J. Phys. Chem B* **101**, 2028 (1997) (b) Zhigilei, L. V., Kodali, P. B. S., and Garrison, B. J., *Appl. Phys. Lett.* **71**, 551 (1997) (c) Zhigilei, L. V., Kodali, P. B. S., and Garrison, B. J., *J. Phys. Chem B* **101**, 8624 (1997)
2. Ichimura, T., Mori, Y., Shinohara, H., Nishi, N., *Chem. Phys.* **189**, 117 (1994).
3. Davidson, R. S., Goodin, J. W., and Kemp, G., *Adv. Phys. Organic Chem.* **20**, 191 (1984) and references therein.
4. (a) Georgiou, S., Koubenakis, A., Kontoleta, P., and Syrrou, M., *Chem. Phys. Lett.* **260**, 166 (1996). (b) Georgiou, S., Koubenakis, A., Kontoleta, P., and Syrrou, M., *Chem. Phys. Lett.* **270**, 491 (1997).
5. Because of the relatively low resolution for masses higher than 100 amu, we can not exclude the possibility that the desorbed species are actually the corresponding ipso-adduct radicals, i.e., $C_6H_5Cl_2$, $C_6H_5Cl-C_6H_5$. We consider, however, more likely the detected species to be the stable phenyl derivatives and we adopt this possibility in the subsequent discussion of this paper.
6. Song, Y., Garder, P., Conrad, H., Bradshaw, A. M., and White, J. M., *Surf. Sci.* **248**, 1279 (1991).
7. (a) Robin, M. B., *Higher Excited States of Polyatomic Molecules*, (Academic Press, New York, 1974) (a) vol. I., pp. 155 - 178 (b) vol II, pp. 230 - 232. 62. (c) Coffman, R., and McClure, D. S., *Can. J. Chem.* **36**, 48 (1958).
8. *Handbook of Chemistry and Physics*, 61 ed., edited by Weast, R. C., and Astle, M. J., (CRC, Boca Raton, FL, 1980)

Electron Configuration Dependence of Kinetic Energy Distributions of Ion-Beam Sputtered Ni Atoms studied by Double Resonant Laser Ionization

V. Philipsen, J. Bastiaansen, E. Vandeweert, P. Lievens, R.E. Silverans

Laboratorium voor Vaste-Stoffysica en Magnetisme, K.U. Leuven
Celestijnenlaan 200 D, B-3001 Leuven, Belgium

Abstract. We report the measurement of state specific kinetic energy distributions of ground and metastable state Ni atoms ejected during 15 keV Ar^+ beam sputtering of clean polycrystalline Ni targets. A very sensitive experimental procedure based on double resonant laser ionization was used to measure a nearly complete set of kinetic energy distributions of the populated states. The state specific kinetic energy distributions and their weighted sum using the measured population partition provide further evidence for resonant electron tunneling during the emission process and its dependence on both the atomic electronic configuration and the bulk electronic structure.

INTRODUCTION

Recent work aiming at identifying the electronic mechanisms involved in the emission of *metastable* atoms following ion-beam sputtering of clean metal surfaces concentrated on the combined measurement of ground and metastable state population partitions and quantum state specific kinetic energy (KE) distributions using resonant laser ionization techniques (1-4). In several cases, the experimental data could not be explained by collision induced excitation and non-radiative deexcitation. The correspondance between the electronic state of the sputtered atoms and the equilibrium electronic structure of the solid as well as resonant electron pick-up have been involved to interpret the observations.

In this contribution we report the measurement of a nearly complete set of KE distributions of ground and metastable state Ni atoms ejected during ion-beam sputtering, and their interpretation in combination with the measured population partition. Ni I exhibits a rich structure of low and high lying metastable states with two types of outer shell electronic configuration ($3d^9 4s^1$ and $3d^8 4s^2$), which allows to monitor the influence of both the electronic configuration and the excitation energy on the velocity dependent population probability of these states.

EXPERIMENTAL PROCEDURE

The sputtering experiments are performed in a UHV chamber with a base pressure of $\sim 5 \times 10^{-10}$ hPa. A plasma ion source produces 15 keV Ar ions directed onto the target foil at 45° incidence. For the measurements of the KE distributions pulsed operation is used (ion current ~0.5 µA/mm^2, pulse duration ~300 ns). Resonant ionization of the sputtered atoms is performed using the overlapping beams of pulsed optical parametric oscillator and dye laser systems: wavelengths from 225 to 1600 nm, pulse duration ~6 ns, repetition rate 10 Hz, band widths ~10 GHz and pulse energies up to 50 mJ. The laser beams have near-Gaussian beam profiles (diameter 1 mm) and cross the plume of sputtered particles at a distance of 5 mm above the surface of the foil. The laser ionized atoms are accelerated into a time-of-flight mass spectrometer through a collimator (diameter 2 mm) and counted.

For the state-selective detection of the sputtered metastable atoms and the quantitative measurement of population partitions, an experimental procedure based on two-step two-colour resonance laser ionization processes is used. The first step sequentially saturates the excitation of atoms, in different metastable states, to *the same intermediate state*. A second independent laser ionization step is applied to ionize the atoms. By this procedure, the relative photoion intensities directly reflect the population partition of the involved metastable states: only saturation of the excitation steps is required, which is already obtained with modest laser pulse energies. More details are given in ref. 5.

To measure the KE distribution of atoms in all states, including the weakly populated high-lying states, very high sensitivity is required. This is obtained by tuning the ionizing laser to resonant transitions into autoionizing states. The high ionization cross sections of such transitions even allow to saturate the ionization step in most cases (6). State specific KE distributions are obtained from measurements of the photoion intensities $S(t)$ as a function of delay time t between the sputter ion pulse and the ionizing laser pulses. From these measured flight-time distributions the kinetic energy flux distributions can be derived (7).

EXPERIMENTAL RESULTS AND DISCUSSION

Systematic differences are observed between the KE distributions of states with $3d^84s^2$ configuration (the ground state multiplet a $^3F_{4,3,2}$ with excitation energies 0 to 0.28 eV, the b $^1D_2$ state at 1.68 eV and the a ^3P_J mutiplet states at ~1.95 eV) and those of the states with $3d^94s^1$ configuration (the a $^3D_{3,2,1}$ multiplet at 0.03 to 0.21 eV, and the a $^1D_2$ state at 0.42 eV). Examples of KE distributions representative for each type of state are presented in FIGURE 1. The maximum in the KE distribution of the $3d^94s^1$ states is shifted to higher kinetic energies with respect to the ones of the $3d^84s^2$ states and the high energy tail of the KE distributions of the former falls off less steeply. This behaviour elucidates that the state specific KE distributions are mainly determined by the electronic configuration of the states.

The KE-integrated population partition of the sputtered atoms over the ground and metastable states has been measured earlier and was reported in ref. 5. The populations of the states of the low-lying multiplets with $3d^94s^1$ configuration are shifted to higher values with respect to the states with comparable excitation energy of the ground state multiplet ($3d^8 4s^2$ configuration). The metastable states of multiplets with excitation energies between 1.6 eV and 2 eV show anomalously high populations of about 10% of the ground state population.

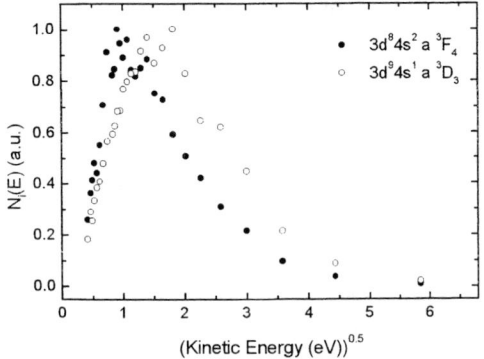

FIGURE 1. Representative examples of state specific KE distributions of sputtered Ni atoms: the $3d^8 4s^2$ a $^3F_4$ state (closed symbols) and the $3d^9 4s^1$ a $^3D_3$ state (open symbols).

The population partition has been interpreted in the frame of the resonant electron transfer (RET) model (5,8), and we show here that the KE-distributions add further support to the applicability of this model. The preferential population of states with $3d^9 4s^1$ outer electron shell configuration has been explained by the fact that their configuration is very similar to the $3d^{9.4} 4s^{0.6}$ character of the conduction band of Ni, causing stronger coupling of these atomic states to the metal states and therefore higher electron tunneling rates.

Multichannel electron transfer into the different atomic states of a particle sputtered with velocity perpendicular to the surface, has been modelled by a set of coupled rate equations describing the competition between the different transfer channels (8). The final occupation probabilities of the states depend on the velocity of the escaping particle. The faster the particle recedes from the surface, the higher the probability of population of the strongly coupled states. This fits to the observation of the differences in the KE distributions of the $3d^9 4s^1$ states compared to those of the $3d^8 4s^2$ states, see FIGURE 1.

Within the collision cascade theory, the KE flux distribution $N_{tot}(E)$ of the *whole ensemble* of sputtered particles is given by the Sigmund-Thompson distribution (9). According to the RET model, the state specific KE distributions $N_i(E)$ will deviate from this distribution due to the KE-dependent RET probability to the different states. But as the ionic fraction of particles sputtered from metals is small for all velocities, the sum of the state specific KE distributions weighted with the populations n_i of the different states should, to a good approximation, correspond to the Sigmund-Thompson distribution:

$$N_{tot}(E) \propto \sum_i n_i N_i(E) \propto \frac{E}{(E+U)^3} \tag{1}$$

with U the surface binding energy for which usually the sublimation energy is taken. This weighted sum, displayed in FIGURE 2, is indeed in agreement with the predictions of collision cascade theory.

The velocity dependence of the RET probability into the two classes of electronic states is further evidenced by FIGURE 3, presenting the KE resolved populations of

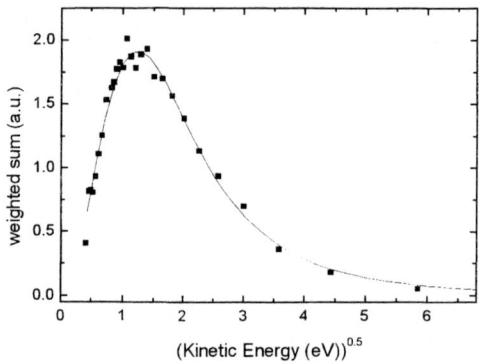

FIGURE 2. Weighted sum of state specific KE distributions (dots) compared to the Sigmund-Thompson KE flux distribution (full line).

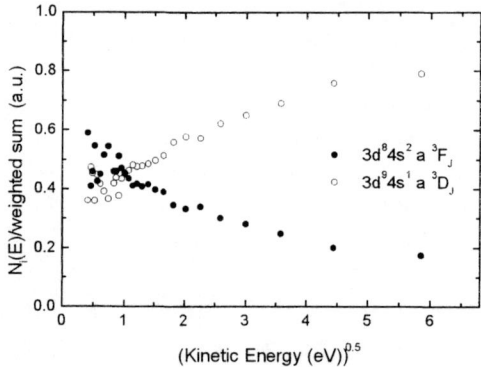

FIGURE 3. Populations of the low-lying a ^3D_J ($3d^94s^1$) multiplet (open symbols) and the a ^3F_J ($3d^84s^2$) ground state multiplet (closed symbols) as function of kinetic energy, relative to the total population of all metastable states.

the low-lying multiplets, obtained by taking the weighted sum of the KE distributions of these multiplets divided by the KE distribution of the whole ensemble. It is clear that the probability of resonant electron transfer into the states with $3d^94s^1$ configuration increases drastically with emission velocity of the sputtered particle.

In conclusion, the state specific KE distributions of a nearly complete set of sputtered Ni atoms in combination with the measured population partition adds further evidence for resonant electron transfer in the emission process and its dependence on the correspondance between the atomic and bulk metal electronic structure.

ACKNOWLEDGMENTS

This work is financially supported by the Fund for Scientific Research - Flanders (F.W.O.), the Flemish Concerted Action (G.O.A.) Research Programme and the Interuniversity Poles of Attraction Programme - Belgian State, Prime Minister's Office - Federal Office for Scientific, Technical and Cultural Affairs. P.L. is a Postdoctoral Research Fellow of the F.W.O.

REFERENCES

1. C. He, Z. Postawa, S.W. Rosencrance, R. Chatterjee, B.J. Garrison, and N. Winograd, Phys. Rev. Lett. **75**, 3950 (1995).
2. G. Nicolussi, W. Husinsky, D. Gruber, and G. Betz, Phys. Rev. B **51**, 8779 (1995) 8779.
3. W. Berthold and A. Wucher, Phys. Rev. Lett. **76**, 2181 (1996).
4. A. Wucher and Z. Sroubek, Phys. Rev. B **55**, 780 (1997).
5. E. Vandeweert, V. Philipsen, W. Bouwen, P. Thoen, H. Weidele, R.E. Silverans, and P. Lievens, Phys. Rev. Lett. **78**, 138 (1997).
6. P. Lievens, E. Vandeweert, P. Thoen, and R.E. Silverans, Phys. Rev. A **54**, 2253 (1996).
7. A. Wucher, M. Wahl, and H. Oechsner, Nucl. Instr. Methods B **82**, 337 (1993).
8. R.E. Silverans and P. Lievens, "Resonance Ionization Spectroscopy Investigations of Electronic Processses during Ion-Beam Sputtering of Metal Atoms", contribution to this conference, *Proceedings RIS-98*.
9. P. Sigmund, Phys. Rev. **184**, 383 (1969).

Development of a Biological Imaging Instrument

N.P. Lockyer, S.C.C. Wong and J.C. Vickerman

Surface Analysis Research Centre, Department of Chemistry,
UMIST, P.O. Box 88, Manchester M60 1QD, U.K.

Abstract. We report on the development of a laser-ionization mass spectrometer designed to have wide applications in biological and medical research. The instrumentation combines chemical specificity with imaging capabilities analogous to the scanning electron microscope. We present some aspects of current research on this instrument, such as the investigation of biomedical implants and molecular nano-engineering.

INTRODUCTION

Scanning electron microscopy has proved a powerful tool in the study of surfaces of biological interest. However, this approach to surface characterisation is limited in that it is not chemically specific. The aim of this project is to develop an instrument analogous to the electron microscope but specifically designed to provide sub-micron surface chemical analysis of complex chemical and biochemical materials.

In collaboration with Pennsylvania State University we have designed and built a new generation laser-ionization mass spectrometer with spatial resolution in the region of 50 nm and the ability to manipulate cryogenic bio-samples.

EXPERIMENTAL

The sample is located in ultra high vacuum on a high stability cold stage that can be maintained at 100 K to reduce evaporation rates. A 25 keV gallium primary ion beam is digitally rastered across the sample to desorb surface molecules. A small fraction of the desorbed material consists of secondary ions that can be mass analysed by a time-of-flight mass spectrometer and detected by a dual microchannel plate detector (Fig. 1). This technique of secondary ion mass spectrometry (SIMS) is a powerful tool for surface analysis, but its application at ultra high spatial resolution is limited by the ionisation probability of ion beam-desorbed material. This is particularly true for high mass organic molecules. For example in a 50 nm x 50 nm pixel there will be only a few thousand surface molecules and typically 0.1% of these will be desorbed as molecular ions in the SIMS process.

CP454, *Resonance Ionization Spectroscopy*
edited by J. C. Vickerman, I. Lyon, N. P. Lockyer, and J. E. Parks
© 1998 The American Institute of Physics   1-56396-810-X/98/$15.00

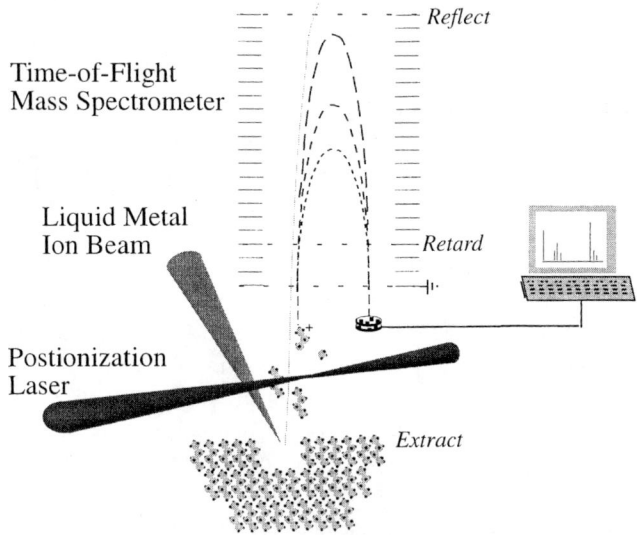

FIGURE 1. Schematic diagram of the experimental technique. Surface molecules are desorbed by a highly focused liquid metal ion beam. Ions formed directly by this process can be discriminated against by adjusting the voltages in the ToF-MS. Species desorbed as neutrals are postionized by laser radiation and detected after transmission through the ToF-MS.

In order to dramatically increase sensitivity laser ionization methods are employed to detect the dominant neutral component of the ion beam-desorbed material. The experimental challenge is to ionize the dominant neutral component of desorbed molecules with uniformly high efficiency and minimum fragmentation. Laser ionization offers the best method of meeting this challenge. This approach is known as laser *postionization* as neutrals are ionized in a separate step after they have been desorbed from the surface.

Multiphoton ionization (MPI) methods using nanosecond laser pulses at above threshold intensity typically result in extensive photofragmentation of biomolecules. Femtosecond pulses have been used in MPI to defeat neutral dissociation mechanisms which operate on slower timescales. This has been shown to increase the ionization efficiency and reduce excessive photofragmentation in many cases (1-4). This is the laser ionization method adopted by the Penn. State Group.

At UMIST we employ a vacuum ultraviolet (VUV) single photon ionisation (SPI) scheme, thereby avoiding dissociative intermediate states altogether. VUV photons of 118 nm are obtained from a third harmonic generation process in xenon/argon mixtures excited with the 355 nm output of a 35 ps modelocked Nd:YAG laser. This approach has been shown to achieve high yields of molecular-type ions from biological samples (5, 6).

RESULTS

The instrument will primarily be used for biological imaging applications. We have already initiated work to this end, with biocompatibility and molecular nano-engineering studies. These projects have to date only utilized the SIMS capability of the instrument. With the installation of the laser ionization facility we expect performance to be enhanced many-fold.

Biocompatability of Implants for Cataract Treatment

In-the-bag implantation of intraocular lenses (IOLs) has become generally accepted as the procedure of choice in cataract patients. This involves surgically removing the cataractous lens and implanting the IOL between the anterior and posterior capsules. However, cataract extraction leaves lens material that can proliferate and migrate on the surface of the IOL leading to Posterior Capsule Opacification (PCO) whereby vision is again impaired (Fig. 2) (7). PCO is the most common postoperative complication of modern cataract surgery, with a reported incidence of 20-50% within 5 years of surgery (8). The laser treatment of PCO is the second most frequently performed surgery in some industrialised countries (9).

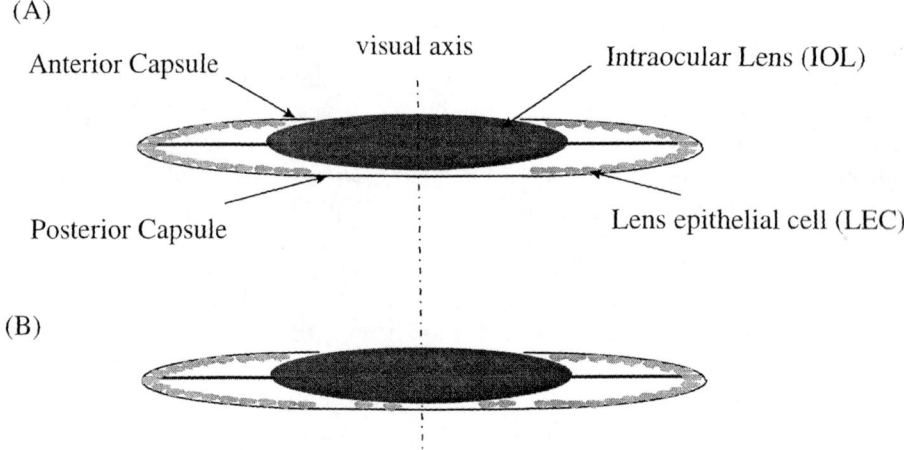

FIGURE 2. Postoperative Complications. (A) a newly implanted IOL between anterior and posterior capsules. LECs remain after removal of the cataractous lens. (B) Proliferation of LECs on the visual axis lead to posterior capsule opacification requiring laser treatment to restore vision.

The incidence of PCO in eyes with IOLs varies with implant design and material suggesting that biocompatibility plays a key role in determining the surgical success rate. The IOL can be biocompatible in a bioinert or bioactive way. An IOL material such as PMMA or silicone which has good biocompatibilty but is bioinert and would allow LECs to proliferate turning the posterior capsule opaque. A bioactive IOL surface could allow a single lens epithelial cell monolayer to form between the IOL and the posterior capsule thereby preventing further PCO.

There is a requirement for evaluation techniques to compare surgical techniques and implant biocompatibility. We are undertaking a study of new and used intraocular lenses of different biomaterials to provide further insights in this area.

Fig. 3 shows SIMS images of sodium and chloride ions on a used PMMA IOL revealing the growth of LECs across the visual axis. With the application of laser postionization methods we hope to increase sensitivity towards protein fragments and organic additives to the IOL thereby providing more information on the interaction of the cells with the IOL surface.

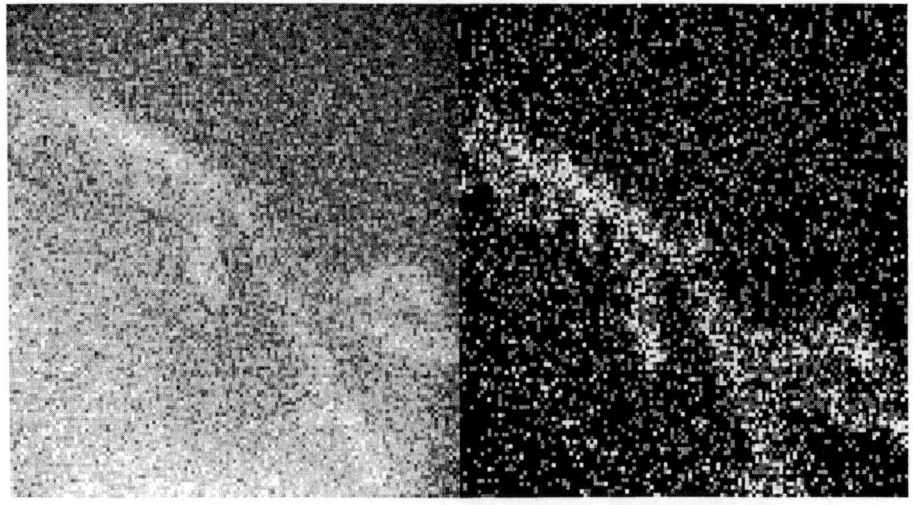

FIGURE 3. Chemical images of cell growth. These images of Na^+ (left) and Cl^- (right) ions reveal the growth of human cells on a poly-methylmethacrylate (PMMA) intraocular lens removed from a patient experiencing visual impairment (image size 250 x 250 μm).

Molecular Nano-Engineering: Self-Assembled Monolayers

Self-assembled monolayers (SAMs) provide idealised model organic surfaces which are thermodynamically stable and mechanically robust. Well defined monolayers may be prepared which are highly ordered, densely packed, and oriented, with a great degree of synthetic flexibility. SAMs have been used to improve the understanding and control of the responses of biological systems to man-made materials such as the inhibition of protein adsorption on contact lenses, and the promotion of cell attachment to artificial implants (10).

Imaging mass spectrometric techniques can be applied to chemically complex systems with high sensitivity and spatial resolution, and are therefore ideal for studying SAMs. Fig. 4 shows the distribution of Au_2SCH_3 and $Au_2SC_2H_5$ ions in a patterned alkanethiol SAM on a gold substrate. From data such as these information can not only be gained on the structure of the SAM pattern, but also on the nature of the bonding of the individual alkanethiol molecules to the substrate.

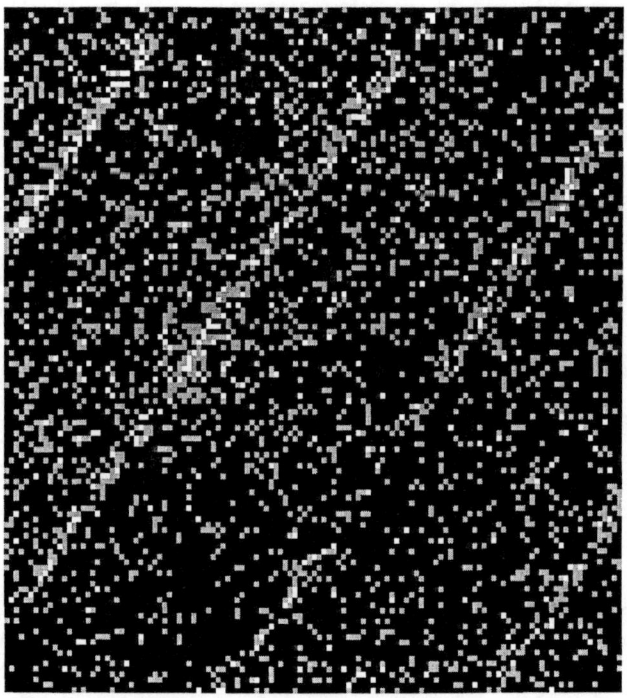

FIGURE 4. Chemical imaging of a SAM pattern. The image is of the Au_2SCH_3 and $Au_2SC_2H_5$ ions which reveal stripes of 3-Mercaptopropanoic acid and 3-Mercaptoundecanoic acid SAMs on a gold substrate (image size 800 μm x 800 μm).

SUMMARY

The application of femtosecond or vacuum ultraviolet photons to ion-beam desorbed organic and bio-organic neutral molecules are exciting new approaches to surface chemical characterisation. Our collaboration with the Penn. State group has allowed us to share our experience in this area and develop a new generation instrument designed specifically for this purpose. With the ability to investigate these two laser ionization methods on essentially identical instruments we are in a unique position to evaluate and compare each approach.

The results presented in this paper illustrate that the instrument is performing well in SIMS mode and already proving a very valuable research tool in biological applications. Laser postionization promises to improve the instrument specification greatly.

Our ultimate goal is to be able to image processes occurring at cell surfaces. The machine will have wide application in other areas where complex chemistry on small area materials (e.g. fibres and particles) is important.

ACKNOWLEDGEMENTS

The authors would like to thank Nick Winograd and Rob Braun of Penn. State University for their collaboration in this project. The roles of KORE Technology Ltd, and Ionoptika Ltd in constructing the key components and turning the ideas into detailed designs are gratefully acknowledged. The UMIST instrument is funded by the EPSRC.

REFERENCES

1. Möllers, R., Terhorst, M., Niehuis, E. and Benninghoven, A., *Surf. Interface Anal.* **18**, 824 (1992).
2. Weinkauf, R., Aicher, P., Wesley, G., Grotemeyer, J. and Schlag, E.W., *J. Phys. Chem.* **98**, 8381 (1994).
3. Brummel, C.L., Willy, K.F., Vickerman, J.C. and Winograd, N., *Int. J. Mass Spectrom. Ion Proc.* **143**, 257 (1995).
4. Ledingham, K.W.D., Kilic, H.S., Kosmidis, C., Deas, R.M., Marshall, A., McCanny, T., Singhal, R.P., Langley, A.J. and Shaikh, W., *Rapid Com. Mass. Spec.* **9**, 1522 (1995).
5. Ayre, C.R., Moro, L. and Becker, C.H., *Anal. Chem.* **66**, 1610 (1994).
6. Lockyer, N.P. and Vickerman, J.C., *Laser Chem.* **17**, 139 (1997).
7. Apple, D.J., Solomon, K.D. and Tetz, M.R., *Surv. Opthalmol.* **37**, 73 (1992).
8. Apple, D.J., Mamalis, N. and Loftfield, K., *Surv. Opthalmol.* **29**, 1 (1984).
9. Sourdille, P., *J. Cataract Refract. Surg.* **23**, 1431 (1997).
10. Mrksich, M. and Whitesides, G.M., *Annu. Rev. Biomol. Struct.* **25**, 55 (1996).

AUTHOR INDEX

A

Abraham, C. J., 53
Allott, R., 229
Amiot, C., 147
Andreyev, A., 191
Arimondo, E., 27

B

Baik, M. G., 261
Bastiaansen, J., 353
Berry, R. S., 253
Bisling, P., 33
Blaum, K., 73, 171, 275
Boesl, U., 59, 305
Bouckaert, S., 243
Bouloufa, N., 125
Bouwen, W., 243
Brault, P., 143
Bruyneel, B., 191
Bushaw, B. A., 73, 171, 177, 275

C

Cabaret, L., 125, 225
Cacciani, P., 125
Camus, P., 125
Cannon, B. D., 177
Cavalieri, S., 19, 249
Cha, B. H., 293, 297
Chakraborty, B. R., 210
Charalambidis, D., 19, 249
Chatterjee, R., 210
Ciampini, D., 27
Clark, E., 229
Codling, K., 155
Comparat, D., 147
Couris, S., 331
Crawford, J. E., 225
Creswell, A., 229
Crubellier, A., 147

D

Dangor, B., 229
Darveau, S. A., 253
Dederichs, J., 33
Deißenberger, R., 183
Donovan, R. J., 103, 325
Dorfner, R., 59, 305, 309
Drag, C., 147
Dulieu, O., 147
Duong, H. T., 225
Dyke, J. M., 81

E

Elizarov, A. Y., 315
Eramo, R., 19, 249
Erdmann, N., 183, 279, 285

F

Fang, X., 229
Faucher, O., 19, 249
Fini, L., 19, 249
Fioretti, A., 147
Fotakis, C., 19, 331
Franchoo, S., 191
Fujii, M., 137
Fuso, F., 27
Fuß, W., 163

G

Gabbanini, C., 27
Gamblin, S. D., 81
Garrison, B. J., 210
Genevey, J., 225
Gentens, J., 191
Georgiou, S., 347
Geppert, C., 275
Gibert-Legrand, T., 143
Gilmour, J. D., 269
Girod, M., 225
Gonthiez, T., 143

Graham, P., 229, 341
Grotemeyer, J., 67
Grun, C., 67
Grüning, C., 279, 285

H

Han, J. M., 257, 261, 297
Heger, H. J., 59, 305
Hernandez, M. B., 269
Hertz, E., 19
Holleman, I., 3
Huber, G., 183, 225, 279, 285
Huyse, M., 191

I

Ibrahim, F., 225
Ishiuchi, S., 137

J

Jeong, D. Y., 289
Jones, O. R., 53

K

Karapanagioti, N. E., 19
Kettrup, A., 59, 305, 309
Kilic, H. S., 341
Kim, J. T., 261
Kim, S. H., 293, 297
Klopp, P., 285
Knippels, G. M. H., 3
Ko, D. K., 293, 297
Köhler, S., 183
Kompa, K. L., 163
Kono, T., 89, 206
Kosmidis, C., 109, 341
Koubenakis, A., 347
Koudoumas, E., 331
Kratz, J. V., 183, 285
Krieg, M., 225
Kruglov, K., 191
Krushelnick, K., 229
Krustev, T., 53

Kudryavtsev, Y., 191
Kunz, P., 279, 285

L

Labrakis, J., 347
Langley, A. J., 341
Lassithiotaki, M., 347
Lawley, K. P., 103, 325
Le Blanc, F., 225
Ledingham, K. W. D., 229, 341
Lee, J., 257, 261, 289, 293, 297
Lee, J. K. P., 225
Lee, S. W., 297
Lehrer, F., 117
Lettry, J., 225
Lievens, P., 243, 353
Lockyer, N. P., 358
Lunney, D., 225

M

Machecek, A., 229
Magill, J., 229
Manson, S. T., 27
Maragò, O., 27
Marangos, J. P., 47
Masnou-Seeuws, F., 147
Materazzi, M., 19, 249
Mazeikis, A., 9
McCanny, T., 229, 341
McMahon, A. W., 269, 301
Meijer, G., 3
Meserole, C. A., 210
Mills, Jr., A. P., 47
Miskinis, J., 9
Miyake, Y., 47
Morris, A., 81
Mueller, W., 191
Müller, P., 73, 171, 275

N

Nagamine, K., 47
Neely, D., 229
Neidhart, B., 33
Noordam, L. D., 95

Norreys, P., 229
Nörtershäuser, W., 73, 171, 275
Nunnemann, M., 183, 279, 285

O

Obert, J., 225
O'Keeffe, P. K., 103
Oms, J., 225

P

Park, H., 257, 261
Passler, G., 183, 279, 285
Péru, S., 225
Peterson, J. R., 183
Petravicius, A., 9
Philipsen, V., 353
Philis, J. G., 109
Pillet, P., 147
Pinard, J., 225
Pitcheev, B., 125
Postawa, Z., 210
Posthumus, J. H., 155
Putaux, J. C., 225

R

Raabe, R., 191
Rateitzak, M., 269
Reusen, I., 191
Rhee, B. K., 289
Rhee, Y. J., 257, 261, 289
Ridley, T., 103, 325
Robino, M., 9
Roussière, B., 225

S

Sanderson, D., 229
Santala, M., 229
Sauvage, J., 225
Schmid, W. E., 163
Schmitt, A., 275
Schofield, P. A., 301
Sebastian, V., 225

Shaw, R. W., 235
Shimomura, K., 47
Silverans, R. E., 197, 243, 353
Singhal, R. P., 229, 341
Smith, D. J., 341
Staudt, C., 217
Stetzer, O., 285

T

Taday, P. F., 341
Tatarakis, M., 229
Telle, H. H., 53
Thoen, P., 243
Trautmann, N., 73, 171, 183, 275, 279
Trushin, S. A., 163
Tzallas, P., 109

V

Vaitkus, J., 9
Van den Bergh, P., 191
van der Meer, A. F. G., 3
Vandeweert, E., 210, 353
Van Duppen, P., 191
Vanhoutte, F., 243
van Roij, A., 3
Van Roosbroeck, J., 191
Vermeeren, L., 191
Verney, D., 225
Vetter, R., 125
Vickerman, J. C., 358
Vivet, L., 143
von Helden, G., 3
Vorsa, V., 89, 206

W

Waldek, A., 183, 279, 285
Wang, S. L., 103, 325, 341
Wark, J., 229
Watts, I., 229
Webb, O. F., 235
Weickhardt, C., 67
Weidele, H., 243
Weinkauf, R., 117
Weissman, L., 191

Weitkamp, C., 33
Wendt, K., 39, 73, 171, 275, 285
West, J. B., 81
Willey, K. F., 89, 206
Winograd, N., 89, 206, 210
Wong, S. C. C., 358
Wright, T. G., 81
Wucher, A., 217

X

Xenakis, D., 19

Y

Yeretzian, C., 309
Yi, J. H., 257, 261
Young, J. P., 235

Z

Zemlyanoi, S., 225
Zepf, M., 229
Zimmermann, R., 59, 131, 305, 309
Zindulis, A., 9